CLASS 7

Pearson IIT Foundation Series

Mathematics

Fifth Edition

CLASS 7

Pearson IIT Foundation Series
Mathematics

Fifth Edition

Trishna Knowledge Systems

Pearson

Photo Credits

All Chapter Openers: Albachiaraa. Shutterstock
Icons of Practice Questions: graphixmania. Shutterstock
Icons of Answer Keys: Viktor88. Shutterstock
Icons of Hints and Explanation: graphixmania. Shutterstock

Editor—Acquisitions: Nandini Basu
Senior Editor—Production: G. Sharmilee

The aim of this publication is to supply information taken from sources believed to be valid and reliable. This is not an attempt to render any type of professional advice or analysis, nor is it to be treated as such. While much care has been taken to ensure the veracity and currency of the information presented within, neither the publisher nor its authors bear any responsibility for any damage arising from inadvertent omissions, negligence or inaccuracies (typographical or factual) that may have found their way into this book.

Copyright © 2017 Trishna Knowledge Systems
Copyright © 2012, 2014, 2015, 2016 Trishna Knowledge Systems

This book is sold subject to the condition that it shall not, by way of trade or otherwise, be lent, resold, hired out, or otherwise circulated without the publisher's prior written consent in any form of binding or cover other than that in which it is published and without a similar condition including this condition being imposed on the subsequent purchaser and without limiting the rights under copyright reserved above, no part of this publication may be reproduced, stored in or introduced into a retrieval system, or transmitted in any form or by any means (electronic, mechanical, photocopying, recording or otherwise), without the prior written permission of both the copyright owner and the publisher of this book.

ISBN 978-93-325-7901-9

First Impression

Published by Pearson India Education Services Pvt. Ltd, CIN: U72200TN2005PTC057128, formerly known as TutorVista Global Pvt. Ltd, licensee of Pearson Education in South Asia.

Head Office: 15th Floor, Tower-B, World Trade Tower, Plot No. 1, Block-C, Sector-16, Noida 201 301, Uttar Pradesh, India.
Registered Office: 4th Floor, Software Block, Elnet Software City, TS-140, Block 2 & 9, Rajiv Gandhi Salai, Taramani, Chennai 600 113, Tamil Nadu, India.
Fax: 080-30461003, Phone: 080-30461060
www.pearson.co.in, Email: companysecretary.india@pearson.com

Compositor: SRS Technologies, Puducherry
Printed in India at Thomson Press India Ltd.

Brief Contents

Preface — ix
Chapter Insights — x
Series Chapter Flow — xii

Chapter 1	Number Systems	1.1
Chapter 2	Expressions and Special Products	2.1
Chapter 3	Ratio and Its Applications	3.1
Chapter 4	Indices	4.1
Chapter 5	Geometry	5.1
Chapter 6	Mensuration	6.1
Chapter 7	Equations and their Applications	7.1
Chapter 8	Formulae	8.1
Chapter 9	Statistics	9.1
Chapter 10	Set Theory	10.1

Contents

Preface	ix
Chapter Insights	x
Series Chapter Flow	xii

CHAPTER 1 NUMBER SYSTEMS — 1.1

Introduction — 1.2
- Natural Numbers — 1.2
- Whole Numbers — 1.2
- Integers — 1.2
- Additive Inverse — 1.4
- Multiplication of Integers — 1.4
- Division of Integers — 1.5
- Fractions — 1.6
- Decimal Numbers — 1.11
- Rational Numbers — 1.15
- Decimal Representation of Rational Numbers — 1.19
- Irrational Numbers — 1.20
- Square Roots — 1.20
- Properties of Square Roots — 1.21

Practice Questions — 1.23
Answer Keys — 1.33
Hints and Explanation — 1.35

CHAPTER 2 EXPRESSIONS AND SPECIAL PRODUCTS — 2.1

Introduction — 2.2
- Constant — 2.2
- Variable — 2.2
- Algebraic Expression — 2.2
- Polynomial — 2.2
- Factorisation — 2.8
- Division of a Polynomial by a Monomial — 2.9
- Division of a Polynomial by a Polynomial — 2.10
- The HCF of the Given Polynomials — 2.10
- Algebraic Expressions—Formulae — 2.12
- Algebraic Expressions–Rules — 2.12

Practice Questions — 2.14
Answer Keys — 2.23
Hints and Explanation — 2.25

CHAPTER 3 RATIO AND ITS APPLICATIONS — 3.1

Introduction — 3.2
- Terms of a Ratio — 3.2
- Properties of a Ratio — 3.2
- Simplest Form of a Ratio — 3.2
- Comparison of Ratios — 3.3
- Proportion — 3.3
- Variation — 3.4
- Percentage — 3.5

Practice Questions — 3.17
Answer Keys — 3.29
Hints and Explanation — 3.31

CHAPTER 4 INDICES — 4.1

Introduction — 4.2
- Laws of Indices — 4.2
- Exponential Equation — 4.4

Practice Questions — 4.8
Answer Keys — 4.17
Hints and Explanation — 4.19

CHAPTER 5 GEOMETRY — 5.1

Introduction — 5.2
- Properties of Parallel Lines — 5.5
- Triangle — 5.7
- Congruence — 5.11
- Quadrilaterals — 5.18
- Construction of Quadrilaterals — 5.22
- Polygons — 5.28
- Circle — 5.29

Practice Questions	5.33
Answer Keys	5.48
Hints and Explanation	5.50

CHAPTER 6 MENSURATION — 6.1

Introduction — 6.2
- Plane Figures — 6.2
- Area of a Rectangle — 6.3
- Area of a Parallelogram — 6.4
- Area of a Triangle — 6.4
- Area of a Trapezium — 6.5
- Area of a Rhombus — 6.5
- Area of a Square — 6.6
- Area of an Equilateral Triangle — 6.6
- Area of a Right-angled Triangle — 6.7
- Isosceles Right Triangle — 6.7
- Circumference of a Circle — 6.7
- Area of a Ring — 6.7
- Solids — 6.9
- Cubes and Cuboids — 6.9

Practice Questions — 6.12
Answer Keys — 6.24
Hints and Explanation — 6.26

CHAPTER 7 EQUATIONS AND THEIR APPLICATIONS — 7.1

Introduction — 7.2
- Numbers and Symbols — 7.2
- Numerical Expressions — 7.2
- Algebraic Expression — 7.2
- Mathematical Sentence — 7.2
- Mathematical Statement — 7.2
- Open Sentences — 7.2
- Equation — 7.3
- Linear Equation — 7.3
- Simple Equation — 7.3

Practice Questions	7.10
Answer Keys	7.18
Hints and Explanation	7.20

CHAPTER 8 FORMULAE — 8.1

Introduction — 8.2
- Subject of a Formula — 8.2
- Framing a Formula — 8.5

Practice Questions — 8.7
Answer Keys — 8.17
Hints and Explanation — 8.19

CHAPTER 9 STATISTICS — 9.1

Introduction — 9.2
- Data — 9.2
- Statistical Graphs — 9.3
- Arithmetic Mean or Mean — 9.8
- Median — 9.10
- Chance and Probability — 9.12

Practice Questions — 9.15
Answer Keys — 9.23
Hints and Explanation — 9.25

CHAPTER 10 SET THEORY — 10.1

Introduction — 10.2
- Set — 10.2
- Operations on Sets — 10.4
- Venn Diagrams — 10.5

Practice Questions — 10.10
Answer Keys — 10.20
Hints and Explanation — 10.22

Preface

Pearson IIT Foundation Series has developed into a trusted resource for scores of students who have aspired to be a part of the elite undergraduate institutions of India. As a result it has become one of the best selling series, providing authentic and class tested content for effective preparation.

The structure of the content is not only student-friendly but is also designed in such a manner that it invigorates the students to go beyond the usual school curriculum and also act as a source of higher learning to strengthen the fundamental concepts of Physics, Chemistry, and Mathematics.

The core objective of the series is to be a one-stop-solution for students preparing for various competitive examinations. Irrespective of the field of study that the student may choose to take up later, it is important to understand that Mathematics and Science form the basis for most modern day activities. Hence, utmost efforts have been made to develop student interest in these basic building blocks via real-life examples and application based problems. Ultimately the aim is to ingrain the art of problem-solving in the minds of the reader.

To ensure high level of accuracy and practicality this series has been authored by a team of highly qualified and experienced faculties involved in grooming the young minds. That said, we believe that there is always scope for doing things in a better manner and hence invite you to provide us with your candid feedback and suggestions on how we can make this series more superior.

Chapter Insights

> **REMEMBER**
> Before beginning this chapter, you should be able to:
> - Know the natural numbers, and whole numbers, etc.
> - Understand terms like fraction, multiples, HCF, etc.

Remember section will help them to memories and review the previous learning on a particular topics

> **KEY IDEAS**
> After completing this chapter, you should be able to:
> - Explain the natural numbers, whole numbers, integers, fractional numbers, decimals, rational and irrational numbers
> - Understand how to obtain LCM and HCF of given numbers

Key points will help the students to identify the essential points in a chapter

Whole Numbers
Zero and natural numbers together are called whole numbers.

Integers
All the numbers .... $-3, -2, -1, 0, 1, 2, 3$ .... are integers.

If a, b, c are natural numbers and $ab = c$, a and b are factors of c. c is a multiple of a as well as of b.

For example, $2 \times 3 = 6$. Therefore, 2 and 3 are factors of 6 and 6 is a multiple of 2 as well as of 3.

Text: concepts are explained in a well structured and lucid manner

Note boxes are some add-on information of related topics

> **Note** When we add two decimal numbers or subtract one decimal number from the other we have to ensure that the digit in the units place of the second number is written exactly below the digit in the units place of the first number. Same is the case with digits in the 100's place, 1000's place, $\frac{1}{10}$ th place, $\frac{1}{100}$ th place and so on.

EXAMPLE 1.12

Find the value of $\sqrt{20\frac{1}{4}} - \sqrt{1\frac{32}{49}}$.

SOLUTION

$$\sqrt{20\frac{1}{4}} = \sqrt{\frac{20 \times 4 + 1}{4}} = \sqrt{\frac{81}{4}} = \frac{9}{2}$$

$$\sqrt{1\frac{32}{49}} = \sqrt{\frac{1 \times 49 + 32}{49}} = \sqrt{\frac{81}{49}} = \frac{9}{7}$$

$$\sqrt{20\frac{1}{4}} - \sqrt{1\frac{32}{49}} = \frac{9}{2} - \frac{9}{7}$$

$$= 9\left(\frac{1}{2} - \frac{1}{7}\right) = 9\left(\frac{5}{14}\right) = \frac{45}{14}.$$

Examples given topic-wise to apply the concepts learned in a particular chapter

Illustrative examples solved in a logical and step-wise manner

Chapter Insights xi

TEST YOUR CONCEPTS
Very Short Answer Type Questions

Directions for questions 1 to 9: State whether the following statements are true or false.

1. The multiplicative inverse of zero does not exist.
2. Division by zero is not defined.
3. Division is commutative for integers.
4. The sum of any two proper fractions is an improper fraction.
5. The product of a proper and an improper fraction is less than the improper fraction.
6. $2.3458 \times 1000 = 234.58$

21. Which of the following sets is not closed under subtraction?
    (a) N (b) Z
    (c) Q (d) R

22. What do you call two fractions, whose product is 1?
    (a) Additive inverse to each other.
    (b) Multiplicative inverse to each other.
    (c) Reciprocals to each other.
    (d) Both (b) and (c)

Different levels of questions have been included in the Test Your Concept as well as on Concept Application which will help students to develop the problem-solving skill

'Test Your Concepts' at the end of the chapter for classroom preparations

CONCEPT APPLICATION
Level 1

Directions for questions 1 to 17: Select the correct alternative from the given choices.

1. What is the additive identity element in the set of whole numbers?
   (a) 0   (b) 1
   (c) –1   (d) None of these

2. Which of the following is/are true?
   (a) $\sqrt{6} \times \sqrt{6}$ is an irrational number.
   (b) $\sqrt{4} \times \sqrt{25}$ is a rational number.
   (c) Both (a) and (b)
   (d) Neither (a) nor (b)

3. M and N are two coprimes. Which of the following is/are true?

5. $\sqrt{X}$ is a perfect square. Which of the following is/are true?
   (a) X is a perfect square.
   (b) X^2 is a perfect square
   (c) Neither (a) nor (b)
   (d) Both (a) and (b)

6. $\sqrt[3]{\dfrac{125}{216}} - \sqrt{\dfrac{25}{36}} =$ _____.
   (a) 5/6   (b) 1/6
   (c) 0   (d) 1

7. The HCF of two numbers is 6 and the product of two numbers is 360. Find the LCM of the numbers.

'Concept Application' section with problems divided as per complexity: Level 1; Level 2; and Level 3

33. The required number is HCF of 81, 144 and 162.

```
    81) 162 (2
        162
        ___
          0
        81) 144 (1
            81
            __
         63 ) 81 (1
              63
              __
           18) 63 (3
               54
               __
                9) 18 (2
                   18
                   __
                    0
```

HCF = 9.

Hence, the correct option is (a)

Hints and Explanation for key questions along with highlights on the common mistakes that students usually make in the examinations

Series Chapter Flow

Class 7

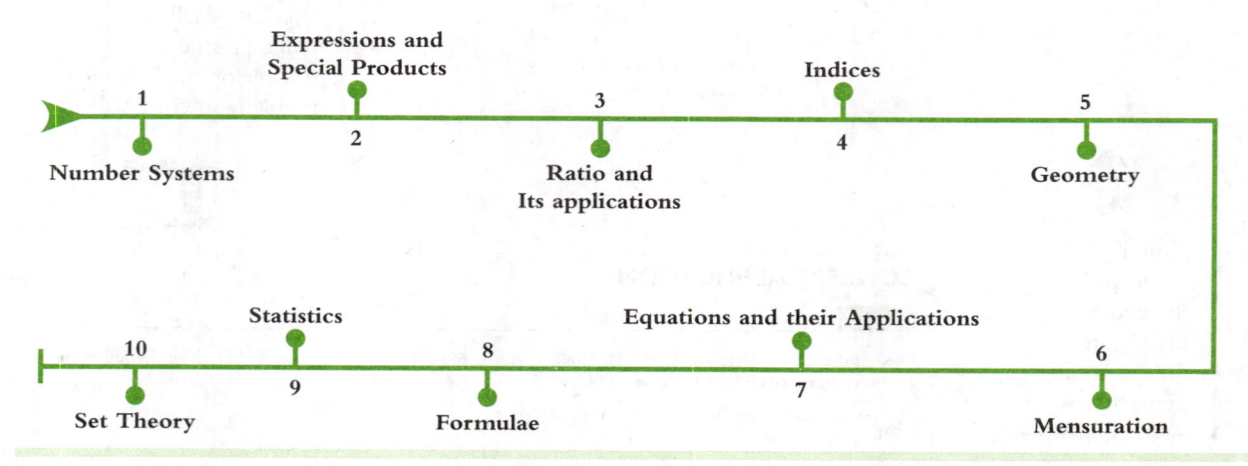

1. Number Systems
2. Expressions and Special Products
3. Ratio and Its applications
4. Indices
5. Geometry
6. Mensuration
7. Equations and their Applications
8. Formulae
9. Statistics
10. Set Theory

Class 9

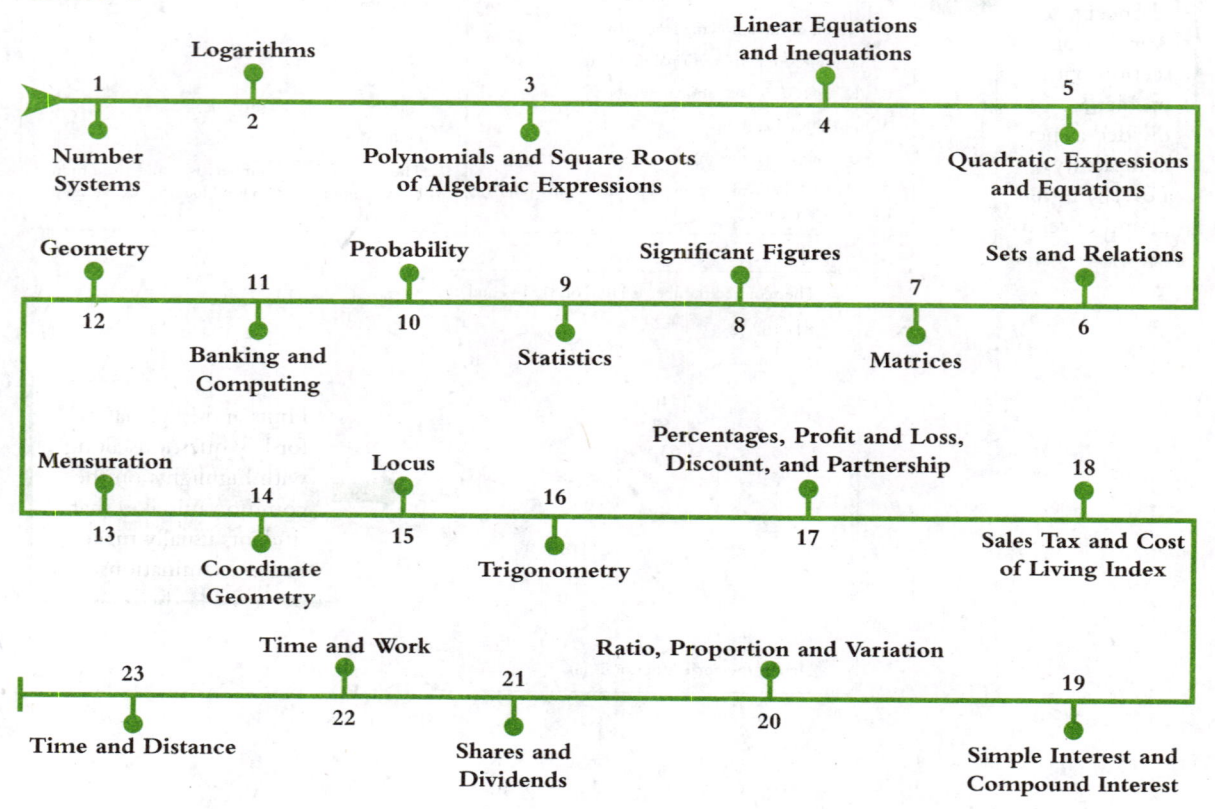

1. Number Systems
2. Logarithms
3. Polynomials and Square Roots of Algebraic Expressions
4. Linear Equations and Inequations
5. Quadratic Expressions and Equations
6. Sets and Relations
7. Matrices
8. Significant Figures
9. Statistics
10. Probability
11. Banking and Computing
12. Geometry
13. Mensuration
14. Coordinate Geometry
15. Locus
16. Trigonometry
17. Percentages, Profit and Loss, Discount, and Partnership
18. Sales Tax and Cost of Living Index
19. Simple Interest and Compound Interest
20. Ratio, Proportion and Variation
21. Shares and Dividends
22. Time and Work
23. Time and Distance

Series Chapter Flow

Class 8

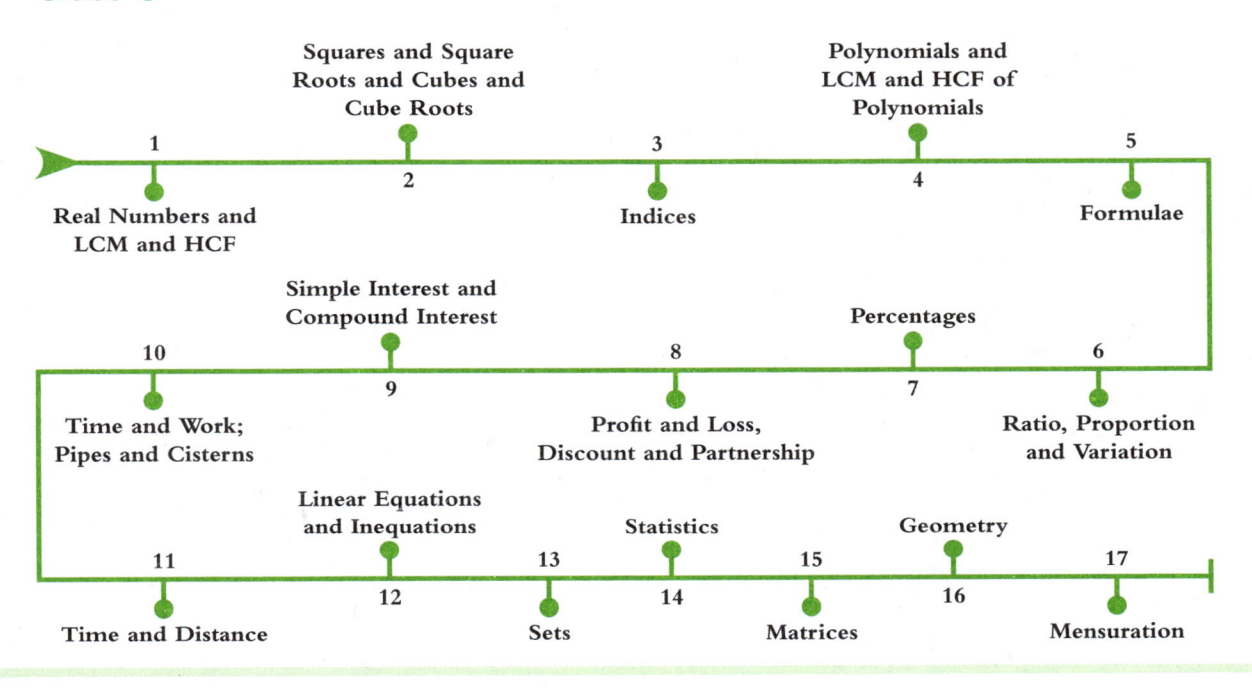

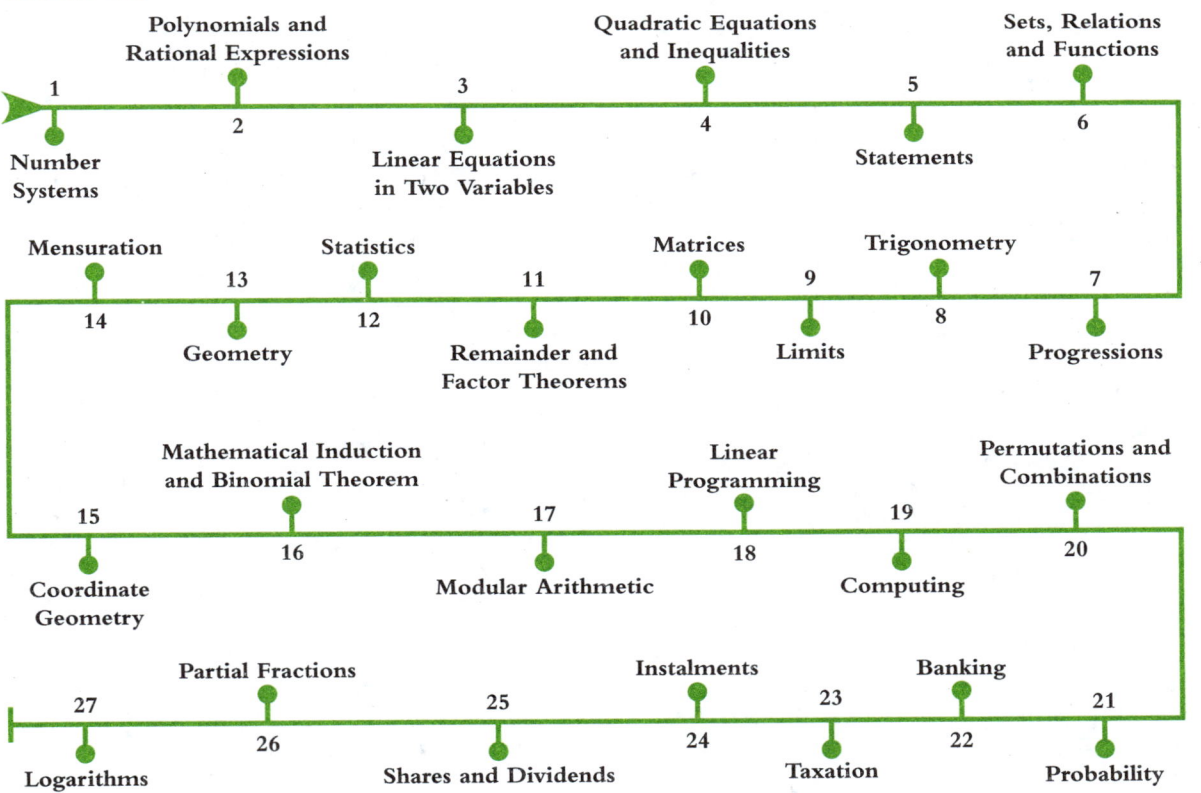

1. Real Numbers and LCM and HCF
2. Squares and Square Roots and Cubes and Cube Roots
3. Indices
4. Polynomials and LCM and HCF of Polynomials
5. Formulae
6. Ratio, Proportion and Variation
7. Percentages
8. Profit and Loss, Discount and Partnership
9. Simple Interest and Compound Interest
10. Time and Work; Pipes and Cisterns
11. Time and Distance
12. Linear Equations and Inequations
13. Sets
14. Statistics
15. Matrices
16. Geometry
17. Mensuration

Class 10

1. Number Systems
2. Polynomials and Rational Expressions
3. Linear Equations in Two Variables
4. Quadratic Equations and Inequalities
5. Statements
6. Sets, Relations and Functions
7. Progressions
8. Trigonometry
9. Limits
10. Matrices
11. Remainder and Factor Theorems
12. Statistics
13. Geometry
14. Mensuration
15. Coordinate Geometry
16. Mathematical Induction and Binomial Theorem
17. Modular Arithmetic
18. Linear Programming
19. Computing
20. Permutations and Combinations
21. Probability
22. Banking
23. Taxation
24. Instalments
25. Shares and Dividends
26. Partial Fractions
27. Logarithms

Chapter 1

Number Systems

$$a)\ 5(3x) - 3(x-4) = 0$$
$$15x - 3x + 12 = 0$$
$$12x + 12 = 0$$
$$12x = -12$$
$$x = -12/12$$
$$x = -1 \qquad x^2 = 1$$

REMEMBER
Before beginning this chapter, you should be able to:
- Know the natural numbers, and whole numbers, etc.
- Understand terms like fraction, multiples, HCF, etc.

KEY IDEAS
After completing this chapter, you should be able to:
- Explain the natural numbers, whole numbers, integers, fractional numbers, decimals, rational and irrational numbers
- Understand how to obtain LCM and HCF of given numbers
- Represent integers and rational numbers on a number line
- Apply operations on integers, fractions, decimals, rational and irrational numbers
- Convert decimal into fraction and vice-versa
- Find the square roots and cube roots of rational numbers

INTRODUCTION

In the lower classes, we have studied natural numbers, whole numbers, integers and fractions. We are also familiar with terms like factors, multiples, primes, coprimes, HCF and LCM. In this chapter, we shall revise these ideas and learn what rational and irrational numbers are.

Natural Numbers

All counting numbers are natural numbers.
1, 2, 3, 4, ... are natural numbers.

Whole Numbers

Zero and natural numbers together are called whole numbers.

Integers

All the numbers .... –3, –2, –1, 0, 1, 2, 3 .... are integers.

If a, b, c are natural numbers and $ab = c$, a and b are factors of c. c is a multiple of a as well as of b.

For example, $2 \times 3 = 6$. Therefore, 2 and 3 are factors of 6 and 6 is a multiple of 2 as well as of 3.

If we take two or more numbers, we can find out the factors of all the numbers and determine the common factors and among these we can determine the highest common factor, i.e., the HCF.

Similarly, we can list the multiples of all the numbers and identify the common multiples and then determine the least common multiple, i.e., the LCM.

For example, $12 = 2 \times 2 \times 3$ and $18 = 2 \times 3 \times 3$.

The HCF = $2 \times 3 = 6$ and the LCM = $2 \times 2 \times 3 \times 3 = 36$.

All natural numbers greater than 1 have at least two factors. If they have exactly 2, they are prime. Otherwise they are composite. Two numbers which have only 1 as a common factor are coprimes.

EXAMPLE 1.1

Find the HCF and the LCM of 300, 360 and 600.

SOLUTION

```
2 | 300, 360, 600
2 | 150, 180, 300
3 |  75,  90, 150
5 |  25,  30,  50
  |   5,   6,  10
```

As no other number is a common factor to 5, 6, 10, the HCF of 300, 360 and 600 is $2^2 \times 3 \times 5 = 60$.

Now we shall get the LCM.

2	300, 360, 600
2	150, 180, 300
2	75, 90, 150
3	75, 45, 75
5	25, 15, 25
5	5, 3, 5
	1, 3, 1

∴ The LCM of 300, 360, and 600 is $2 \times 2 \times 2 \times 3 \times 5 \times 5 \times 3 = 1800$.

Representation of Integers on the Number Line

A line on which numbers are represented is known as the number line.

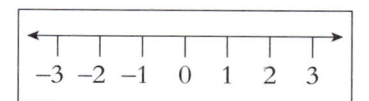

Some integers are represented on the number line given below.

We observe the following about the number line:

1. As we move from left to right, the integers are in the ascending order.

    $-3 < -2 < -1 < 0 < 1 < 2$ ....

    As we move from right to left, the integers are in the descending order.

2. When we add a positive integer, the result obtained is to the right of the given integer.

    For example, $0 + 1 = 1$, 1 is to the right of 0.

    $1 + 1 = 2$, 2 is to the right of 1.

    $1 + 3 = 4$, 4 is to the right of 1.

3. When we add a negative integer to a given integer, the result obtained is to the left of the given integer.

    For example, $2 + (-1) = 1$, 1 is to the left of 2.

    $4 + (-2) = 2$, 2 is to the left of 4.

4. When we subtract a positive integer, from a given integer, the result obtained is to the left of the given integer.

    For example, $3 - 1 = 2$, 2 is to the left of 3.

    $2 - 2 = 0$, 0 is the left of 2.

5. When we subtract a negative integer from a given integer, the result obtained is to the right of the given integer.

    For example, $3-(-2) = 5$ and 5 is to the right of 3.

6. The sum of two positive integers is a positive integer.

    For example, $2 + 5 = 7$; $6 + 4 = 10$.

7. The sum of two negative integers is a negative integer.

    For example, $(-3) + (-4) = -7$; $(-5) + (-4) = -9$.

8. The sum of a positive integer and a negative integer may be positive or negative.

    For example, $5 + (-3) = 2$; $4 + (-7) = -3$.

Additive Inverse

When the sum of two integers is zero, then each integer is said to be the additive inverse of the other.

For example, $2 + (-2) = 0$.

-2 is the additive inverse of 2 and 2 is the additive inverse of -2.

5 is the additive inverse of -5.

-20 is the additive inverse of 20.

Properties of Integers under Addition

1. **Closure property:** For any two integers p and q, $(p + q)$ is an integer.

   When we add two integers, their sum is also an integer.

   For example, $15 + 12 = 27$, $20 + (-11) = 9$.

2. **Commutative property:** For any two integers p and q, $p + q = q + p$.

   For example, $3 + 2 = 2 + 3$ and $5 + (-16) = (-16) + 5$.

   > **Note** Subtraction is not commutative for integers. For example, $1 - 3 \neq 3 - 1$.

3. **Associate property:** For any three integers p, q and r.

   $(p + q) + r = p + (q + r)$.

   For example, $(1 + 2) + 3 = 1 + (2 + 3)$.

4. **Additive Identity:** There is an integer 0, such that for any integer p, $p + 0 = 0 + p = p$.

   If we add zero to any integer, we get the same integer.

   For example, $2 + 0 = 2$.

Multiplication of Integers

1. The product of two positive integers p and q is also a positive integer.

   For example, $3 \times 5 = 15$ and $6 \times 8 = 48$.

2. The product of a positive integer and a negative integer is negative.

   For example, $4 \times (-6) = -24$ and $(-8) \times 5 = -40$.

3. The product of two negative integers is a positive integer.

   For example, $(-7) \times (-4) = +28$ and $(-6) \times (-5) = +30$.

4. The product of three negative integers is a negative integer.

   For example, $(-3) \times (-4) \times (-2) = (+12 \times -2) = -24$.

5. The product of four negative integers is a positive integer.

   For example, $(-1) \times (-2) \times (-3) \times (-4) = 24$.

We may observe that the product of an odd number of negative integers is a negative integer and the product of an even number of integers is a positive integer.

However, the product of any number of positive integers is always a positive integer.

Properties of Integers under Multiplication

1. **Closure property:** For any two integers p and q, $p \times q$ is an integer.
   For example, $2 \times 3 = 6$, $5 \times 10 = 50$ and $8 \times 0 = 0$.

2. **Commutative property:** For any two integers p and q, $p \times q = q \times p$.
   For example, $2 \times 4 = 4 \times 2$ and $-5 \times 6 = 6 \times -5$.

3. **Associative property:** For any three integers, p, q and r, $(p \times q) \times r = p \times (q \times r)$.
   $(1 \times 2) \times 3 = 1 \times (2 \times 3)$ and $(4 \times 5) \times -6 = 4 \times [5 \times (-6)]$

4. **Multiplication by zero:** For any integer p, $p \times 0 = 0$.
   For example, $5 \times 0 = 0$ and $(-4) \times 0 = 0$.

5. **Multiplicative identity:** There is an integer 1 such that for any integer p,
   $p \times 1 = 1 \times p = p$. For example, $2 \times 1 = 1 \times 2 = 2$ and $(-10) \times 1 = 1 \times (-10) = -10$. If we multiply any integer by 1, we get the same integer.

6. **Distributive property:** For any three integers p, q and r, $p \times (q + r) = (p \times q) + (p \times r)$.
   This is known as distributive property of multiplication over addition.
   For example, $2 \times (3 + 2) = (2 \times 3) + (2 \times 2)$.
   LHS = $2 \times (3 + 2) = 2 \times 5 = 10$ and RHS = $(2 \times 3) + (2 \times 2) = 6 + 4 = 10$.

Division of Integers

We know that $2(3) = 6$. We can write that $6 \div 2 = 3$ or $6 \div 3 = 2$.

Thus, we can say that division is an inverse operation of multiplication.

1. When we divide a positive integer by another positive integer, we get a positive number, which need not be an integer.
   For example, $6 \div 5 = \dfrac{6}{5}$ or $1\dfrac{1}{5}$.
   We observe that division is not closed for integers.

2. When a negative integer is divided by a positive integer, we get a negative number.
   For example, $-12 \div 3 = -\dfrac{12}{3} = -4$.
   When 12 is divided by 3, the quotient is 4 and we get a negative sign before 4.

3. When a positive integer is divided by a negative integer, we get a negative number.
   For example, $20 \div (-10)$ can be written as $-\left(\dfrac{20}{10}\right) = -2$.

4. When a negative integer is divided by a negative integer, we get a positive number.
   For example, $(-40) \div (-8)$ can be written as $\left(\dfrac{40}{8}\right) = 5$.

5. For two integers p and q, $(p \div q) \neq (q \div p)$. For example, $2 \div 1 = 2$ and $1 \div 2 = \dfrac{1}{2}$.
   $2 \neq \dfrac{1}{2}$.
   We observe that division is not commutative for integers.

6. When a positive integer p is divided by 1, the result is p. $p \div 1 = p$.

For example, $2 \div 1 = 2$, $15 \div 1 = 15$ and $0 \div 1 = 0$.

However, $1 \div 0$ is not defined. There is no integer which when multiplied by zero gives one.

It is important to note that $0 \div 0$ is not defined. Any integer (or any number) when multiplied by zero gives zero.

Fractions

We have learnt what fractions are in the previous classes. In this section, we shall learn about the types of fractions, representation of fractions on the number line, ordering of fractions and the basic arithmetic operations on fractions, etc.

Imagine that you have invited five friends to your birthday party. You have cut the cake into six equal pieces. Each person gets one piece out of a total of six pieces. In other words each person gets one-sixth of the cake. 1/6 is a fraction. Here cake is the whole and a piece is a part.

3/4, 4/5, 4/9 and 21/2 are some examples of fractions.

A number of the form a/b, where a, b are whole numbers and $b \neq 0$ is a fraction. The number a is the numerator and b is the denominator.

There are three types of fractions:

1. Proper fraction

2. Improper fraction and

3. Mixed fraction

1. *Proper fraction:* A fraction that represents a part of whole is a proper fraction.

1/2, 3/4, 5/6, 0/7 are some examples of proper fractions. In a proper fraction, numerator is less than the denominator. The whole number 0 can be expressed as a proper fraction.

2. *Improper fraction:* A fraction that represents a combination of a whole and a part of the whole is an improper fraction. 5/3, 8/5, 20/9, 10/10 are some examples of improper fractions. In an improper fraction, the numerator is greater than or equal to its denominator. All natural numbers can be expressed as improper fractions.

3. *Mixed Fraction:* When an improper fraction is written as an integer followed by a proper fraction, it is a mixed fraction.

$\dfrac{5}{3}$ can be written as $1 + \dfrac{2}{3}$ or $1\dfrac{2}{3}$. Here, 1 is a whole and $\dfrac{2}{3}$ is a part.

$\dfrac{20}{9}$ can be written as $2 + \dfrac{2}{9}$ or $2\dfrac{2}{9}$. Here, 2 is a whole and $\dfrac{2}{9}$ is a part.

We observe that an improper fraction is the sum of an integer and a proper fraction.

Comparison of Fractions

Let us learn this with the help of an example.

EXAMPLE 1.2

Compare $\dfrac{2}{5}$ and $\dfrac{3}{7}$.

SOLUTION

$\dfrac{2}{5}$ can be written as $\dfrac{2 \times 7}{5 \times 7} = \dfrac{14}{35}$.

$\dfrac{3}{7}$ can be written as $\dfrac{3 \times 5}{7 \times 5} = \dfrac{15}{35}$.

$\dfrac{15}{35} > \dfrac{14}{35}$

$\therefore \dfrac{3}{7} > \dfrac{2}{5}$

EXAMPLE 1.3

Compare $\dfrac{5}{6}$ and $\dfrac{7}{9}$.

SOLUTION

$\dfrac{5}{6}$ can be written as $\dfrac{5 \times 3}{6 \times 3} = \dfrac{15}{18}$.

$\dfrac{7}{9}$ can be written as $\dfrac{7 \times 2}{9 \times 2} = \dfrac{14}{18}$.

$\dfrac{15}{18} > \dfrac{14}{18} \Rightarrow \dfrac{5}{6} > \dfrac{7}{9}$

Addition of Fractions

1. **A fraction and an Integer:** The integer is expressed as a fraction with the same denominator as the given fraction and the numerators are added.

    For example,

    (a) $\dfrac{5}{8} + 1 = \dfrac{5}{8} + \dfrac{8}{8} = \dfrac{13}{8}$

    (b) $\dfrac{5}{8} + 0 = \dfrac{5}{8} + \dfrac{0}{8} = \dfrac{5}{8}$

    (c) $\dfrac{5}{8} + (-2) = \dfrac{5}{8} + \dfrac{-16}{8} = \dfrac{-11}{8}$

2. **Two or more fractions:** All the fractions in the expression are expressed with a common denominator and the numerators are added.

    For example,

    (a) $\dfrac{2}{3} + \dfrac{3}{4} = \dfrac{2(4)}{3(4)} + \dfrac{3(3)}{4(3)} = \dfrac{8}{12} + \dfrac{9}{12} = \dfrac{17}{12}$

    (b) $\dfrac{5}{6} + \dfrac{7}{9} - \dfrac{1}{2} = \dfrac{15}{18} + \dfrac{14}{18} - \dfrac{9}{18} = \dfrac{20}{18} = \dfrac{10}{9}$

Multiplication of Fractions

We shall now learn how to multiply fractions.

1. *A fraction and an integer:* First let us consider multiplying a proper fraction with a whole number.

$$2\left(\frac{1}{2}\right) = \frac{2(1)}{2} = 1.$$

Suppose you have half a laddu and your friend has another half of the same laddu. It can be diagrammatically represented as follows:

The shaded portions of the two circles on the left hand side of the equality sign represent two equal halves. The circle on the right hand side represents the complete circle.

We can write the same as $\frac{1}{2} + \frac{1}{2} = 1$ or $2\left(\frac{1}{2}\right) = 1$.

Similarly, $6 \times \left(\frac{1}{3}\right) = \frac{6(1)}{3} = 2.$

$$12\left(\frac{3}{5}\right) = \frac{12(3)}{5} = \frac{36}{5} = 7\frac{1}{5}$$

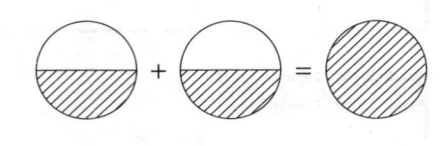

So far we have seen multiplication of a proper fraction and a whole number.

Let us consider multiplying an improper fraction and a whole number.

$$5\left(\frac{8}{3}\right) = \frac{5(8)}{3} = \frac{40}{3} = 13\frac{1}{3}$$

$$3\left(\frac{5}{4}\right) = \frac{3(5)}{4} = \frac{15}{4} = 3\frac{3}{4}$$

We observe that when an improper fraction or a proper fraction is multiplied with a whole number, the numerator of the fraction is multiplied with the whole number and the denominator is retained as it is.

2. *Multiplication of a mixed fraction with a whole number:*

For example, $9\left(5\frac{2}{3}\right) = \frac{9(17)}{3} = 3(17) = 51$ and $3\left(2\frac{5}{12}\right) = 3\left(\frac{29}{12}\right) = \frac{3(29)}{12} = \frac{29}{4} = 7\frac{1}{4}.$

We can convert the mixed fraction to an improper fraction and then multiply as shown in the above example.

EXAMPLE 1.4

In a class there are 60 students. One third of them are girls. How many girls are there in the class?

SOLUTION

It is given that $\frac{1}{3}$ of the total students are girls.

$\therefore$ The number of girls $\frac{1}{3}$ of 60 = $\frac{1}{3}$ (60) = 20.

We observe that multiplication is also used as an operator 'of'.

$$\frac{1}{4} \text{ of } 20 = \frac{1}{4}(20) = 5.$$

$$\frac{1}{5} \text{ of } 80 = \frac{1}{5}(80) = 16.$$

3. **Multiplication of a fraction by a fraction:**

EXAMPLE 1.5

Suppose your father has brought a big bar of chocolates. He divided it into two equal pieces and gave one piece to you. You wanted to share it with two of your friends. You divided your piece of chocolate into three equal pieces. How much did each of you get?

SOLUTION

You have $\frac{1}{2}$ a chocolate bar.

You made 3 pieces of it where each piece is $\frac{1}{3}$ rd of the half.

$$\frac{1}{3} \times \left(\frac{1}{2}\right) = \frac{1}{6}$$

Each of the three got $\frac{1}{6}$ th of the chocolate bar.

Thus, we observe that the product of two fractions = $\dfrac{\text{The product of their numerators}}{\text{The product of their denominators}}$.

We also make the following observations while multiplying fractions.

1. The product of two proper fractions is less than one.

   For example, $\frac{1}{2}\left(\frac{1}{3}\right) = \frac{1}{6} < 1$ and $\frac{2}{3}\left(\frac{3}{4}\right) = \frac{1}{2} < 1$.

2. The product of two proper fractions is less than each of the two fractions.

   For example, $\frac{3}{4}\left(\frac{1}{2}\right) = \frac{3}{8}$. We see that $\frac{3}{8} < \frac{3}{4}$ and $\frac{3}{8} < \frac{1}{2}$.

   $\frac{1}{2}\left(\frac{1}{3}\right) = \frac{1}{6} \Rightarrow \frac{1}{6}$ is less than $\frac{1}{2}$ as well as $\frac{1}{3}$.

3. The product of two improper fractions is more than one.

   For example, $\frac{3}{2}\left(\frac{5}{4}\right) = \frac{15}{8} > 1$.

   $\frac{5}{4}\left(\frac{7}{6}\right) = \frac{35}{24} > 1$

4. The product of two improper fractions is greater than each of the fractions.

   For example, $\frac{5}{2}\left(\frac{7}{3}\right) = \frac{35}{6}$ and we note that $\frac{35}{6} > \frac{5}{2}$ also $\frac{35}{6} > \frac{7}{3}$.

5. When a proper fraction and an improper fraction are multiplied, the product obtained is greater than the proper fraction but less than the improper fraction.

For example,

(a) $\dfrac{1}{2}\left(\dfrac{5}{4}\right) = \dfrac{5}{18}$. We see that $\dfrac{1}{2} < \dfrac{5}{8} < \dfrac{5}{4}$.

(b) $\dfrac{2}{3}\left(\dfrac{7}{5}\right) = \dfrac{14}{15}$. We see that $\dfrac{2}{3} < \dfrac{14}{15} < \dfrac{7}{5}$.

Division of Fractions

1. **Division of whole number by a fraction:** Suppose you have ₹4. You have to give half-a-rupee to each of some children. Can you guess to how many children the rupee can be given?

   ₹4 ÷ ₹$\dfrac{1}{2}$ or $4 \div \dfrac{1}{2}$ gives you the answer.

   We observe that there are two half parts in one, four half parts in two, six half parts in three and eight half parts in 4.

   Therefore, $4 \div \dfrac{1}{2} = 8$.

   We can also write the above as:

   $4 \div \dfrac{1}{2}$ as $4\left(\dfrac{2}{1}\right) = 8$.

   Thus, we find that the division sign is replaced by a multiplication sign and the fraction $\dfrac{a}{b}$ is replaced by the fraction $\dfrac{b}{a}$.

   For example, $6 \div \dfrac{1}{3} = 6\left(\dfrac{3}{1}\right) = 18$.

   $12 \div \dfrac{3}{4} = 12\left(\dfrac{4}{3}\right) = 16$.

   Multiplicative inverse of a fraction: Thus, the multiplicative inverse of $\dfrac{1}{2}$ is 2 and that of $\dfrac{1}{3}$ is 3.

   Multiplicative inverse of a number is also known as its reciprocal. The product of a number and its reciprocal is one.

   For example, $4\left(\dfrac{1}{4}\right) = 1$. As the multiplicative inverse of $\dfrac{a}{b}$ is $\dfrac{b}{a}$, that of $\dfrac{b}{a}$ is $\dfrac{a}{b}$, i.e., if the multiplicative inverse of x is y and that of y is x.

   When the product of two non-zero numbers is one, then each of them is the reciprocal of the other.

2. **Division of a fraction by a whole number:** We multiply the fraction by the multiplicative inverse of the whole number.

   For example, $\dfrac{1}{2} \div 3$ can be written as $\dfrac{1}{2} \div \dfrac{3}{1} = \dfrac{1}{2}\left(\dfrac{1}{3}\right) = \dfrac{1}{6}$.

3. Division of a fraction by another fraction: We multiply the first fraction by the multiplicative inverse of the second.

For example,

(i) $\dfrac{2}{9} \div \dfrac{3}{4} = \dfrac{2}{9}\left(\dfrac{4}{3}\right) = \dfrac{8}{27}$

(ii) $\dfrac{5}{6} \div \dfrac{7}{9} = \dfrac{5}{6}\left(\dfrac{9}{7}\right) = \dfrac{45}{42} = \dfrac{15}{14}$

Decimal Numbers

Consider the fractions $\dfrac{1}{2}, \dfrac{1}{4}$ and $\dfrac{1}{8}$.

$\dfrac{1}{2}$ can be written as $\dfrac{1(5)}{2(5)} = \dfrac{5}{10}$ and $\dfrac{5}{10}$ can be written as 0.5.

Similarly, $\dfrac{1}{4} = \dfrac{1(25)}{4(25)} = \dfrac{25}{100}$. This can be written as 0.25.

$\dfrac{3}{100}$ can be written as 0.03 and $\dfrac{35}{10}$ can be written as 3.5.

0.5, 0.25, 0.03 and 3.5 are some examples of decimal numbers. A fraction in which the denominator is 10, 100, 1000, … (i.e., a power of 10) is called a *decimal fraction*. When the denominator is indicated by placing a point in the numerator, the expression is called a decimal.

In lower classes, we learnt about place value, units, tens, hundreds, etc. The place value of the first digit after the decimal point is $\dfrac{1}{10}$ and that of the second digit is $\dfrac{1}{100}$ and so on.

For example, 254.36 can be written in the expanded form as

$$2(100) + 5(10) + 4(1) + 3\left(\dfrac{1}{10}\right) + 6\left(\dfrac{1}{100}\right).$$

In the number 254.36,

2 is in the 100's place,

5 is in the 10's place,

4 is in the units place,

3 is in the $\dfrac{1}{10}$ th place

and 6 is in the $\dfrac{1}{100}$ th place.

The number 254.36 is read as two hundred and fifty four point three, six. All the digits after the decimal point are read as one digit at a time.

Similarly, $48.352 = 4(100) + 8(10) + 3\left(\dfrac{1}{10}\right) + 5\left(\dfrac{1}{100}\right) + 2\left(\dfrac{1}{1000}\right)$. This is read as forty eight point three, five, two.

Comparison of decimal numbers

Consider the decimal numbers 12.34 and 12.41. Which of the two is greater?

We have to first compare the digits on the left hand side of the decimal point, starting from the left most digit.

In the given two numbers, 12 is common. Now the first digit after the decimal has to be considered. In 12.41, 4 is the first digit to the right of the decimal (the digit in the tenth place) and in 12.34, 3 is the first digit to the right of the decimal (the digit in the tenth place).

4 > 3

∴ 12.41 > 12.34

In the same way,

32.45 < 32.54

24.01 < 24.10

Addition of decimals

Decimal numbers are added in the following way.

For example, the sum of 25.36 and 34.52 can be written as:

25.36
34.52
59.88

The sum of 123.56 and 38.23 can be written as:

123.56
38.23
161.79

Note When we add two decimal numbers or subtract one decimal number from the other we have to ensure that the digit in the units place of the second number is written exactly below the digit in the units place of the first number. Same is the case with digits in the 100's place, 1000's place, $\frac{1}{10}$ th place, $\frac{1}{100}$ th place and so on.

For example, subtract 35.006 from 123.41.

123.410
− 35.006
   88.404

Multiplication of Decimal Numbers

Suppose you go to a market to buy 2.50 litres of milk at the rate of ₹22.50 per litre, then the total cost of milk = 2.50 (22.50). This is a situation where we need to find the product of two decimal numbers. We shall begin with a simpler example.

$$0.2 = \frac{2}{10} \text{ and } 0.3 = \frac{3}{10}$$

$$0.2 \times 0.3 = \frac{2}{10}\left(\frac{3}{10}\right) = \frac{6}{100}$$

∴ $\frac{6}{100} = 0.06$

0.2(0.3) = 0.06.

Let us try to understand the method using diagrams.

In the following diagram there are ten squares of equal area. Two of the ten squares are shaded.

We can express the shaded part as $\dfrac{2}{10}$ of the total area.

Similarly, $\dfrac{4}{10}$ of the total area can be represented as follows.

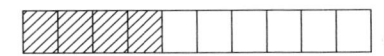

We know that $10 \times 10 = 100$ and $2 \times 4 = 8$.

In the following figure there are 100 equal squares, out of which eight are shaded.

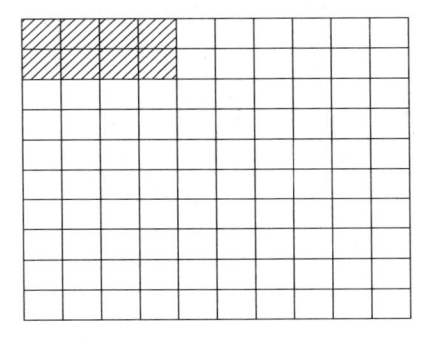

The shaded region is $\dfrac{8}{100}$ th part.

In 0.2 or 0.4 there is one digit to the right of the decimal point. In the product of these two numbers, the decimal should come before two digits from the right.

$\therefore \qquad\qquad\qquad 0.2\,(0.4) = 0.08.$

Let us consider some more examples.

$0.5\,(0.07) = 0.035$

$0.3\,(0.004) = 0.0012$

$1.3\,(1.2) = 1.56$

$1.5\,(1.2) = 1.80$

$2.34\,(10) = \dfrac{234}{100}(10) = \dfrac{234}{10} = 23.4$

$12.31\,(100) = \dfrac{1231}{100}(100) = 1231$

$0.23\,(1000) = \dfrac{23}{100}(1000) = 230$

From the above, we observe that when a decimal number is multiplied by 10, the decimal point is moved to right by one place. When a decimal number is multiplied by 100, the decimal point is shifted by two places to the right. In general, we can state that when a decimal number is multiplied by 10, 100, 1000 or any higher power of 10, the decimal point is shifted to the right, by n places, where n is the number of zeroes in the power of 10, i.e., for 10, 100, 1000 ; n is 1, 2, 3 respectively.

Division of Decimal Numbers

Your brother tells you that he purchased some onions for ₹52.50 at the rate of ₹10.50 per kg. He asks you to find out the weight of onions he purchased.

You know that for ₹10.50 he gets 1 kg. Therefore, for ₹52.50, he gets 52.50 ÷ 10.50 kg.

Here, we have come across a situation where a decimal number has to be divided by another decimal number.

Let us consider some easier examples.

1. $12.6 \div 0.6$ can be written as $\left(\dfrac{126}{10}\right) \div \left(\dfrac{6}{10}\right) = \dfrac{126}{10} \times \dfrac{10}{6} = \dfrac{126}{6} = 21.$

2. $15.5 \div 0.05 = \dfrac{155}{10} \div \dfrac{5}{100} = \dfrac{155}{10}\left(\dfrac{100}{5}\right) = \dfrac{155}{5}\left(\dfrac{100}{10}\right) = 31(10) = 310.$

3. $0.4 \div 0.004 = \dfrac{4}{10} \div \dfrac{4}{1000} = \dfrac{4}{10}\left(\dfrac{1000}{4}\right) = 100.$

Decimal numbers are divided by natural numbers in the following way.

EXAMPLE 1.6

(a) Divide 8.4 by 4.

(b) Divide 36.48 by 8.

SOLUTION

(a) $8.4 \div 4 = \dfrac{84}{10} \div 4 = \dfrac{84}{10}\left(\dfrac{1}{4}\right) = \dfrac{84}{4}\left(\dfrac{1}{10}\right) = \dfrac{21}{10} = 2.10$

(b) $\dfrac{36.48}{8} = \dfrac{3648}{100}\left(\dfrac{1}{8}\right) = \dfrac{3648}{8}\left(\dfrac{1}{100}\right) = 456\left(\dfrac{1}{100}\right) = 4.56$

Decimal numbers are divided by 10, 100, 1000, etc., in the following way.

EXAMPLE 1.7

(a) Divide 23.4 by 10.

(b) Divide 23.4 by 100.

SOLUTION

(a) $\dfrac{23.4}{10} = \dfrac{234}{10}\left(\dfrac{1}{10}\right) = \dfrac{234}{100} = 2.34$

(b) $\dfrac{23.4}{100} = \dfrac{234}{10}\left(\dfrac{1}{100}\right) = \dfrac{234}{1000} = 0.234$

Note When we divide a decimal number by 10, 100 or 1000, the decimal point in the quotient shift towards left by n places, where n is the number of zeroes in the divisor, i.e., 10 or 100 or 1000.

EXAMPLE 1.8

A fruit vendor purchased 125 oranges for ₹165. Find the cost of one orange.

SOLUTION

Cost of 125 oranges = ₹165

Cost of 1 orange = ₹ $\dfrac{165}{125}$

∴ Cost of one orange = ₹1.32.

Rational Numbers

So far we have discussed integers, fractions, decimals and the basic arithmetic operations on these numbers.

Sometimes we face certain situations, where the understanding of numbers we have already studied may not be sufficient.

For example, the temperature at Jammu on a certain day is $5\dfrac{1}{2}$ below zero degrees. The temperature there is $-5\dfrac{1}{2}$. This number is neither an integer nor a fraction. It is a rational number.

A number that can be expressed in the form of $\dfrac{p}{q}$, where $q \neq 0$ and p and q are integers is a rational number.

$\dfrac{1}{2}, \dfrac{3}{4}, \dfrac{5}{1}, \dfrac{0}{1}, \dfrac{-7}{2}$ are some examples of rational numbers.

We observe that all integers can be expressed in the form of $\dfrac{p}{q}$ ($q \neq 0$).

If p is an integer, it can be written as $\dfrac{p}{1}$.

We observe the following:

1. All integers are rational numbers.
2. All fractions are rational numbers.
3. Decimal numbers that can be written in the form of $\dfrac{p}{q}$ are rational numbers.

For example, 0.3, 0.7, 0.25, 0.125 can be written as $\dfrac{3}{10}, \dfrac{7}{10}, \dfrac{25}{100}, \dfrac{125}{1000}$ respectively.

Equivalent Rational Numbers

A rational number $\dfrac{2}{3}$ can be written as $\dfrac{2(2)}{3(2)} = \dfrac{4}{6}$ or $\dfrac{2(3)}{3(3)} = \dfrac{6}{9}$ or $\dfrac{2(4)}{3(4)} = \dfrac{8}{12}$ and so on.

The numbers $\dfrac{2}{3}, \dfrac{4}{6}, \dfrac{6}{9}, \dfrac{8}{12}$ are examples of equivalent rational numbers.

When the numerator and the denominator of a rational number is multiplied by the same non-zero integer, we get an equivalent rational number. A rational number has infinite equivalent rational numbers.

Standard form of a Rational Number

1. A rational number is expressed in its simplest form.

   For example, $\frac{15}{20}$ is expressed as $\frac{3}{4}$.

   Similarly, $\frac{20}{60} = \frac{1}{3}$; $\frac{45}{55} = \frac{9}{11}$, etc.

2. A rational number $\frac{p}{-q}$ is expressed as $\frac{-p}{q}$.

   For example, $\frac{2}{-3}$ is written as $\frac{-2}{3}$.

   Similarly, $\frac{2}{-3} = \frac{2(-1)}{-3(-1)} = \frac{-2}{3}$; $\frac{5}{-6} = \frac{5(-1)}{-6(-1)} = \frac{-5}{6}$, etc.

3. A rational number whose numerator as well as denominator is negative is expressed as a positive rational number.

   For example, $\frac{-5}{-8} = \frac{-5(-1)}{-8(-1)} = \frac{5}{8}$.

Representation of a Rational Number on the Number Line

Earlier, we have discussed representing integers on a number line.

All positive numbers are represented on the right side of zero and all negative numbers are represented on the left side of the zero on the number line. Let us say 2 is represented on the right side of zero and 2 units away from zero.

Then −2 is represented on the left side of zero but 2 units away from zero. Therefore, 2 and −2 are equidistant from zero. In the same way, $\frac{3}{2}$ and $\frac{-3}{2}$ are equidistant from zero.

We can understand this by studying the following number line.

```
        −3/2                    3/2
         |                       |
◄──┬────┬────┬────┬────┬────┬────┬──►
  −3   −2   −1    0    1    2    3
```

Comparison of Rational Numbers

We already have an idea on comparison of integers, fractions and decimal numbers. We shall discuss comparison of rational numbers.

1. Any positive rational number is greater than any negative rational number.

   For example, $\frac{1}{2} > \frac{-3}{2}$; $\frac{3}{4} > \frac{-50}{4}$, etc.

2. Zero is less than any positive rational number but greater than any negative rational number.

   For example, −3/4 < 0 < 3/4, − 10 < 0 < 1/2, etc.

3. We can compare two positive rational numbers because we have already learned to compare fractions like 5/12 and 7/18.

4. Comparison of two negative rational numbers −3/4 and −1/2.

We know that $\dfrac{3}{4} > \dfrac{1}{2}$.

$\therefore \dfrac{-3}{4} < \dfrac{-1}{2}$.

Inserting Rational Numbers Between Two Given Rational Numbers

Have you noticed how many integers are between 10 and 15? 11, 12, 13 and 14, i.e., four integers are between 10 and 15.

Integers between −9 and −5 are −8, −7 and −6.

The number of integers between any two integers is finite.

Between any two rational numbers, there are infinite rational numbers.

Let us consider two rational numbers 1/2 and 1/5.

$$\dfrac{1}{2} = \dfrac{1(5)}{2(5)} = \dfrac{5}{10}$$

$$\dfrac{1}{5} = \dfrac{1(2)}{5(2)} = \dfrac{2}{10}$$

3/10, 4/10 are two rational numbers between 2/10 and 5/10, i.e., 1/5 and 1/2.

We can insert rational numbers between 3/10 and 4/10.

$$\dfrac{3}{10} = \dfrac{3(10)}{10(10)} = \dfrac{30}{100}$$

$$\dfrac{4}{10} = \dfrac{4(10)}{10(10)} = \dfrac{40}{100}$$

$\dfrac{31}{100}, \dfrac{32}{100} \ldots \ldots \dfrac{39}{100}$ are some rational numbers between $\dfrac{3}{10}$ and $\dfrac{4}{10}$.

In this way, we observe that infinite rational numbers can be inserted between any two rational numbers.

Addition of Rational Numbers

Let us consider some rational numbers having the same denominator.

For example,

1. $\dfrac{-3}{8} + \dfrac{5}{8} = \dfrac{-3+5}{8} = \dfrac{2}{8} = \dfrac{1}{4};$

2. $\dfrac{8}{11} + \dfrac{(-3)}{11} = \dfrac{8+(-3)}{11} = \dfrac{5}{11};$

3. $\left(\dfrac{-2}{7}\right) + \left(\dfrac{-3}{7}\right) = \dfrac{(-2)+(-3)}{7} = \dfrac{-5}{7}$

Let us now consider the rational numbers with different denominators.

For example,

1. $\dfrac{3}{4} + \left(\dfrac{-5}{8}\right) = \dfrac{6}{8} + \left(\dfrac{-5}{8}\right) = \dfrac{6-5}{8} = \dfrac{1}{8}$

Since, the denominator of −5/8 is 8 and the denominator of 3/4 is 4, we try to write $\frac{3}{4}$ as an equivalent to a fraction whose denominator is 8.

That is why we write $\frac{3}{4}$ as $\frac{3(2)}{4(2)} = \frac{6}{8}$. We observe that the LCM of 4 and 8 is 8.

2. $\frac{5}{12} = \frac{5(3)}{12(3)} = \frac{15}{36}$ and $\frac{-7}{18} = \frac{-7(2)}{18(2)} = \frac{-14}{36}$ (∵ The LCM of 12 and 18 is 36)

$\frac{5}{12} + \frac{-7}{18} = \frac{15}{36} + \frac{-14}{36} = \frac{15-14}{36} = \frac{1}{36}$

A rational number can be subtracted from another in the same way.

For example,

1. $\frac{3}{22} - \frac{5}{11} = \frac{3}{22} - \frac{10}{22} = \frac{3-10}{22} = \frac{-7}{22}$;

2. $\frac{-3}{4} - \frac{7}{20} = \frac{-3 \times 5}{4 \times 5} - \frac{7}{20} = \frac{-15}{20} - \frac{7}{20} = \frac{-15-7}{20} = \frac{-22}{20} = \frac{-11}{10}$;

3. $\frac{2}{5} - \left(\frac{-8}{15}\right) = \frac{2}{5} + \frac{8}{15} = \frac{6}{15} + \frac{8}{15} = \frac{6+8}{15} = \frac{14}{15}$.

Multiplication of Rational Numbers

Rational numbers are multiplied in the following way.

1. The numerators of the rational numbers are multiplied. Let it be *a*.
2. The denominators of the rational numbers are multiplied. Let it be *b*.
3. The product of the rational numbers is *a/b*.

For example,

1. The product of −7/12 and 5/6 is −35/72   (∵ −7 (5) = −35 and 12 (6) = 72)

2. The product of −15/49 and 7/5 is $\frac{-105}{245} = \frac{-3}{7}$ (∵ −15 (7) = −105 and 49 (5) = 245)

Division of Rational Numbers

Let us recall the division of fractions. Dividing a fraction *a* by another fraction *b* is the same as multiplying *a* and 1/*b*.

1/*b* is the reciprocal of *b* or the multiplicative inverse of *b*.

For example,

1. $\frac{3}{4} \div \left(\frac{-2}{3}\right)$

$= \frac{3}{4}\left(\frac{3}{-2}\right) = \frac{3(3)}{4(-2)} = \frac{9}{-8} = \frac{-9}{8}$

2. $\frac{-15}{7} \div \frac{2}{9} = \frac{-15}{7}\left(\frac{9}{2}\right) = \frac{-15(9)}{7(2)} = \frac{-135}{14}$

Decimal Representation of Rational Numbers

Every rational number can be expressed as a terminating decimal or a repeating decimal.

A terminating decimal is a decimal that contains finite number of digits.

0.5, 0.25, 0.4 are some examples of terminating decimals.

Repeating Decimal

A repeating decimal (or a recurring) is a decimal in which one digit or a block of digits repeat (or occur again and again) indefinitely.

0.333………., 0.1666………..

0.454545 ……….. are some examples of repeating decimals.

In 0.333 ……….., only one digit, i.e., 3 is repeating indefinitely.

In 0.454545 ……….., a block of digits, i.e., 45 is repeating indefinitely.

Conversion of a Decimal Number into a Fraction

Every terminating decimal and every non-terminating repeating decimal can be expressed in the form of p/q.

For example,

1. $0.37 = 37/100$
2. $1.436 = 1436/1000 = 359/250$
3. $0.333……… = 0.\overline{3}$

   Since, only one digit is repeating after the decimal, we say that the periodicity is one.

   Let $x = 0.333 ………$ (1)

   Multiply both sides with 10.

   $10x = (0.333…)10 = 3.333 …$ (2)

   $(2) - (1) \Rightarrow$

   $(10x - x) = (3.33…………….) - (0.33 ……………..)$

   $\Rightarrow 9x = 3 \Rightarrow x = 3/9 = 1/3$

4. Express $1.231231………$ in the form of p/q.

   Let $x = 1.231231$.

   $1000x = 1231.231$

   $x = 1.231 ………$

   $(1000x - x) = 1230 \Rightarrow 999x = 1230 \Rightarrow x = 1230/999 = 410/333$.

   Note   Recurring decimal $= \dfrac{A - B}{10^C - 10^D}$

Where A = The whole number obtained by writing digits in order.

        B = The whole number made by non–recurring digits in order.

        C = The number of digits after the decimal point.

        D = The number of digits after decimal point that are not recurring.

EXAMPLE 1.9

Express $2.3\overline{45}$ in the form of p/q where, p and q are coprimes.

SOLUTION

$2.3\overline{45} = \dfrac{2345 - 23}{10^3 - 10^1} = \dfrac{2322}{990} = \dfrac{129}{55}.$

Irrational Numbers

There are some decimal numbers which are neither terminating nor recurring. They are irrational numbers.

1.2354 ....,

3.1415 ....

1.414 ...., 1.732 .... are some examples of irrational numbers.

Square Roots

When a whole number is multiplied with itself once, then the product is the square of the number.

$a(a) = a^2$. Here a^2 is the square of a and a is the square root of a^2.

$2(2) = 4$. The number 4 is the square of 2 and 2 is the square root of 4.

$3(3) = 9$. The number 9 is the square of 3 and 3 is the square root of 9. Also $0(0) = 0$ and $1(1) = 1$.

We write $\sqrt{4} = 2$, $\sqrt{9} = 3$, $\sqrt{16} = 4$, $\sqrt{0} = 0$ and $\sqrt{1} = 1$.

The symbol $\sqrt{}$ is used to denote square root.

We know that $1^2 = 1$ and $2^2 = 4$. $\therefore$ 1 and 2 are the square roots of 1 and 4 respectively.

The square roots of 2 and 3 are not integers.

$1 < \sqrt{2} < 2$ and $1 < \sqrt{3} < 2$. Since, there are no integers between 1 and 2, the square root of 2 and 3 must lie between 1 and 2. Let us try to find the square root of 2 by the method of approximation.

Let us take the average of 1 and 2.

$1 + 2/2 = 1.5$ and $(1.5)^2 = 2.25$.

While $(1.4)^2 = 1.96$.

As $(1.4)^2 < 2 < (1.5)^2$, it follows that $1.4 < \sqrt{2} < 1.5$.

Next, by trial, we see that $1.41^2 = 1.9881$ and $1.42^2 = 2.0614$.

$\therefore 1.41 < \sqrt{2} < 1.42$.

Again by trial, $1.414^2 = 1.999396$ and $1.415^2 = 2.002225$.

$\therefore 1.414 < \sqrt{2} < 1.415$.

It turns out that the decimal number which is approximately equal to the value of $\sqrt{2}$ is neither terminating nor repeating. It is irrational.

Similarly, $\sqrt{3}, \sqrt{5}, \sqrt{6}$, all these numbers are irrational numbers.

All rational numbers and all irrational numbers are together known as real numbers.

Properties of Square Roots

1. A positive number has two square roots, one positive and the other negative. Their magnitudes are equal but signs are different. For example, square root of 4 = +2 or −2 and square root of 9 = +3 or −3.

2. The square root of a negative number is not a real number.

3. For all $a > 0$, $(\sqrt{a})^2 = a$.

   For example, $\sqrt{-4}$ is not a real number. It cannot be represented on the number line.

4. If a and b are two real numbers, $\sqrt{ab} = \sqrt{a}\sqrt{b}$ and $\sqrt{a/b} = \sqrt{a}/\sqrt{b}$.

   Note If $x = a \times a \times a = a^3$ (where, a is a whole number), then x is called a perfect cube and a is called the cube root of x.

   For example, $8 = 2 \times 2 \times 2 = 2^3$, therefore 8 is a perfect cube.

   2 is the cube root of 8.

   Note The sign of cube root is $\sqrt[3]{}$.

EXAMPLE 1.10

Find the cube root of 216.

SOLUTION

$216 = 2^3 \times 3^3 = (2 \times 3)^3 = 6^3$.

∴ Cube root of 216 is $\sqrt[3]{6^3} = 6$.

EXAMPLE 1.11

What is the least positive integer by which 50 should be multiplied so that the product is a perfect square?

SOLUTION

$$50 = 2 \times 5^2$$

∴ The required number is 2.

$$(2 \times 50 = 100 = 10^2)$$

EXAMPLE 1.12

Find the value of $\sqrt{20\frac{1}{4}} - \sqrt{1\frac{32}{49}}$.

SOLUTION

$$\sqrt{20\frac{1}{4}} = \sqrt{\frac{20 \times 4 + 1}{4}} = \sqrt{\frac{81}{4}} = \frac{9}{2}$$

$$\sqrt{1\frac{32}{49}} = \sqrt{\frac{1 \times 49 + 32}{49}} = \sqrt{\frac{81}{49}} = \frac{9}{7}$$

$$\sqrt{20\frac{1}{4}} - \sqrt{1\frac{32}{49}} = \frac{9}{2} - \frac{9}{7}$$

$$= 9\left(\frac{1}{2} - \frac{1}{7}\right) = 9\left(\frac{5}{14}\right) = \frac{45}{14}.$$

Number Systems 1.23

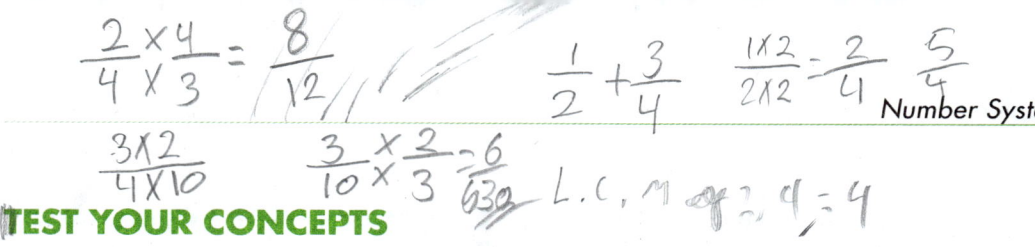

TEST YOUR CONCEPTS

Very Short Answer Type Questions

Directions for questions 1 to 9: State whether the following statements are true or false.

1. The multiplicative inverse of zero does not exist.
2. Division by zero is not defined.
3. Division is commutative for integers.
4. The sum of any two proper fractions is an improper fraction.
5. The product of a proper and an improper fraction is less than the improper fraction.
6. $2.3458 \times 1000 = 234.58$
7. $34/10000 = 0.0034$
8. Irrational numbers cannot be represented on the number line.
9. The square root of a negative number is irrational.

Directions for questions 10 to 19: Fill in the blanks.

10. The additive identity element in the set of natural numbers _____.
11. A fraction in which the numerator is greater than the denominator is called _____.
12. The sum of an integer and a proper fraction is _____ fraction (proper/mixed).
13. The product of two proper fractions is always _____ fraction.
14. $0.2 \times 0.3 =$ _____.
15. The decimal form of a proper fraction whose denominator is a multiple of 3 is a _____ decimal.
16. $1.999999 \ldots\ldots\ldots =$ _____.
17. Non-repeating and non-terminating decimals are called _____.
18. $\sqrt{2} =$ _____. (up to 3 places of decimals)
19. $\sqrt{3} =$ _____. (up to 3 places of decimals)

Directions for questions 20 to 30: Select the correct alternative from the given choices.

20. What is the multiplicative identity element in the set of whole numbers?
    (a) 0 (b) –1
    (c) 1 (d) None

21. Which of the following sets is not closed under subtraction?
    (a) N (b) Z
    (c) Q (d) R

22. What do you call two fractions, whose product is 1?
    (a) Additive inverse to each other.
    (b) Multiplicative inverse to each other.
    (c) Reciprocals to each other.
    (d) Both (b) and (c)

23. Period of the decimal $2.3636363\ldots\ldots$ is ____.
    (a) 2 (b) 3
    (c) 36 (d) 63

24. Periodicity of $981.7\overline{836}$ is _____.
    (a) 8 (b) 836
    (c) 3 (d) 6

25. What is the smallest positive irrational number?
    (a) $\sqrt{2}$ (b) $\sqrt{3}$
    (c) $\sqrt{\dfrac{1}{2}}$ (d) Not defined

26. What is the multiplicative identity element in the set of integers?
    (a) 0 (b) 1
    (c) –1 (d) None of these

27. Find the HCF of 17 and 19.
    (a) 1 (b) 2
    (c) 17 (d) 19

28. $\sqrt{225} + \sqrt[3]{\dfrac{1}{64}} =$ _____.
    (a) $15\dfrac{1}{4}$ (b) $15\dfrac{1}{8}$
    (c) $15\dfrac{1}{2}$ (d) $15\dfrac{1}{16}$

Chapter 1

29. The LCM of two numbers is 30 and the product of two numbers is 150. Find the HCF.
    (a) 3 (b) 5
    (c) 10 (d) 15

30. If $m = (-1)^{2000}$ and $n = (-1)^{2002}$, then find the value of $\dfrac{m}{n}$.
    (a) −1 (b) 1
    (c) 2000 (d) 2002

Short Answer Type Questions

31. Arrange the following decimal numbers in the ascending order.
    0.52314, 0.52313, 0.53201 and 0.52321

32. Find the product of $3\dfrac{1}{7} \times 1\dfrac{5}{6} \times 1\dfrac{2}{5} \times 1\dfrac{1}{11}$.

33. Convert the following terminating decimals as fractions.
    (i) 2.3675 (ii) 54.26
    (iii) 75.35 (iv) 0.7575

34. Simplify:
    (i) 22.308 ÷ 7.436 (ii) $\dfrac{837}{125} \div \dfrac{558}{4750}$

35. Add the following rational numbers.
    $\dfrac{3}{10}, \dfrac{0.4}{100}, \dfrac{-21}{1000}, \dfrac{7}{10000}$

36. Which of the following fractions are recurring decimals?
    (i) $\dfrac{237}{625}$ (ii) $\dfrac{125}{192}$
    (iii) $\dfrac{1000}{1024}$ (iv) $\dfrac{2008}{2560}$

37. Find the value of 30.32 × 4.5.

38. Convert the following improper fractions into mixed fractions.
    (i) 23/12 (ii) 37/8 (iii) 108/52

39. Find the HCF and the LCM for the following sets of numbers.
    (i) 24, 36 (ii) 16, 20, 48 (iii) 25, 35, 40

40. The LCM and the HCF of two numbers are 625 and 5 respectively. Find the product of the two numbers.

41. Convert the following decimals into p/q form. ($p, q \in Z$)
    (i) 2.345 (ii) 3.5 $\overline{2}$
    (iii) 281.$\overline{31}$ (iv) 108.$\overline{001}$

42. Find any three rational numbers between 5/8 and 7/12.

43. Arrange the following fractions in the descending order.
    10/12, 13/15, 21/25 and 43/45

44. Find the difference between 3.47777…… and 2.85888…….

45. Show that $2.\overline{9} = 3$.

46. Find the value of $3.\overline{4} \times 5.\overline{6}$.

47. The LCM and the HCF of two numbers are 96 and 12 respectively. If one of the numbers is 24, then find the other number.

48. Find the least positive integer that should be multiplied to 720 so that the product obtained is a perfect square.

49. Krishna covers a certain distance in 150 minutes. He covers half of the distance in 4/15 of the time. Find the time taken to cover the remaining distance.

50. A family requires 2.2 litres of milk every day. Find the total quantity of milk required in a month of 31 days.

51. A is the smallest four-digit number formed by using all the digits 0, 1, 2 and 3. B is the greatest four-digit number formed by using all the digits 0, 1, 2 and 3. Find B − A.

52. Find the greatest number that can divide 76 and 60 leaving remainders of 4 and 6 respectively.

53. Simplify: 44.4444…… − 27.2727…. =

54. Simplify: $\sqrt{2\dfrac{14}{25}} - \sqrt{1\dfrac{7}{9}}$.

55. Simplify: $\sqrt{2.42} \times \sqrt{2.88}$.

Essay Type Questions

56. Find the least three-digit number which when divided by 20, 30, 40 and 50 leaves a remainder 10 in each case.

57. Find the value of $\sqrt{5}$ up to 2 places of decimals.

58. In a class, there are 72 boys and 64 girls. If the class is to be divided into the least number of groups such that each group contains either only boys or only girls, then how many groups will be formed?

59. Find the largest number that divides 59 and 54 leaving remainders 3 and 5 respectively.

60. Two bells toll at intervals of 120 sec and 180 sec respectively. If they toll together at 12 noon, what is the earliest time at which they toll together?

61. A man purchased a plot which is in the shape of a square. The area of the plot is 12 hectares 3201 m². Find the length of each side of the plot.

62. Find the divisor, given that the dividend is 2200, remainder is 13 and the divisor is one-third of the quotient.

63. A certain number of men went to a hotel. Each man spent as many rupees as one-fourth of the men. If the total bill paid was ₹20449, then how many men visited the hotel?

64. Find the value of $\sqrt[3]{288} \times \sqrt[3]{432} \times \sqrt[3]{648}$.

65. Volume of a cube is given by the formula, $V = s^3$, where s is the length of the side. Find the side of the cube if its volume is 10.648 cubic units.

CONCEPT APPLICATION

Level 1

Directions for questions 1 to 17: Select the correct alternative from the given choices.

1. What is the additive identity element in the set of whole numbers?
   (a) 0
   (b) 1
   (c) −1
   (d) None of these

2. Which of the following is/are true?
   (a) $\sqrt{6} \times \sqrt{6}$ is an irrational number.
   (b) $\sqrt{4} \times \sqrt{25}$ is a rational number.
   (c) Both (a) and (b)
   (d) Neither (a) nor (b)

3. M and N are two coprimes. Which of the following is/are true?
   (a) LCM (M, N) = M × N
   (b) HCF (M, N) = 1
   (c) Both (a) and (b)
   (d) Neither (a) nor (b)

4. If $a = (-1)^{2009}$ and $b = (-1)^{2010}$, then find the value of ab.
   (a) 1
   (b) −1
   (c) 2009
   (d) 2010

5. $\sqrt{X}$ is a perfect square. Which of the following is/are true?
   (a) X is a perfect square.
   (b) X^2 is a perfect square
   (c) Neither (a) nor (b)
   (d) Both (a) and (b)

6. $\sqrt[3]{\dfrac{125}{216}} - \sqrt{\dfrac{25}{36}} = $ _____.
   (a) 5/6
   (b) 1/6
   (c) 0
   (d) 1

7. The HCF of two numbers is 6 and the product of two numbers is 360. Find the LCM of the numbers.
   (a) 60
   (b) 30
   (c) 12
   (d) 6

8. If M and N are two natural numbers, then $\dfrac{LCM(M,N)}{HCF(M,N)}$ is
   (a) An integer.
   (b) A rational number which is not necessarily an integer.

(c) A real number which is not necessarily an integer.

(d) None of these.

9. X is the smallest four-digit number formed by all the digits 0, 1, 2 and 3. Find X.

(a) 123
(b) 1023
(c) 1000
(d) 102

10. Find the least natural number which when divided by 6, 9, 12 and 18 leaves no remainders.

(a) 36
(b) 72
(c) 12
(d) 18

11. If $r^3 = 1728$, then find r.

(a) 12
(b) 14
(c) 16
(d) 18

12. If two numbers are equal, then

(a) Their LCM is equal to their HCF.
(b) Their LCM is less than their HCF.
(c) Their LCM is equal to two times their HCF.
(d) None of these

13. X is a perfect square. Which of the following is necessarily being true?

(a) X^2 is a perfect square
(b) $\sqrt{X}$ is a perfect square
(c) Both (a) and (b)
(d) Neither (a) nor (b)

14. The following are the steps involved in converting $0.\overline{23}$ into p/q form where $q \neq 0$. Arrange them in sequential order from the first to the last.

(A) $\therefore 100x = 23.23\ 23\ 23.$ (2)
(B) As periodicity is 2, multiply the equation (1) with 100.
(C) $\therefore 99x = 23 \Rightarrow x = \dfrac{23}{99}.$
(D) Let $x = 0.\overline{23} = 0.23\ 23\ 23.$ (1)
(E) Subtract equation (1) from equation (2).

(a) DBEAC
(b) DBAEC
(c) DBCAE
(d) DABEC

15. The following are the steps involved in finding the positive value of x from the equation $x^2 = 12.96$. Arrange them in sequential order from the first to the last.

(A) $x^2 = \dfrac{(36)^2}{10^2}$
(B) $\therefore x = 3.6$
(C) $x^2 = 12.96 = \dfrac{1296}{100}$
(D) $x = \dfrac{36}{10}$

(a) CDAB
(b) CABD
(c) CADB
(d) CDBA

16. The following are the steps involved in finding the positive value of x from the equation $x^2 = \dfrac{8.1}{36.1}$. Arrange them in sequential order from the first to the last.

(A) $x = \dfrac{9}{19}$
(B) $x^2 = \dfrac{9^2}{(19)^2}$
(C) $x^2 = \dfrac{8.1}{36.1} = \dfrac{81}{361}$

(a) ACB
(b) BAC
(c) ABC
(d) CBA

17. The following are the steps involved in converting $1.\overline{52}$ into p/q form where, $q \neq 0$. Arrange them in sequential order from the first to last.

(A) $\therefore 10x = 15.2222.$ (2)
(B) Let $x = 1.\overline{52} = 1.5\ 2222.$ (1)
(C) $9x = 13.7 \Rightarrow x = \dfrac{13.7}{9} = \dfrac{137}{90}.$
(D) As periodicity is 1, multiply the equation (1) with 10.
(E) Subtract equation (1) from the equation (2).

(a) BDAEC
(b) DBAEC
(c) DBCEA
(d) BDCEA

Directions for questions 18 to 21: Match Column A with Column B.

Column A		Column B
18. $(-22) + 21 + (-22) + 21 +$ ------(40 terms) = ___ ( )		(a) 75.6
19. $0.756 \times 100 =$ ___ ( )		(b) 20
20. $75.6 \div 10 =$ ___ ( )		(c) −1
21. $(9)(-1/3)(-3)(-1/9) =$ ___		(d) 1
		(e) −20
		(f) 7.56

Number Systems 1.27

Directions for questions 22 to 25: Match Column A with Column B.

Column A		Column B
22. (−2) (−3) (6) (−1)	( )	(a) 2.5
23. 0.25 ÷ 100 =	( )	(b) 36
24. 0.025 × 100 =	( )	(c) −36
25. 86 + (−28) + 12 + (−34)	( )	(d) 0.0025

Level 2

Directions for questions 26 to 65: Select the correct alternative from the given choices.

26. If the sum of two integers is −26 and one of them is 14, then find the other integer.
    (a) −12 (b) 12
    (c) −40 (d) 40

27. Which of the following pairs of integers have 5 as a difference?
    (a) 10, 5 (b) −10, −5
    (c) 15, −20 (d) Both (a) and (b)

28. If the product of two integers is 72 and one of them is −9, then find the other integer.
    (a) 8 (b) −8
    (c) 81 (d) 63

29. Simplify: 29 × 76 − 71 × 29
    (a) 148 (b) 147
    (c) 146 (d) 145

30. A has to travel a certain distance. If he travels three-fifth of the distance on a day and the rest the next day, then what part of the distance has he travelled on the second day?
    (a) 3/5 (b) 2/5
    (c) 1/5 (d) 4/5

31. If the lengths of two poles P_1 and P_2 are 26.79 m and 29.34 m respectively, then how much longer is P_2 than P_1?
    (a) 2.45 m (b) 2.35 m
    (c) 2.55 m (d) 2.65 m

32. Which of the following fractions lies between 2/3 and 5/7?
    (a) 3/4 (b) 4/5
    (c) 5/6 (d) None of these

33. Find the greatest number that exactly divides 81, 144 and 162.
    (a) 9 (b) 27
    (c) 3 (d) 81

34. Find the least number which when divided by 24, 36 and 48 leaves zero as its remainder.
    (a) 124 (b) 144
    (c) 164 (d) 224

35. If $p = (-1)^{205}$ and $q = (-1)^{202}$, then $p + q$ is
    (a) $(-1)^{407}$ (b) $(-1)^4$
    (c) 0 (d) None of these

36. $\sqrt[3]{\dfrac{27}{125}} + \sqrt{\dfrac{4}{25}} = $ _____.
    (a) 4 (b) 3
    (c) 2 (d) 1

37. Which of the following is true?
    (a) $\sqrt{a} + \sqrt{b} = \sqrt{a+b}$ (b) $\sqrt{a}.\sqrt{b} = \sqrt{ab}$
    (c) $\sqrt{a} - \sqrt{b} = \sqrt{a-b}$ (d) None of these

38. The period of the decimal number $10.23\overline{46}$ is _____.
    (a) 23 (b) 46
    (c) 10 (d) 1023

39. Which of the following is/are recurring decimals?
    (a) 1/3 (b) 2/7
    (c) 1/5 (d) Both (a) and (b)

1.28 Chapter 1

40. If $x = \sqrt{3}$, $y = \sqrt{27}$ and $z = \sqrt{243}$, then which of the following is/are rational numbers?
    (a) xy
    (b) xz
    (c) yz
    (d) All of these

41. $0.\overline{3} + 0.\overline{45} =$ _____.
    (a) $0.\overline{75}$
    (b) $0.\overline{48}$
    (c) $0.\overline{76}$
    (d) $0.\overline{78}$

42. $0.1\overline{2} - 0.1\overline{2} =$ _____.
    (a) $0.\overline{011}$
    (b) $0.00\overline{1}$
    (c) $0.0\overline{01}$
    (d) $0.\overline{001}$

43. $\sqrt{1\frac{9}{16}} - \sqrt{1\frac{7}{9}} =$ _____.
    (a) 1/12
    (b) 2/3
    (c) −1/12
    (d) −2/3

44. $\sqrt{3.38} \times \sqrt{3.92} =$ _____.
    (a) 1.82
    (b) 1.72
    (c) 3.64
    (d) 3.44

45. X is the smallest four-digit number formed by all the digits 0, 7, 8 and 9. Y is the greatest four-digit number formed by all the digits 0, 7, 8 and 9. Find $Y - X$.
    (a) 9081
    (b) 2781
    (c) 2777
    (d) 1890

46. $\sqrt{6.05} \times \sqrt{8.45} =$
    (a) 6.95
    (b) 7.35
    (c) 7.55
    (d) 7.15

47. Find the greatest number that can divide 101 and 115 leaving remainders 5 and 7 respectively.
    (a) 6
    (b) 9
    (c) 12
    (d) 18

48. In a school, the number of students in each section is equal to the number of sections. If the total number of students is 625, then find the number of sections.
    (a) 10
    (b) 20
    (c) 15
    (d) 25

Level 3

49. Arrange 3/4, 9/13, 12/17 and 1/2 in the ascending order.
    (a) 1/2, 9/13, 3/4, 12/17
    (b) 3/4, 9/13, 1/2, 12/17
    (c) 1/2, 3/4, 9/13, 12/17
    (d) 1/2, 9/13, 12/17, 3/4

50. The smallest of the fractions 2/3, 4/7, 8/11 and 5/9 is _____.
    (a) 2/3
    (b) 4/7
    (c) 8/11
    (d) 5/9

51. A, B and C shared a total of ₹6024. Share of A is one-third of the total money and share of B is half of the total money. Find the share of C.
    (a) ₹1004
    (b) ₹104
    (c) ₹208
    (d) ₹2008

52. The average weight of each student of a class is 32¾ kg. If there are 24 students in the class, then find the total weight of the class.
    (a) 768 kg
    (b) 786 kg
    (c) 867 kg
    (d) 876 kg

53. The HCF and the LCM of two numbers are 24 and 1008. If one of the numbers is 168, then find the other number.
    (a) 336
    (b) 252
    (c) 148
    (d) 144

54. Find the greatest number that can divide 246 and 279 by leaving remainders 2 and 3 respectively.
    (a) 4
    (b) 40
    (c) 6
    (d) 60

55. Find the greatest possible quantity which can be used to measure exactly the quantities $3\ell\ 250m\ell$, $3\ell\ 500m\ell$ and 4ℓ.
    (a) $25m\ell$
    (b) $125m\ell$
    (c) $250m\ell$
    (d) $500m\ell$

56. What is the least positive integer by which 4500 should be divided so that the quotient is a perfect cube?

(a) 6	(b) 36	(a) 40	(b) 10
(c) 2	(d) 3	(c) 20	(d) 30

57. A total of 1152 students were assembled in rows and columns. If there are n rows and $n/2$ columns, then find the number of students in each row.

    (a) 36   (b) 42
    (c) 48   (d) 34

58. The LCM of two numbers is 26, then which of the following can be their HCF?

    (a) 1   (b) 2
    (c) 13   (d) All of these

59. In a computer game, if we hit a balloon, we get 500 points and if we miss the balloon, we lose 300 points. Raj hits 20 balloons and misses 40 balloons. Find his net score.

    (a) 2000   (b) −2000
    (c) 1000   (d) −1000

60. Krishna purchased 20 pencils for his two sons Akhil and Nikhil. Akhil took two-fifth of the total number of these pencils and Nikhil took the remaining pencils. Find the number of pencils taken by Nikhil.

    (a) 8   (b) 12
    (c) 6   (d) 14

61. If $x = (-23) + 22 + (-23) + 22\ldots\ldots$ (40 terms) and $y = 11 + (-10) + 11 + (-10) + \ldots$ (20 terms), then $y - x$ is _____.

62. Ram went to a shop to purchase an article. But he had an amount which is equal to [13/27] of the cost of the article. If the cost of the article was ₹540, then find the amount with him.

    (a) ₹240   (b) ₹250   (c) ₹260   (d) ₹270

63. If x and y are the smallest and the greatest four-digit numbers formed by using 1, 3, 5 and 9, then find $y - x$.

    (a) 5940   (b) 8172
    (c) 3600   (d) 6336

64. If $x = (-1)^1 + (-1)^2 + \ldots\ldots + (-1)^{2009}$ and $y = (-1)^1 - (-1)^2 + (-1)^3 - (-1)^4 + \ldots\ldots + (-1)^{2009}$, then find $x - y$.

    (a) 2009   (b) 2008
    (c) 0   (d) 1004

65. Raj went to a market to buy a radio. But he had an amount which is equal to $\left[\dfrac{15}{28}\right]$ of the cost of the radio. If the cost of the radio is ₹560, then find the amount with him.

    (a) ₹280   (b) ₹300
    (c) ₹240   (d) ₹320

ASSESSMENT TESTS

Test 1

Directions for questions 1 to 11: Select the correct alternative from the given choices.

1. Find the divisor, given that the dividend is 789, the remainder is 5 and the quotient is one-fourth of the divisor.

   The following are the steps involved in solving the above problem. Arrange them in sequential order.

   (A) $789 = (4x \times x) + 5$.

   (B) We have, dividend = (divisor × quotient) + remainder.

   (C) Let the quotient be $x \Rightarrow$ divisor = $4x$.

   (D) $x = \sqrt{\dfrac{784}{4}} = 14$.

   (E) ∴ The divisor = $4x = 4 \times 14 = 56$.

   (a) BCDAE   (b) CBADE
   (c) BCAED   (d) CBDAE

2. In a class, each student contributed as many rupees as the number of students in the class, to the flood relief fund. If the total amount collected is ₹841.

1.30 Chapter 1

Find the number of students in the class.

The following are the steps involved in solving the above problem. Arrange them in sequential order.

(A) The total amount contributed (in ₹) = $x \times x = x^2$.

(B) Given, $x^2 = 841 \Rightarrow x = \sqrt{841} = 29$.

(C) Let the number of students in the class be x.

(D) ∴ The number of students in the class = 29.

(a) CABD   (b) BCAD
(c) BCDA   (d) CBAD

3. The area of a square field is 1444 m². Find the cost of fencing the field at the rate of ₹5 per metre.
   (a) ₹760   (b) ₹720
   (c) ₹680   (d) ₹640

4. What is the least natural number that should be added to the result of 88 × 89, so that the sum obtained is a perfect square?
   (a) 1   (b) 8
   (c) 88 (d) 89

5. Find the least number which when divided by 5, 7 and 8 leaves 3 as the remainder in each case.
   (a) 283   (b) 78
   (c) 578   (d) 57

6. $2.\overline{3} + 3.\overline{4} - 4.\overline{8} = $ _____.
   (a) $0.\overline{7}$ (b) $1.\overline{2}$

(c) $0.\overline{8}$ (d) $1.\overline{9}$

7. Find the value of $\sqrt[3]{27} \times \sqrt[3]{216} \times \sqrt[3]{64}$.
   (a) 24   (b) 45
   (c) 72   (d) 96

8. Find the largest number that divides 92 and 75 and leave the remainders 2 and 5 respectively.
   (a) 10   (b) 15
   (c) 25   (d) 30

9. In a game, if we hit a balloon, we get 300 points and if we miss the balloon, we lose 100 points. Raj hits 15 balloons and misses 40 balloons. Find his net score.
   (a) 500   (b) 400
   (c) 300   (d) 200

10. The LCM of two numbers is 420. Which of the following cannot be the HCF of the two numbers?
    (a) 70   (b) 60
    (c) 210 (d) 80

11. Which of the following fractions represent a non-terminating decimal?
    (a) $\dfrac{63}{24}$ (b) $\dfrac{18}{15}$
    (c) $\dfrac{14}{21}$ (d) $\dfrac{33}{44}$

Directions for questions 12 to 15: Match Column A with Column B.

Column A		Column B
12. The period of $3.5\overline{31}$ is a ( )		(a) Terminating decimal
13. $\dfrac{51}{96}$ can be expressed as a ( )		(b) Non-terminating and repeating decimal
14. $\dfrac{5}{24}$ can be expressed as a ( )		(c) 31
15. 3.101100111000111100011111......... is ( )		(d) Non-terminating and non-repeating decimal
		(e) 2

PRACTICE QUESTIONS

Test 2

Directions for questions 16 to 26: Select the correct alternative from the given choices.

16. Find the least possible remainder, given that the dividend is 967 and the quotient is equal to the divisor.

    The following are the steps involved in solving the above problem. Arrange them in sequential order.

    (A) $967 = x \times x + R$.

    (B) Let each of the divisor and the quotient be x. Let the remainder be R.

    (C) We have, dividend = divisor × quotient + remainder.

    (D) $x^2 \leq 967 \Rightarrow x^2$ is 961 $\therefore$ Remainder is 6.

    (a) BCDA (b) ABCD
    (c) BADC (d) BCAD

17. Convert $2.\overline{45}$ into $\dfrac{p}{q}$ form, where p and q are coprimes.

    The following are the steps involved in solving the above problem. Arrange them in sequential order.

    (A) $100x - x = 245.4545\ldots - 2.4545\ldots$

    (B) $100x = 245.4545\ldots$

    (C) Let $x = 2.\overline{45} = 2.454545\ldots$

    (D) $99x = 243$ $x = \dfrac{243}{99} = \dfrac{27}{11}$

    (a) BDCA (b) CBAD
    (c) BCDA (d) CBDA

18. The area of a square field is 1681 m². Find the cost of fencing the field at the rate of ₹3 per metre.

    (a) ₹672 (b) ₹564
    (c) ₹492 (d) ₹372

19. What is the least natural number that should be added to the product of 30 and 31, so that the sum obtained is a perfect square?

    (a) 10 (b) 3
    (c) 30 (d) 31

20. Find the least number which when divided by 9, 12 and 15, leaves 5 as the remainder in each case.

    (a) 180 (b) 50
    (c) 185 (d) 77

21. $3.\overline{4} + 5.\overline{8} - 7.\overline{9} =$ ____.

    (a) 4/3 (b) 8/5
    (c) 7/9 (d) 5/4

22. Find the value of $\sqrt[3]{512} + \sqrt[3]{343} + \sqrt[3]{729}$.

    (a) 26 (b) 24
    (c) 56 (d) 48

23. Find the largest number that divides 64 and 72 and leaves remainders 12 and 7 respectively.

    (a) 17 (b) 13
    (c) 14 (d) 18

24. In a game, if we reach a level we get 400 points and if we miss any level, we lose 300 points. Rohit reaches 20 levels and misses 30 levels. Find his net score.

    (a) 1000 (b) 500
    (c) −500 (d) −1000

25. The HCF of two numbers is 18. Which of the following cannot be their LCM?

    (a) 324 (b) 260
    (c) 648 (d) 360

26. Which of the following fractions represent a non-terminating repeating decimal?

    (a) $\dfrac{625}{128}$ (b) $\dfrac{273}{250}$
    (c) $\dfrac{750}{216}$ (d) $\dfrac{150}{300}$

Directions for questions 27 to 30: Match the Column A with Column B.

Column A		Column B
27. $\frac{3}{7}$ can be expressed as a	( )	(a) Terminating decimal.
28. $\frac{5}{2}$ can be expressed as a	( )	(b) Non-terminating and non-repeating decimal
29. 2.012341234523456… is a	( )	(c) Non-terminating and repeating decimal
30. The periodicity of $3.\overline{01}$ is	( )	(d) 2
		(e) 01

Number Systems

TEST YOUR CONCEPTS

Very Short Answer Type Questions

1. True
2. True
3. False
4. False
5. True
6. False
7. True
8. False
9. False
10. does not exist.
11. improper fraction
12. mixed
13. proper
14. 0.06
15. recurring
16. 2
17. irrational numbers
18. 1.414
19. 1.732
20. (c)
21. (a)
22. (d)
23. (c)
24. (c)
25. (d)
26. (b)
27. (a)
28. (a)
29. (b)
30. (b)

Short Answer Type Questions

31. $b < a < d < c$
32. $8\frac{4}{5}$
33. (i) $\frac{947}{400}$ (ii) $\frac{2713}{50}$
    (iii) $\frac{1507}{20}$ (iv) $\frac{303}{400}$
34. (i) 3   (ii) 57
35. 0.2837
36. $\frac{125}{192}$
37. 136.44
38. (i) $1\frac{11}{12}$ (ii) $4\frac{5}{8}$ (iii) $2\frac{1}{13}$
39. (i) 12; 72   (ii) 4; 240   (iii) 5; 1400
40. 3125
41. (i) $\frac{469}{200}$ (ii) $\frac{317}{90}$
    (iii) $\frac{27850}{99}$ (iv) $\frac{106921}{990}$
42. $\frac{59}{96}, \frac{29}{48}, \frac{57}{96}$
43. $\frac{43}{45} > \frac{23}{25} > \frac{13}{15} > \frac{10}{12}$
44. $\frac{557}{900}$
45. $19\frac{14}{27}$
47. 48
48. 5
49. 110 minutes
50. 68.21
51. 2187

Chapter 1

52. 18
53. $\dfrac{1700}{99}$
54. 4/15
55. 2.64

Essay Type Questions

56. 610
57. ~ 2.24
58. 17
59. 7
60. 12:06
61. 351 m
62. 27
63. 286
64. 432
65. 2.2 units

CONCEPT APPLICATION

Level 1

1. (a)	2. (b)	3. (c)	4. (b)	5. (d)	6. (c)	7. (a)	8. (a)	9. (b)	10. (a)
11. (a)	12. (a)	13. (a)	14. (b)	15. (c)	16. (d)	17. (a)	18. (e)	19. (a)	20. (f)
21. (c)	22. (c)	23. (d)	24. (a)	25. (b)					

Level 2

26. (c)	27. (d)	28. (b)	29. (d)	30. (b)	31. (c)	32. (d)	33. (a)	34. (b)	35. (c)
36. (d)	37. (b)	38. (b)	39. (d)	40. (d)	41. (d)	42. (c)	43. (c)	44. (c)	45. (b)
46. (d)	47. (c)	48. (d)							

Level 3

| 49. (d) | 50. (d) | 51. (a) | 52. (b) | 53. (d) | 54. (a) | 55. (c) | 56. (b) | 57. (c) | 58. (d) |
| 59. (b) | 60. (b) | 61. (d) | 62. (c) | 63. (b) | 64. (b) | 65. (b) | | | |

ASSESSMENT TESTS

Test 1

| 1. (b) | 2. (a) | 3. (a) | 4. (d) | 5. (a) | 6. (c) | 7. (c) | 8. (a) | 9. (a) | 10. (d) |
| 11. (c) | 12. (c) | 13. (a) | 14. (b) | 15. (d) | | | | | |

Test 2

| 16. (d) | 17. (b) | 18. (c) | 19. (d) | 20. (c) | 21. (a) | 22. (b) | 23. (b) | 24. (d) | 25. (b) |
| 26. (c) | 27. (c) | 28. (a) | 29. (b) | 30. (d) | | | | | |

Number Systems 1.35

CONCEPT APPLICATION
Level 1

1. 0 is the additive identity element in the set of whole numbers.
   Hence, the correct option is (a)

2. (a) $\sqrt{6} \times \sqrt{6} = (\sqrt{6})^2 = 6$ is a rational number.
   ∴ Given statement is false.
   (b) $\sqrt{4} \times \sqrt{25} = \sqrt{100} = 10$ is a rational number.
   ∴ Given statement is true.
   Hence, the correct option is (b)

3. LCM of two coprimes is their product.
   HCF of two coprimes is 1.
   ∴ Both (a) and (b) are true.
   Hence, the correct option is (c)

4. Given, $a = (-1)^{2009}$
   $\Rightarrow a = -1$
   Given, $b = (-1)^{2010}$
   $\Rightarrow b = 1$
   ∴ $ab = (-1)(1) = -1$
   Hence, the correct option is (b)

5. Given, $\sqrt{x}$ is a perfect square.
   $\Rightarrow x$ is a perfect square.
   $\Rightarrow x^2$ is also perfect square.
   For example, $\sqrt{16}$, i.e., 4 is a perfect square.
   $\Rightarrow 16$ and $(16)^2$ are also perfect squares.
   Hence, the correct option is (d)

6. $\sqrt[3]{\dfrac{125}{216}} - \sqrt{\dfrac{25}{36}} = \sqrt[3]{\dfrac{5^3}{6^3}} - \sqrt{\dfrac{5^2}{6^2}} = \dfrac{5}{6} - \dfrac{5}{6} = 0$
   Hence, the correct option is (c)

7. The product of two numbers = (Their LCM) × (Their HCF)
   $\Rightarrow 360 = \text{LCM} \times 6$
   $\Rightarrow \text{LCM} = \dfrac{360}{6} = 60$
   Hence, the correct option is (a)

8. LCM of any two numbers is a multiple of their HCF.
   ∴ $\dfrac{\text{LCM}(M,N)}{\text{HCF}(M,N)}$ must be an integer.
   Hence, the correct option is (a)

9. Given digits are 0, 1, 2 and 3.
   X is the smallest possible number formed using the digits.
   $X = 1023$
   Hence, the correct option is (b)

10. The required number is the LCM of 6, 9, 12 and 18 is 36.
    Hence, the correct option is (a)

11. Given, $r^3 = 1728$.
    $r = \sqrt[3]{1728} = \sqrt[3]{(12)^3} = 12$
    Hence, the correct option is (a)

12. If $P = Q$, LCM (P, Q) = HCF (P, Q)
    For example, LCM of $(4, 4) = 4$
    HCF of $(4, 4) = 4$
    Hence, the correct option is (a)

13. Given, X is a perfect square.
    ∴ X^2 is also a perfect square.
    But $\sqrt{x}$ need not be perfect square.
    If $x = 4$, $\sqrt{x} = 2$ which is not a perfect square.
    Hence, the correct option is (a)

14. (D), (B), (A), (E) and (C) is the sequential order from the first to the last.
    Hence, the correct option is (b)

15. (C), (A), (D) and (B) is the sequential order from the first to the last.
    Hence, the correct option is (c)

16. (C), (B) and (A) is the sequential order from the first to the last.
    Hence, the correct option is (d)

1.36 Chapter 1

17. (B), (D), (A), (E) and (C) is the sequential order from the first to the last.

    Hence, the correct option is (a)

18. (−22) + 21 + (−22) + 21 + ------ (40 terms)

    = (−1) + (−1) + ------ (20 terms) = −20

    Hence, the correct option is (e)

19. 0.756 × 100 = 75.6

    Hence, the correct option is (a)

20. 75.6 ÷ 10 = 7.56

    Hence, the correct option is (f)

21. (9) (−1/3) (−3) (−1/9) = −1

    Hence, the correct option is (c)

22. → c: (−2) (−3) (6) (−1) = −36

23. → d: 0.25 ÷ 100 = 0.0025

24. → a: 0.025 × 100 = 2.5

25. → b: 86 + (−28) + 12 + (−34) = 98 − 62 = 36

Level 2

26. Sum of two integers = −26.

    One of the integers = 14.

    Let the other integer be x.

    $x + 14 = −26$

    $x = −26 − 14 = −40$

    Hence, the correct option is (c)

27. 10 − 5 = 5

    −5 − (−10) = 5

    15 − (−20) = 35

    Option (d) follows

    Hence, the correct option is (d)

28. The product of the two integers = 72.

    One of the integers = −9.

    Let the other integer be x.

    $−9 × x = 72$

    $X = \dfrac{72}{-9} = -8$

    Hence, the correct option is (b)

29. 29 × 76 − 71 × 29

    = 29 (76 − 71)

    = 29 (5) = 145

    Hence, the correct option is (d)

30. The distance travelled on first day = 3/5 of the total distance.

    The distance travelled on second day = $\left(1 - \dfrac{3}{5}\right)$ of total distance. = $\dfrac{2}{5}$ of total distance.

    Hence, the correct option is (b)

31. Length of the pole P_1 = 26.79 m.

    Length of the pole P_2 = 29.34 m.

    Difference = 29.34 − 26.79

    = 2.55 m.

    Hence, the correct option is (c)

32. $2/3 = 0.\overline{6}$ and 5/7 0.71

    (a) 3/4 = 0.75

    (b) 4/5 = 0.8

    (c) $5/6 = 0.8\overline{3}$

    No fraction among the given fractions lies between $\dfrac{2}{3}$ and $\dfrac{5}{7}$.

    Hence, the correct option is (d)

33. The required number is HCF of 81, 144 and 162.

    ```
    81) 162 (2
        162
        ───
          0

        81) 144(1
            81
            ──
            63 )81(1
                63
                ──
                18)63(3
                   54
                   ──
                   9)18(2
                     18
                     ──
                      0
    ```

    HCF = 9.

    Hence, the correct option is (a)

34. The required number is LCM of 24, 36 and 48.

$$\begin{array}{r|l} 2 & 24-36-48 \\ \hline 2 & 12\text{-}18\text{-}24 \\ \hline 2 & 6\text{-}9\text{-}12 \\ \hline 3 & 3\text{-}9\text{-}6 \\ \hline & 1\text{-}3\text{-}2 \end{array}$$

LCM $= 2^4 \times 3^2 = 144$.

Hence, the correct option is (b)

35. $p = (-1)^{205} = -1$

$q = (-1)^{202} = +1$

$p + q = 0$

Hence, the correct option is (c)

36. $\sqrt[3]{\dfrac{27}{125}} + \sqrt{\dfrac{4}{25}}$

$= \sqrt[3]{\left(\dfrac{3}{5}\right)^3} + \sqrt{\left(\dfrac{2}{5}\right)^2}$

$= \dfrac{3}{5} + \dfrac{2}{5} = 1$

Hence, the correct option is (d)

37. In general,

$\sqrt{a} + \sqrt{b} \neq \sqrt{a+b}$

$\sqrt{a} \cdot \sqrt{b} = \sqrt{ab}$

$\sqrt{a} - \sqrt{b} \neq \sqrt{a-b}$

Hence, the correct option is (b)

38. The recurring part of the decimal is called period.

Period of $10.23\overline{46}$ is 46.

Hence, the correct option is (b)

39. (a) $\dfrac{1}{3} = 0.\overline{3}$

(b) $\dfrac{2}{7} = 0.\overline{285714}$

(c) $\dfrac{1}{5} = 0.2$

Both (a) and (b) are recurring decimals.

Hence, the correct option is (d)

40. $x = \sqrt{3}$

$y = \sqrt{27} = \sqrt{3^3}$

$z = \sqrt{243} = \sqrt{3^5}$

(i) $xy = \sqrt{3^4} = 3^2 = 9$

(a rational number)

(ii) $xz = \sqrt{3^6} = 3^3 = 27$

(a rational number)

(iii) $yz = \sqrt{3^8} = 3^4 = 81$

(a rational number)

Hence, the correct option is (d)

41. $0.\overline{3} + 0.\overline{45}$

$= \dfrac{3}{9} + \dfrac{45}{99} = \dfrac{33+45}{99} = \dfrac{78}{99} = 0.\overline{78}$

Hence, the correct option is (d)

42. $0.1\overline{2} + 0.\overline{12}$

$= \dfrac{12-1}{90} - \dfrac{12}{99} = \dfrac{11}{99} - \dfrac{12}{99}$

$= \dfrac{121-120}{9 \times 10 \times 11} = \dfrac{1}{990} = 0.00\overline{1}$

Hence, the correct option is (c)

43. $\sqrt{1\dfrac{9}{16}} - \sqrt{1\dfrac{7}{9}}$

$= \sqrt{\dfrac{25}{16}} - \sqrt{\dfrac{16}{9}}$

$= \sqrt{\left(\dfrac{5}{4}\right)^2} - \sqrt{\left(\dfrac{4}{3}\right)^2}$

$= \dfrac{5}{4} - \dfrac{4}{3} = \dfrac{15-16}{12} = -\dfrac{1}{12}$

Hence, the correct option is (c)

44. $\sqrt{3.38 \times 3.92}$

$= \sqrt{1.69 \times 1.96 \times 2^2}$

$= \sqrt{(1.3)^2 \times (1.4)^2 \times 2^2}$

$= 1.3 \times 1.4 \times 2$

$= 3.64$

Hence, the correct option is (c)

Chapter 1

45. Given digits are 0, 7, 8 and 9.

X is the smallest possible number formed using the digits 0, 7, 8, 9.

∴ X = 7089.

Y is the greatest possible number formed using the digits 0, 7, 8, 9

∴ Y = 9870.

⇒ Y − X = 2781

Hence, the correct option is (b)

46. $\sqrt{6.05} \times \sqrt{8.45}$

$= \sqrt{5 \times 1.21} \times \sqrt{5 \times 1.69}$

$= \sqrt{5^2 \times \dfrac{121}{100} \times \dfrac{169}{100}}$

$= \sqrt{5^2 \times \dfrac{11}{10} \times \dfrac{13}{10}}$

$= 5 \times \dfrac{11}{10} \times \dfrac{13}{10}$

$= \dfrac{715}{100} = 7.15$

Hence, the correct option is (d)

47. Let the required number be N.

Given that 101 when divided by N leaves a remainder of 5.

∴ N must divide 101 − 5, i.e., 96.

Similarly that N must also divide 115 − 7, i.e., 108 exactly.

∴ N must be the HCF of 96 and 108, i.e., 12.

Hence, the correct option is (c)

48. Let the number of sections be x.

∴ The number of students in each section = x

Total number of students = 625

⇒ x × x = 625 ⇒ $x^2 = 25^2$ ⇒ x = 25

∴ The number of sections = 25.

Hence, the correct option is (d)

Level 3

49. 3/4 = 0.75

9/13 ≃ 0.69

12/17 ≃ 0.71

1/2 = 0.5

1/2, 9/13, 12/17 and 3/4 are in the ascending order.

Hence, the correct option is (d)

50. 2/3 = $0.\overline{6}$

4/7 = $0.\overline{571428}$

8/11 = $0.\overline{72}$

$\dfrac{5}{9} = 0.\overline{5}$

The smallest fraction is $\dfrac{5}{9}$.

Hence, the correct option is (d)

51. Total money = ₹6024

Share of C = $(1 - \dfrac{1}{3} - \dfrac{1}{2})$ 6024

$= \left(\dfrac{6 - 2 - 3}{6}\right) 6024$

$= \left(\dfrac{1}{6}\right) 6024 = ₹1004$

Hence, the correct option is (a)

52. The average weight of a student = 32 ¾ kg.

Total weight of 24 students = (32 ¾) = 24.

$= \left(\dfrac{131}{4}\right) 24 = 131 \times 6$

= 786 kg

Hence, the correct option is (b)

53. Given, LCM = 1008.

HCF = 24.

One of the two numbers = 168.

Let the other number be x.

We have, the product of two numbers = LCM × HCF.

$168 \times 2 = 1008 \times 24$

$x = \dfrac{1008 \times 24}{168}$

$= \dfrac{1008}{7} = 144$

Hence, the correct option is (d)

54. The required number is HCF of (246 − 2 and 279 − 3).

    i.e., HCF of 244 and 276.

    ```
    244)276(1
        244
        ───
        32)244(7
           224
           ───
           20)32(1
              20
              ──
              12)20(1
                 12
                 ──
                 8)12(1
                   8
                   ─
                   4)8(2
                     8
                     ─
                     0
    ```

    HCF = 4.

    The required number is 4.

    Hence, the correct option is (a)

55. The required quantity is the HCF of 3ℓ 250 mℓ, 3ℓ 500 mℓ and 4ℓ.

    ```
    3250) 4000 (1
          3250
          ────
          750 ) 3250 (4
                3000
                ────
                250)750(3
                    750
                    ───
                     0
    ```

    ```
    250) 3500 (14
         3500
         ────
           0
    ```

    HCF = 250.

    The required quantity = 250 mℓ.

    Hence, the correct option is (c)

56. $4500 = 6^2 \times 5^3$

    The required number is 62. $(\because 6^2 \times \dfrac{5^3}{6^2} = 5^3)$

    Hence, the correct option is (b)

57. Total number of students = 1152.

    Let $n \times \dfrac{n}{2} = 1152$

    $n^2 = 2304$

    $n = \sqrt{(48)^2} = 48$.

    Hence, the correct option is (c)

58. The HCF is a factor of the LCM.

    As 1, 2 and 13 are factors of 26, HCF can be 1 or 2 or 13.

    Hence, the correct option is (d)

59. Net score = (20 × 500) − (40 × 300)

    = 10000 − 12000

    = −2000

    Hence, the correct option is (b)

60. Total number of pencils = 20.

    Number of pencils taken by Nikhil = $(1 - \dfrac{2}{5})$ 20.

    $\left(1 - \dfrac{2}{5}\right) 20$

    $\left[\dfrac{3}{5}\right] 20 = 12$

    Hence, the correct option is (b)

61. $x = (-23) + 22 + (-23) + 22 + \ldots\ldots$ (40 terms).

    $= (-1) + (-1) + \ldots\ldots (20 \text{ terms}) = -20$

    $y = 11 + (-10) + 1 + (-10) + \ldots\ldots (20 \text{ terms})$.

    $= 1 + 1 + \ldots\ldots (10 \text{ terms})$

    $= 10$

    $y - x = 10 - (-20) = 30$.

    Hence, the correct option is (d)

62. Cost of the article = ₹540

    The amount with Ram = $\left(\dfrac{13}{27}\right) 540$

    $= 13 \times 20 = ₹260$

    Hence, the correct option is (c)

1.40 Chapter 1

63. Given digits are 1, 3, 5 and 9.
    X = 1359
    Y = 9531
    Y − X = 9531 − 1359 = 8172.
    Hence, the correct option is (b)

64. $x = (-1)^1 + (-1)^2 + \ldots (-1)^{2009}$
    $= -1 + 1 - 1 + 1 - 1 - 1 = (-1)$
    $y = (-1)^1 - (-1)^2 + (-1)^3 - (-1)^4 + \ldots (-1)^{2009}$
    $= -1 - 1 - 1 - 1 - 1 \ldots 2009$ times
    $= -2009$
    $= x - y = -1 - (-2009) = 2008.$
    Hence, the correct option is (b)

65. Let the money Raj has be ₹ x.
    $x = \dfrac{15}{28} \times 560$ (given) = ₹ 300.
    Hence, the correct option is (b)

ASSESSMENT TESTS

Test 1

Solutions for questions 1 to 11:

1. (C), (B), (A), (D) and (E) is the required sequential order.
   Hence, the correct option is (b)

2. (C), (A), (B) and (D) is the required sequential order.
   Hence, the correct option is (a)

3. Side of the square field = $\sqrt{1444}$ m = 38 m.
   Perimeter of the field = 4 × 38 m = 152 m.
   The cost of fencing = ₹ (152 × 5) = ₹760.
   Hence, the correct option is (a)

4. ∴ 88 × 89 + x = 89² (∴ 88² < 88 × 89 < 89²)
   x = 89² − 88 × 89
   = 89 (89 − 88)
   x = 89 × 1
   ∴ x = 89
   Hence, the correct option is (d)

5. ∴ The LCM of 5, 7 and 8 = 5 × 7 × 8 = 280.
   The required number is (280 + 3) = 283.
   Hence, the correct option is (a)

6. $2.\overline{3} + 3.\overline{4} - 4.\overline{8}$
   $= \dfrac{23-2}{9} + \dfrac{34-3}{9} - \dfrac{48-4}{9} = \dfrac{21}{9} + \dfrac{31}{9} - \dfrac{44}{9}$
   $= \dfrac{52-44}{9} = \dfrac{8}{9} = 0.\overline{8}$
   Hence, the correct option is (c)

7. $\sqrt[3]{27} \times \sqrt[3]{216} \times \sqrt[3]{64}$
   $= \sqrt[3]{3^3} \times \sqrt[3]{6^3} \times \sqrt[3]{4^3} = 3 \times 6 \times 4 = 72$
   Hence, the correct option is (c)

8. The HCF of 92 − 2 = 90 and 75 − 5 = 70 is 10.
   ∴ 10 is the required number.
   Hence, the correct option is (a)

9. Net score = 15 × 300 − 40 × 100
   = 4500 − 4000 = 500
   Hence, the correct option is (a)

10. The HCF must be a factor of LCM.
    70, 60, 210 are some of the factors of 420 but 80 is not a factor of 420.
    Hence, the correct option is (d)

11. $\dfrac{63}{24} = \dfrac{21}{8}$, $\dfrac{21}{2^3} = \dfrac{18}{15} = \dfrac{6}{5}$, $\dfrac{33}{44} = \dfrac{3}{4} = \dfrac{3}{2^2}$ are terminating and $\dfrac{14}{21} = \dfrac{2}{3}$ is non-terminating.
    Hence, the correct option is (c)

Solutions for questions 12 to 15:

12. → c: $3.5\overline{31}$, the period is 31.

13. → a: $\dfrac{51}{96} = \dfrac{17}{32} = \dfrac{17}{2^5}$ can be expressed as a terminating decimal.

14. → b: $= \dfrac{5}{24} = \dfrac{5}{2^3 \times 3}$ can be expressed as a non-terminating repeating decimal.

15. → d: Given number is a non-terminating and non-repeating decimal.

Test 2

Solutions for questions 16 to 26:

16. (B), (C), (A) and (D) is the required sequential order.

    Hence, the correct option is (d)

17. (C), (B), (A) and (D) is the required sequential order.

    Hence, the correct option is (b)

18. Side of the square field = $\sqrt{1681}$ m = 41 m

    Perimeter of the field = 4 × 41 m = 164 m

    The cost of fencing = ₹(164 × 3) = ₹492

    Hence, the correct option is (c)

19. Let 30 × 31 + x = 31². (∵ 30² < 30 × 31 < 31²)

    x = 31² − 30 × 31

    x = 31(31 − 30)

    x = 31

    ∴ 31 should be added to the result of 30 × 31 so that the sum obtained is a perfect square.

    Hence, the correct option is (d)

20. Required number is LCM of (9, 12 and 15) + 5.

    The LCM of 9, 12 and 15 = 180.

    ∴ The required number = 180 + 5 = 185.

    Hence, the correct option is (c)

21. $3.\overline{4} + 5.\overline{8} - 7.\overline{9} = \dfrac{34-3}{9} + \dfrac{58-5}{9} - \dfrac{79-7}{9}$

    $= \dfrac{31}{9} + \dfrac{53}{9} - \dfrac{72}{9}$

    $= \dfrac{84}{9} - \dfrac{72}{9} = \dfrac{12}{9} = \dfrac{4}{3}$

    Hence, the correct option is (a)

22. $\sqrt[3]{512} = \sqrt[3]{8^3} = 8$

    $\sqrt[3]{343} = \sqrt[3]{7^3} = 7$

    $\sqrt[3]{729} = \sqrt[3]{9^3} = 9$

    ∴ $\sqrt[3]{512} + \sqrt[3]{343} + \sqrt[3]{729}$

    = 8 + 7 + 9 = 24

    Hence, the correct option is (b)

23. Required number is the HCF of (64 − 12) = 52 and (72 − 7) = 65

    ```
    52)65(1
       52
       ──
       13)52(4
          52
          ──
           0
    ```

    ∴ 13 divides 64 and 72 by leaving remainders 12 and 7 respectively.

    Hence, the correct option is (b)

24. Net score = 20 × 400 − 30 × 300

    = 8000 − 9000 = −1000.

    Hence, the correct option is (d)

25. The HCF must be a factor of LCM.

    Going by options, 18 is not a factor of 260.

    Hence, the correct option is (b)

26. $\dfrac{750}{216}$, i.e., $\dfrac{125}{36}$ is a non-terminating repeating decimal, since its denominator has 3 as a factor.

    Hence, the correct option is (c)

Solutions for questions 27 to 30:

27. → c: $\dfrac{3}{7}$ can be expressed as a non-terminating and repeating decimal.

28. → a: $\dfrac{5}{2}$ can be expressed as a terminating decimal.

29. → b: Given decimal is a non-terminating and non-repeating decimal.

30. → d: periodicity = 2

Chapter 2

Expressions and Special Products

a) $5(3x) - 3(x-4) = 0$
$15x - 3x + 12 = 0$
$12x + 12 = 0$
$12x = -12$
$x = -12/12$
$x = -1 \qquad x^2 = 1$

REMEMBER

Before beginning this chapter, you should be able to:

- Clear on the four fundamental operations on numbers, i.e., addition, subtraction, multiplication and division
- Have idea about on variable numbers like x, y, etc.

KEY IDEAS

After completing this chapter, you should be able to:

- Understand the concept of algebraic expression and polynomial
- Categorize the polynomials based on degree of polynomial and number of polynomial
- Apply operations on polynomials
- Obtain factorization and HCF of polynomials
- Learn about value of an expression and zero of polynomials

INTRODUCTION

In this chapter, we will look at the definitions of terms, like terms, constant, variable, algebraic expression, application of four fundamental operations, i.e., addition, subtraction, multiplication and division. Besides, we shall discuss some special products and simple factorisation which are related to polynomials.

Constant

A number having a fixed numerical value is called a constant.

For example, 5, $\frac{3}{4}$, 3.6, $2.\overline{3}$, etc.

Variable

A number which can take various numerical values is known as a variable.

For example, x, y, z, etc.

A number which is a power of another variable where the power is not zero is also a variable.

For example, x^3, $y^{\frac{2}{3}}$, $z^{0.5}$, etc.

A number which is the product of a constant and a variable is also a variable.

For example, $5x^6$, $6x^2$, $-3x$, etc.

A combination of two or more variables separated by a + sign or a − sign is also a variable.

For example, $x^2 + y + z^5$, $x^3 - y^3 - z^2$, etc.

Algebraic Expression

A combination of constants and variables connected by + or − or × or ÷ signs is known as an algebraic expression.

For example, $3x^2 - 5$, $x^3 - 2xy + 3xy^2 + 6$, etc.

Terms: Several parts of an algebraic expression, separated by + or − signs are called the terms of the expression.

For example, in the expression $3x - 4y + 7$, we say $3x$, $-4y$, and 7 are terms.

Coefficient of a term: Consider the term $7x^3$. In this case, 7 is called the numerical coefficient of x^3 and x^3 is said to be the literal coefficient of 7.

In case of $-6xy$, the numerical coefficient is -6 and the literal coefficient is xy.

Like terms: Terms having the same literal coefficients are called like terms.

For example, (1) $3x^3$, $-5x^3$ and $4x^3$ are like terms. (2) $3x^2y$, $-5x^2y$ and $7x^2y$ are like terms

Unlike terms: Terms having different literal coefficients are called unlike terms.

For example, $-3x^2$, $7x^3$, $4x^4$ are unlike terms.

Polynomial

An algebraic expression in which the variables involved have only non-negative integer powers is called a polynomial.

For example, $5x^2 - x + 1$, $7x^3 - 4x^2 - 2x - 7$, $2z^2 - 5y - 7$, etc.

Expressions and Special Products 2.3

The expression $3x^3 + 4x + 5/x + 6x^{3/2}$ is not a polynomial since it has powers of x which are negative and fractions.

A polynomial that contains only one variable is known as a polynomial in that variable.

For example, $3x^3 + 2x^2 - 8x + 5$ is a polynomial in the variable x.

$5y^3 + 2y^2 - 2y - 3$ is a polynomial in the variable y.

A polynomial that contains two variables say x and y is known as a polynomial in two variables x and y.

For example, $x^3y^2 - 2x^4y^3 + 5xy$ is a polynomial in x and y.

Degree of a polynomial in one variable: The highest exponent of the variable in a polynomial of one variable is called the degree of the polynomial.

For example, 1. $x^4 - 5x^2 - x + 7$ is a polynomial of degree 4.

2. $7x^9 - 3x^6 + 7x^3 - x^2 - 5$ is a polynomial of degree 9.

Types of Polynomials with Respect to Degree

1. **Linear polynomial:** A polynomial of degree one is called a linear polynomial.

   For example, $3x - 5$, $2x + 3$, $8y + 1$, etc.

2. **Quadratic polynomial:** A polynomial of degree two is called a quadratic polynomial.

   For example, $3x^2 - 5x + 7$, $5 - 2x - x^2$, $y^2 - 5y - 6$, etc.

3. **Cubic polynomial:** A polynomial of degree three is called a cubic polynomial.

   For example, $2x^3 + 5x^2 - 3x + 7$, $8y^3 - 2y - 1$, etc.

4. **Biquadratic polynomial:** A polynomial of degree four is called a biquadratic polynomial

   For example, $2x^4 - x^3 + 4x - 3$, $5x^4 - x^2 - 7$, etc.

5. **Constant polynomial:** A polynomial having only one term which is a constant is called a constant polynomial. Degree of a constant polynomial is 0.

   For example, $7, -2, 5$ are constant polynomials.

Types of Polynomials with Respect to Number of Terms

1. **Monomial:** An expression containing only one term is called a monomial.

   For example, $3x$, $5x^2y$, $6xyz$, etc.

2. **Binomial:** An expression containing two terms is called a binomial.

   For example, $2x - 3y$, $5x^2y^2 - 7$, $xy - 3y$.

3. **Trinomial:** An expression containing three terms is called a trinomial.

   For example, $2x - 3y + 4z$, $3xy^2 - xy + 5$.

Addition of polynomials: The sum of two or more polynomials can be obtained by arranging the terms and then adding the numerical coefficients of like terms.

EXAMPLE 2.1

Add $2x - 3y + z$, $5y - x + 7z$ and $3x - y - 6z$.

SOLUTION

$$\begin{array}{r} 2x - 3y + z \\ -x + 5y + 7z \\ 3x - y - 6z \\ \hline 4x + y + 2z \end{array}$$

The required sum is $4x + y + 2z$.

Subtraction of polynomials: The difference of two polynomials can be obtained by arranging the terms and subtracting the numerical coefficients of the like terms.

EXAMPLE 2.2

Subtract $8p - 5q + 7r$ from $5p - 9q + 3r$.

SOLUTION

$$\begin{array}{r} 5p - 9q + 3r \\ -(8p - 5q + 7r) \\ \hline -3p - 4q - 4r \end{array}$$

∴ The required difference is $-3p - 4q - 4r$.

EXAMPLE 2.3

What should be subtracted from $x^3 - 7x^2 + 17x + 17$ so that the difference is a multiple of $x - 3$?

SOLUTION

Let k be subtracted.
$\Rightarrow 32 - k = 0$
$\Rightarrow k = 32$.

Multiplication of two polynomials: The result of multiplication of two polynomials is obtained by multiplying each term of the polynomial by each term of the other polynomial and then taking the algebraic sum of these products.

EXAMPLE 2.4

Multiply $(5 - 3x - 4x^2)$ with $2x - 3$.

SOLUTION

$$\begin{array}{r} 5 - 3x - 4x^2 \\ \times (2x - 3) \\ \hline 10x - 6x^2 - 8x^3 \\ -15 + 9x + 12x^2 \\ \hline -15 + 19x + 6x^2 - 8x^3 \end{array}$$

Expressions and Special Products 2.5

∴ Required product = $-8x^3 + 6x^2 + 19x - 15$.

Thus, the sum or product of two or more polynomials is also a polynomial.

Now, we shall learn some **algebraic identities** (relations that are true for all values of the symbols that appear in them) and use these identities to find the product of different algebraic expressions.

Identity 1: $(x + a)(x + b) = x^2 + (a + b)x + ab$

EXAMPLE 2.5

(a) Expand $(x + 4)(x + 7)$.

(b) Expand $(x + 2)(x - 5)$

SOLUTION

(a) Putting $a = 4$ and $b = 7$, we have $(x + 4)(x + 7) = x^2 + (4 + 7)x + 4 \times 7 = x^2 + 11x + 28$.

(b) Putting $a = 2$ and $b = -5$, we have
$(x + 2)(x - 5) = x^2 + (2 - 5)x + 2 \times (-5) = x^2 - 3x - 10$.

Identity 2: $(x + a)(x + b)(x + c) = x^3 + (a + b + c)x^2 + (ab + bc + ca)x + abc$

EXAMPLE 2.6

Expand $(x + 3)(x - 2)(x + 6)$.

SOLUTION

Putting $a = 3$, $b = -2$ and $c = 6$, we have
$(x + 3)(x - 2)(x + 6)$
$= x^3 + (3 - 2 + 6)x^2 + [3 \times (-2) + (-2)(6) + 3 \times 6]x + 3 \times (-2) \times 6$
$= x^3 + 7x^2 + (0)x - 36 = x^3 + 7x^2 - 36$.

Identity 3: $(a + b)^2 = a^2 + 2ab + b^2$

EXAMPLE 2.7

Expand $(3x + 4y)^2$.

SOLUTION

Here $a = 3x$ and $b = 4y$, using the identity $(a + b)^2 = a^2 + 2ab + b^2$.
We have $(3x + 4y)^2 = (3x)^2 + 2 \cdot 3x \cdot 4y + (4y)^2 = 9x^2 + 24xy + 16y^2$.

Note $\left(a + \dfrac{1}{a}\right)^2 = a^2 + \dfrac{1}{a^2} + 2$

EXAMPLE 2.8

If $x + \dfrac{1}{x} = 3$, then find the value of $x^2 + \dfrac{1}{x^2}$.

SOLUTION

We know $\left(x + \dfrac{1}{x}\right)^2 = x^2 + \dfrac{1}{x^2} + 2$, substituting the value of $x + \dfrac{1}{x} = 3$.

We have $(3)^2 = x^2 + \dfrac{1}{x^2} + 2$

$\Rightarrow x^2 + \dfrac{1}{x^2} = 7$.

Identity 4: $(a - b)^2 = a^2 - 2ab + b^2$

EXAMPLE 2.9

Expand $(8x - 7y)^2$.

SOLUTION

Here, $a = 8x$ and $b = 7y$ using the identity $(a - b)^2 = a^2 - 2ab + b^2$.
We have $(8x - 7y)^2 = (8x)^2 - 2.8x.7y + (7y)^2 = 64x^2 - 112xy + 49y^2$.

Note $\left(a - \dfrac{1}{a}\right)^2 = a^2 + \dfrac{1}{a^2} - 2$

EXAMPLE 2.10

If $x - \dfrac{1}{x} = 5$, then find the value of $x^2 + \dfrac{1}{x^2}$.

SOLUTION

We know $\left(x - \dfrac{1}{x}\right)^2 = x^2 + \dfrac{1}{x^2} - 2$, substituting the value of $x - \dfrac{1}{x} = 5$ we have

$(5)^2 = x^2 + \dfrac{1}{x^2} - 2$

$\Rightarrow x^2 + \dfrac{1}{x^2} = 27$.

EXAMPLE 2.11

If $y + \dfrac{1}{y} = -2$, then find the value of $y^{50} - \dfrac{1}{y^{50}}$.

Expressions and Special Products

SOLUTION

$y + \dfrac{1}{y} = -2$

$y^2 + 1 = -2y$

$y^2 + 2y + 1 = 0$

$(y)^2 + 2(y)(1) + 1 = 0$

$(y + 1)^2 = 0$

$y = -1$

$\Rightarrow y^{50} - \dfrac{1}{y^{50}}$

$\Rightarrow (-1)^{50} - \dfrac{1}{(-1)^{50}}$

$= 1 - \dfrac{1}{1} = 1 - 1 = 0.$

EXAMPLE 2.12

If $x^2 - y^2 = 12xy$, then find the value of $\dfrac{x^2}{y^2} + \dfrac{y^2}{x^2}$.

SOLUTION

$x^2 - y^2 = 12xy$

Dividing with xy,

$\dfrac{x}{y} - \dfrac{y}{x} = 12$

Taking squares on both the sides

$\left(\dfrac{x}{y} - \dfrac{y}{x}\right)^2 = 12^2$

$\Rightarrow \dfrac{x^2}{y^2} + \dfrac{y^2}{x^2} - 2 = 144$

$\Rightarrow \dfrac{x^2}{y^2} + \dfrac{y^2}{x^2} = 146.$

Identity 5: $(a + b)^2 + (a - b)^2 = 2(a^2 + b^2)$

EXAMPLE 2.13

Simplify $(3x + 4y)^2 + (3x - 4y)^2$.

SOLUTION

Here, $a = 3x$ and $b = 4y$.

Using identity $(a + b)^2 + (a - b)^2 = 2(a^2 + b^2)$.

We have $(3x + 4y)^2 + (3x - 4y)^2 = 2[(3x)^2 + (4y)^2] = 2(9x^2 + 16y^2) = 18x^2 + 32y^2.$

Identity 6: $(a + b)^2 - (a - b)^2 = 4ab$

EXAMPLE 2.14

Simplify $(11x + 3y)^2 - (11x - 3y)^2$.

SOLUTION

Here, $a = 11x$ and $b = 3y$, using the identity $(a + b)^2 - (a - b)^2 = 4ab$, we have
$(11x + 3y)^2 - (11x - 3y)^2 = 4(11x)(3y) = 132xy$.

Identity 7: $(a + b)(a - b) = a^2 - b^2$

EXAMPLE 2.15

Simplify $(13x - 9y)(13x + 9y)$

SOLUTION

Here, $a = 13x$ and $b = 9y$, using the identify $(a + b)(a - b) = a^2 - b^2$, we have
$(13x + 9y)(13x - 9y)$.
$= (13x)^2 - (9y)^2 = 169x^2 - 81y^2$.

Factorisation

Factorisation can be defined as the method of expressing a given polynomial as a product of two or more polynomials. In some of the cases, we use algebraic identities in factorisation.

For example, $2x^2 - 6x = 2x(x - 3)$ as $2x^2 - 6x$ is expressed as the product of $2x$ and $x - 3$.

Hence, $2x$ and $x - 3$ are the factors of $2x^2 - 6x$.

1. **Factorization of polynomials of the form $x^2 - y^2$:** $x^2 - y^2 = (x + y)(x - y) \Rightarrow x + y$ and $x - y$ are the factors of $x^2 - y^2$.

EXAMPLE 2.16

Factorise $49x^2 - 16y^2$.

SOLUTION

Let $a = 7x$ and $b = 4y$.
Using the identity $a^2 - b^2 = (a - b)(a + b)$, we have
$49x^2 - 16y^2 = (7x)^2 - (4y)^2 = (7x - 4y)(7x + 4y)$.

2. **Factorisation of polynomials by regrouping of terms:** In this method, we regroup the terms of the polynomials in such a way that we get a common factor out of them.

EXAMPLE 2.17

(a) Factorise: $x^2 - (z - 5)x - 5z$

(b) Factorise: $x^2 + x - y + y^2 - 2xy$

SOLUTION

(a) $x^2 - (z - 5)x - 5z$
$= x^2 - xz + 5x - 5z = x(x - z) + 5(x - z) = (x - z)(x + 5)$
$\therefore x^2 - (z - 5)x - 5z = (x - z)(x + 5)$.

(b) $x^2 + x - y + y^2 - 2xy$
$= x^2 + y^2 - 2xy + x - y = (x - y)^2 + 1(x - y) = (x - y)(x - y + 1)$
$\therefore x^2 + x - y + y^2 - 2xy = (x - y)(x - y + 1)$.

3. **Factorisation of a perfect square trinomial:** A trinomial of the form, $x^2 \pm 2xy + y^2$ is equivalent to $(x \pm y)^2$.

   This identity can be used to factorise perfect square trinomials.

EXAMPLE 2.18

(a) Factorise: $9x^2 - 24xy + 16y^2$

(b) Factorise: $25x^2 + \dfrac{1}{25x^2} + 2$

SOLUTION

(a) $9x^2 - 24xy + 16y^2 = (3x)^2 - 2.3x.4y + (4y)^2 = (3x - 4y)^2$.

(b) $25x^2 + \dfrac{1}{25x^2} + 2$

$= (5x)^2 + \left(\dfrac{1}{5x}\right)^2 + 2.5x.\dfrac{1}{5x} = \left(5x + \dfrac{1}{5x}\right)^2$

$\therefore 25x^2 + \dfrac{1}{25x^2} + 2 = \left(5x + \dfrac{1}{5x}\right)^2$.

Division of a Polynomial by a Monomial

To divide a polynomial by a monomial, we need to divide each term of the polynomial by the monomial.

EXAMPLE 2.19

Divide $18x^4 - 27x^3 + 6x$ by $3x$.

SOLUTION

$\dfrac{18x^4 - 27x^3 + 6x}{3x} = \dfrac{18x^4}{3x} - \dfrac{27x^3}{3x} + \dfrac{6x}{3x} = 6x^3 - 9x^2 + 2$

$\therefore$ The required result $= 6x^3 - 9x^2 + 2$.

Division of a Polynomial by a Polynomial

Long Division Method

Step 1: First arrange the terms of the dividend and the divisor in the descending order of their degrees.
Step 2: Now, the first term of the quotient is obtained by dividing the first term of the dividend by the first term of the divisor.
Step 3: Then multiply all the terms of the divisor by the first term of the quotient and subtract the result from the dividend.
Step 4: Consider the remainder as the new dividend and proceed as before.
Step 5: Repeat this process till we obtain a remainder which is either 0 or a polynomial of a degree less than that of the divisor.

EXAMPLE 2.20

Divide $10x^3 + 41x^2 - 4x - 15$ by $5x + 3$.

SOLUTION

$$
\begin{array}{r}
5x + 3 \overline{\smash{\big)}\ 10x^3 + 41x^2 - 4x - 15} \ (2x^2 + 7x - 5 \\
\underline{-10x^3 \pm 6x^2 } \\
35x^2 - 4x \\
\underline{-35x^2 \pm 21x } \\
-25x - 15 \\
\underline{+25x + 15} \\
0
\end{array}
$$

$\therefore (10x^3 + 41x^2 - 4x - 15) \div (5x + 3) = 2x^2 + 7x - 5.$

The HCF of the Given Polynomials

For two given polynomials, $f(x)$ and $g(x)$, $r(x)$ can be taken as the highest common factor, if

1. $r(x)$ is a common factor of $f(x)$ and $g(x)$ and
2. every common factor of $f(x)$ and $g(x)$ is also a factor of $r(x)$.

Highest common factor is generally referred to as HCF.

Method for Finding the HCF of the Given Polynomials

Step 1: Express each polynomial as a product of the powers of irreducible factors which also requires the numerical factors to be expressed as the product of the powers of primes.
Step 2: If there is no common factor, then the HCF is 1 and if there are common irreducible factors, we find the least exponent of these irreducible factors in the factorised form of the given polynomials.
Step 3: Raise the common irreducible factors to the smallest or the least exponents found in step 2 and take their product to get the HCF.

Expressions and Special Products 2.11

EXAMPLE 2.21

(a) Find the HCF of $28x^4$ and $70x^6$.

(b) Find the HCF of $48x^2(x+3)^2(2x-1)^3(x+1)$ and $60x^3(x+3)(2x-1)^2(x+2)$.

SOLUTION

(a) Let $f(x) = 28x^4$ and $g(x) = 70x^6$.

Writing $f(x)$ and $g(x)$ as a product of powers of irreducible factors.

We have $f(x) = 2^2 \times 7 \times x^4$, $g(x) = 2 \times 5 \times 7 \times x^6$.

$\therefore$ The common factors with the least exponents are 2, 7 and x^4.

$\Rightarrow$ HCF $= 2 \times 7 \times x^4 = 14x^4$.

(b) Let $f(x) = 48x^2(x+3)^2(2x-1)^3(x+1)$.

And $g(x) = 60x^3(x+3)(2x-1)^2(x+2)$.

Writing $f(x)$ and $g(x)$ as the product of the powers of irreducible factors we have

$f(x) = 2^4 \times 3 \times x^2 (x+3)^2 (2x-1)^3 (x+1)$

$g(x) = 2^2 \times 3 \times 5 \times x^3 (x+3) (2x-1)^2 (x+2)$.

The common factors with least exponents are 2^2, 3, x^2, $x+3$, $(2x-1)^2$.

$\therefore$ H.C.F. of the given polynomials

$= 2^2 \times 3 \times x^2 (x+3) (2x-1)^2 = 12x^2(x+3)(2x-1)^2$.

Value of an expression: In mathematics, we come across many situations in which we need to find out the value of an expression. The value of an expression purely depends upon the values of the variables that form the expression.

For example, when $x = 5$ the value of the expression $3x + 2$ is $3 \times 5 + 2 = 17$.

And the value of $3x + 2y$ when $x = 2$ and $y = -1$ is $3 \times 2 + (-1) = 6 - 2 = 4$.

Zero of a polynomial: The number for which value of the variable, the value of the polynomial is zero is called zero of the polynomial.

For example, when $x = -2$ the value of the expression $x + 2$ is zero.

$\Rightarrow x = -2$ is the zero of $x + 2$.

To find zero of a polynomial equate the given polynomial to zero and solve it for the value of the variable.

EXAMPLE 2.22

Find the zero of the polynomial $3x + 5$.

SOLUTION

Equating the given polynomial $3x + 5$ to zero we have $3x + 5 = 0$.

$\Rightarrow 3x = -5$ and $x = -\dfrac{5}{3}$

$\therefore x = -\dfrac{5}{3}$ is the zero of the given polynomial.

Algebraic Expressions—Formulae

In mathematics, we use many formulae to find perimeter, area, etc., which are expressed as algebraic expressions.

1. For example, to find the perimeter of a rectangle, we use the formula, $P = 2(l + b)$.

   If the values of l and b are given, we can find the perimeter of a rectangle.

2. To find the area of a triangle, we use the formula, $A = \frac{1}{2} \times$ base $\times$ height which is generally written as $A = \frac{1}{2} b \times h$, if the values of b and h are known, we can find the area of the triangle.

EXAMPLE 2.23

Find the perimeter of a rectangle whose length (l) = 15 cm and breadth (b) = 12 cm.

SOLUTION

Here, l = 15 cm and b = 12 cm.
$P = 2(15 + 12) = 54$ cm.
∴ Perimeter of rectangle is 54 cm.

EXAMPLE 24

Find the area of the triangle with base (b) = 12 cm and the corresponding height (h) = 8 cm.

SOLUTION

Area of a triangle = $\frac{1}{2} b \times h$.
Here b = 12 cm and h = 8 cm.
∴ Required area = $\frac{1}{2} \times 12 \times 8 = 48$ cm^2.

Algebraic Expressions–Rules

Consider the following number patterns.

1. 2, 4, 6, 8, ...
2. 1, 3, 5, 7..........

In the first pattern, we see all even numbers. We know, that all even numbers are divisible by 2 hence, every even number can be expressed as $2n$ (algebraic expression) by substituting n = 1, 2, 3,....., we will get different even numbers. Here, the even numbers are expressed as an algebraic expression which can be called as a rule.

Similarly, the second pattern can be expressed as $2n + 1$ for n = 0, 1, 2, 3, ...

∴ $2n + 1$ is a rule to obtain odd numbers.

Points to Remember

- Every polynomial is an expression but every expression need not be a polynomial.
- The degree of a polynomial is the highest degree of a monomial which is present in the polynomial.
- If the degree of polynomials A and B are m and n respectively, then the degree of AB is $m + n$.
- The number for which the value of a polynomial is zero is called zero of the polynomial.
- If a trinomial is a perfect square, then it is the square of a binomial.

Important Formulae

- $(a + b)^2 = a^2 + 2ab + b^2$
- $(a - b)^2 = a^2 - 2ab + b^2$
- $(a + b)(a - b) = a^2 - b^2$
- $(x + a)(x + b) = x^2 + x(a + b) + ab$
- $(x + a)(x + b)(x + c) = x^3 + (a + b + c)x^2 + (ab + bc + ac)x + abc$
- $(a + b)^2 + (a - b)^2 = 2(a^2 + b^2)$
- $(a + b)^2 - (a - b)^2 = 4ab$

Chapter 2

TEST YOUR CONCEPTS

Very Short Answer Type Questions

Directions for questions 1 to 10: State whether the following statements are true or false.

1. $5x + 10y - 15z = 5(x + 2y - 3z)$.
2. $(x + 2)(x - 5) = x^2 - 3x + 10$.
3. $(a + 2b)^2 - (a - 2b)^2 = 8ab$.
4. $9.2 \times 8.8 = 72.96$.
5. An equation which is true for all real values of its variables is called an identity.
6. HCF of $12x^2yz$ and $16xyz^2$ is $4xyz$.
7. $\sqrt{\dfrac{x^2}{y^2 z^6}} = xy^{-1}z^{-3}$
8. $(a^2 - 1) \div (a - 1) = a + 1$.
9. If $a + b = 6$ and $a^2 - b^2 = 24$, then $a - b = 4$.
10. The constant term in the product of $(5x^2 - 7x + 4)(7x - 8)$ is 12.

Directions for questions 11 to 15: Fill in the blanks.

11. $5x^2(x - y) = $ _____ .
12. $4x^2 - 4x + 1 = ($ _____ $)^2$.
13. $(x + a)(x + b) = $ _____ .
14. $x^2 - y^2 = $ _____ .
15. $(a + 3b)^2 = $ _____ .

Directions for questions 16 to 30: Select the correct alternative from the given choices.

16. If $\left(x + \dfrac{1}{x}\right) = 2$, then $x^2 + \dfrac{1}{x^2} = $ _____ .
    (a) 2 (b) 4
    (c) 0 (d) 3

17. $(a - 1)(a^2 - 2a + 1) = $ _____ .
    (a) $(a - 1)^2$ (b) $(a - 1)^3$
    (c) $a^2 - 1$ (d) 1

18. $(a + b)(\sqrt{a} + \sqrt{b})(\sqrt{a} - \sqrt{b}) = $ _____ .
    (a) $(a + b)^2$ (b) $a^2 - b^2$
    (c) 1   (d) $(a - b)^2$

19. If $a = 2, b = -1$, then $a^2 + b^2 + 2ab = $ _____
    (a) 9 (b) 4
    (c) 2 (d) 1

20. $(x + a)(x - b) = $ _____ .
    (a) $x^2 - (a + b)x + ab$ (b) $x^2 + (a - b)x + ab$
    (c) $x^2 - (a - b)x + ab$ (d) $x^2 + (a - b)x - ab$

Directions for questions 21 to 24: Find the degree of the following expressions.

21. The degree of $4x^2$ is _____ .
    (a) 1 (b) 2
    (3) 3 (d) 4

22. The degree of $-3x^3 + 5x^2 + 4$ is _____ .
    (a) 1 (b) 2
    (c) 3 (d) 4

23. The degree of xyz is _____ .
    (a) 1 (b) 2
    (c) 3 (d) 4

24. The degree of $x + y + z$ is _____ .
    (a) 1 (b) 2
    (c) 3 (d) 4

25. The zero of $x + 2$ is ____ .
    (a) 0   (b) 2
    (c) -2 (d) 1

26. The zero of $x - 3$ is ____ .
    (a) 0   (b) 3
    (c) -3 (d) 4

27. The zero of $3x + 2$ is ____ .
    (a) 0   (b) $\dfrac{2}{3}$
    (c) $\dfrac{-2}{3}$ (d) $\dfrac{-1}{3}$

28. The zero of $5x - 3$ is ____
    (a) 3   (b) 5
    (c) $\dfrac{3}{5}$ (d) $\dfrac{5}{3}$

29. The expanded form of $(x + y)(x - y)$ is a _____
    (a) Monomial   (b) Binomial
    (c) Trinomial   (d) None of these

30. Find the degree of $(x^2 - x)^2$.
    (a) 3   (b) 4
    (c) 5   (d) 6

Short Answer Type Questions

Directions for questions 31 to 34: If $A = 3x^2 + 2x - 7$ and $B = 7x^3 - 3x + 4$, then find

31. $A + B$
32. $A - B$
33. $2A + 3B$
34. $2A - 3B$

Directions for questions 35 and 36: Find the following products.

35. $(3x + 2)(x\ 3)$
36. $(x - 5)(x + 6)$
37. Find the product of $5x^2 - 7x + 6$ and $3x$ and verify it when $x = -2$.
38. Find the product of $(3x - 7)(4x + 6)$ and verify it when $x = 1$.

Directions for questions 39 to 44: By using an appropriate identity, expand the following:

39. $(3x + 4y)^2$
40. $\left(5x + \dfrac{1}{5x}\right)^2$
41. $(x - 2y)^2$
42. $\left(3x - \dfrac{1}{3x}\right)^2$
43. $(x + 2y)(x - 2y)$
44. $(x + y)(x - y)(x^2 + y^2)$

Directions for questions 45 to 52: By using an appropriate identity, find the following.

45. $(102)^2$
46. $(54)^2$
47. $\left(10\dfrac{1}{4}\right)^2$
48. $(95)^2$
49. $(28)^2$
50. $\left(9\dfrac{1}{2}\right)^2$
51. 89×111
52. 92×108

Directions for questions 53 and 54: Find the HCF of the following.

53. $26x^2y^2z^2$ and $39x^3y^2z$
54. $50ab$ and $60bc$

Directions for questions 55 and 56: Factorise the following.

55. $x^2(x + y) + y^2(x + y)$
56. $(3x + 2y)(a - b) + (3x - 2y)(a - b)$
57. Find the value of 994×1006 by using an appropriate identity.
58. Check whether the following are perfect squares or not?
    (a) $64x^2 + 81y^2 - 144xy$.
    (b) $4x^2 + 8y^2 + 16xy$.
59. Divide $x^2 + 6x + 8$ by $x + 2$ and find the quotient.
60. Factorise: $a^3 - ab^2 + a^2b - b^3$.
61. Find the HCF of $24abc^3$, $36ab^3c$ and $48a^3bc$.
62. Factorise: $x^4 + y^4 - 2x^2y^2$.
63. If $y + \dfrac{1}{2y} = 4$, then find $y^2 + \dfrac{1}{4y^2}$.
64. Simplify: $\sqrt{\dfrac{169p^3q^2}{225pq^4}}$
65. Simplify: $\dfrac{x^4 - y^4}{x^2 + y^2}$

Essay Type Questions

66. Find the product of $(3a + 4b)$ and $(3a - 4b)$ and verify it when $a = -1$ and $b = 1$.

67. If $x + \dfrac{1}{2x} = 4$, then find $x^2 + \dfrac{1}{4x^2}$.

68. If $5x - \dfrac{1}{2x} = 3$, then find $25x^2 + \dfrac{1}{4x^2}$.

69. If $a + \dfrac{1}{a} = 6$, then find $a^4 + \dfrac{1}{a^4}$.

70. Divide $(a^4 - b^4)$ by $a - b$ and find the quotient and remainder.

71. Divide $4x^3 + 8x^2 - 9x + 6$ by $2x + 3$ and verify the division rule.

72. Divide $x^3 + y^3$ by $x + y$ and verify the division rule.

Directions for questions 73 to 75: Factorise the following.

73. $4x^2 + 8y^2 + 12xy$

74. $(3x + 2y)^2 - (5x - 3y)^2$

75. $a^4 - a - a^2 - 1$

76. If $x - \dfrac{1}{x} = 2$, then find $x^4 + \dfrac{1}{x^4}$.

77. If $36a^2 + \dfrac{1}{4a^2} = 31$, then find the value(s) of $6a - \dfrac{1}{2a}$.

78. If $m + n = 14$ and $mn = 48$, then find the value(s) of $m - n$.

CONCEPT APPLICATION

Level 1

Directions for questions 1 to 13: Select the correct alternative from the given choices.

1. Which of the following pairs is/are like terms?
   (A) x (B) x^2
   (C) $3x^3$ (D) $4x^3$
   (a) A, B   (b) B, C
   (c) C, D   (d) None of these

2. Degree of $5x^2y + 3xy$ is _____.
   (a) 3   (b) 2
   (c) 4   (d) 5

3. Zero of $3x - \dfrac{3}{2}$ is _____.
   (a) $\dfrac{3}{2}$ (b) $\dfrac{2}{3}$
   (c) $\dfrac{1}{2}$ (d) $\dfrac{1}{3}$

4. If x, y and z are variables, then $x + y + z$ is a _____.
   (a) Monomial   (b) Binomial
   (c) Trinomial (d) None of these

5. Which of the following is not an identity?
   (a) $a^2 + 2ab + b^2 = (a + b)(a + b)$.
   (b) $(x - y)^2 = x^2 - 2xy + y^2$.
   (c) $(p + q)(p - q) = p^2 - q^2$.
   (d) $x + 2 = 3$.

6. The zero of $2x + 3$ is _____.
   (a) -2 (b) -3
   (c) $-\dfrac{3}{2}$ (d) $-\dfrac{2}{3}$

7. The expanded form of $(x + y)^2$ is a _____.
   (a) Monomial
   (b) Binomial
   (c) Trinomial
   (d) None of these

8. $\sqrt{\dfrac{256a^4b^4}{625a^2b^2}} = $ _____.
   (a) $\dfrac{16b}{25a^2}$ (b) $\dfrac{16b}{25a}$
   (c) $\dfrac{4b}{25a}$ (d) $\dfrac{4b^2}{25a}$

9. $\dfrac{a^2 - b^2}{a - b} = $ _____.

   (a) $a + b$
   (b) $a - b$
   (c) $a^2 + b^2$
   (d) $(a - b)^2$

10. The following are the steps involved in expanding $(x + 3y)^2$. Arrange them in sequential order from the first to the last.

    (A) $(x + 3y)^2 = x^2 + 6xy + 9y^2$.
    (B) $(x + 3y)^2 = (x)^2 + 2(x)(3y) + (3y)^2$.
    (C) Using identity, $(a + b)^2 = a^2 + 2ab + b^2$, where $a = x$ and $b = 3y$.

    (a) ACB
    (b) CAB
    (c) CBA
    (d) ABC

11. The following are the steps involved in finding the value of $(97)^2$ by using a suitable identity. Arrange them in sequential order from the first to the last.

    (A) $10000 - 600 + 9 = 9409$
    (B) Using identity, $(a - b)^2 = a^2 - 2ab + b^2$, where $a = 100$ and $b = 3$.
    (C) $(97)^2 = (100 - 3)^2$
    (D) $(100 - 3)^2 = (100)^2 - 2(100)(3) + 3^2$

    (a) CABD
    (b) CADB
    (c) CBAD
    (d) CBDA

12. The following are the steps involved in finding the value of $a^4 + \dfrac{1}{a^4}$ when $a + \dfrac{1}{a} = 1$. Arrange them in sequential order from the first to the last.

    (A) $a^2 + \dfrac{1}{a^2} + 2 = 1 \Rightarrow a^2 + \dfrac{1}{a^2} = -1$
    (B) $(a^2)^2 + \left(\dfrac{1}{a^2}\right)^2 + 2 = 1$
    (C) $\left(a + \dfrac{1}{a}\right)^2 = 1^2$
    (D) $\left(a^2 + \dfrac{1}{a^2}\right)^2 = (-1)^2$
    (E) $a^4 + \dfrac{1}{a^4} = -1$

    (a) CADBE
    (b) CDBAE
    (c) CBADE
    (d) CEDAB

13. The following are the steps involved in the factorisation of $x^2(x - y) + (y - x)y^2$. Arrange them in sequential order from the first to the last.

    (A) $(x - y)[x^2 - y^2]$
    (B) $x^2(x - y) - (x - y)y^2$
    (C) $(x - y)^2(x + y)$
    (D) $(x - y)[(x + y)(x - y)]$

    (a) ABDC
    (b) BADC
    (c) ABCD
    (d) CDAB

Directions for questions 14 to 17: Match Column A with Column B.

Column A		Column B
14. If $a^2 - b^2 = 16$ and $a - b = 2$, then $a + b$	( )	(a) $3x^2 - x - 11$
15. The degree of $(x - a)(x - b)(x - c)(x - d)$	( )	(b) $3x^2 - x + 5$
16. $4x^2 + 20xy + 25y^2$	( )	(c) 8
17. $(5x^2 + 7x - 3) - (2x^2 + 8x - 8)$	( )	(d) 4
		(e) $(4x + 5y)(x + 5y)$
		(f) $(2x + 5y)^2$

Directions for questions 18 to 21: Match Column A with Column B.

Column A		Column B
18. $(3x^2 - 5) - (2x^2 - 5 + y^2)$	( )	(a) $x^2 + xy + y^2$
19. $9x^2 - 16y^2$	( )	(b) 2
20. $\dfrac{x^3 - y^3}{x - y}$	( )	(c) $(9x + 16y)(9x - 16y)$
21. The degree of $(x + 2)(x + 3)$	( )	(d) $x^2 - y^2$
		(e) 1
		(f) $(3x + 4y)(3x - 4y)$

2.18 Chapter 2

Level 2

Directions for questions 22 to 40: Select the correct alternative from the given choices.

22. If $a^2 - b^2 = 36$ and $a + b = 4$, then $(a - b)^2 =$ _____.
    (a) 36   (b) 9
    (c) 81   (d) 144

23. If $X = 3x^3 + 3x^2 + 3x + 3$ and $Y = 3x^2 - 3x + 3$, then $X - Y =$ _____.
    (a) $3x^3$
    (b) $3x^3 + 6x^2 + 6x + 6$
    (c) $6x^2 + 6x + 6$
    (d) $3x^3 + 6x$

24. The sum of the values of the expression $2x^2 - 2x + 2$ when $x = -1$ and $x = 1$ is _____.
    (a) 6   (b) 8
    (c) 4   (d) 2

25. If $X = 2x^2$, $Y = 4x^6 - 6x^4$, then find the value of $\dfrac{Y}{X}$ when $x = 1$.
    (a) −4   (b) 2
    (c) −1   (d) 3

26. If $x + \dfrac{1}{x} = 6$, then find $x^2 + \dfrac{1}{x^2}$.
    (a) 34   (b) 36
    (c) 32   (d) 38

27. If $x = -2$ and $x^2 + y^2 + 3xy = -5$, then find y.
    (a) −2   (b) 3
    (c) −4   (d) 9

28. Number of terms in the expansion $(a + b)(c + d)$ is _____.
    (a) 1   (b) 2
    (c) 3   (d) 4

29. $46 \times 46 + 54 \times 54 + 2 \times 46 \times 54 =$ _____.
    (a) 9996   (b) 10004
    (c) 9800   (d) 10000

30. $\dfrac{79^2 - 9^2}{89^2 - 9^2} =$ _____.
    (a) $\dfrac{5}{8}$ (b) $\dfrac{13}{15}$
    (c) $\dfrac{6}{9}$ (d) $\dfrac{11}{14}$

31. $\sqrt{\dfrac{81(x+y)^2}{144(x-y)^2}} =$ _____.
    (a) $\dfrac{9(x+y)^{\sqrt{2}}}{12(x-y)^{\sqrt{2}}}$ (b) $\dfrac{9(x+y)^2}{12(x-y)^2}$
    (c) $\dfrac{3(x+y)}{4(x-y)}$ (d) None of these

32. If $A = (3x + 6)$ and $B = 2x^2 + 3x + 4$, then the degree of AB is _____.
    (a) 4   (b) 3
    (c) 2   (d) 1

33. Find the HCF of $18p^2qr$, $24pq^2r$ and $27pqr^2$.
    (a) $216pqr$ (b) $3pqr$
    (c) $216(pqr)^2$ (d) $3(pqr)^2$

34. Find the degree of $(x^3 - x^2)^2$.
    (a) 12   (b) 4
    (c) 6   (d) 9

35. The value of 998^2 is
    (a) 996064 (b) 996004
    (c) 998004 (d) 998064

36. $x^4 + y^4 + 2x^2y^2 =$ _____.
    (a) $(x + y)^4$
    (b) $(x^2 - y^2)^2$
    (c) $(x^2 + y^2)(x^2 - y^2)$
    (d) $(x^2 + y^2)^2$

37. If $3x + \dfrac{1}{x} = 6$, then find $9x^2 + \dfrac{1}{x^2}$.
    (a) 24   (b) 27
    (c) 30   (d) 33

38. Which of the following is/are perfect squares?
    (a) $16a^2 + 36b^2 - 48ab$
    (b) $9x^2 + 18xy + 9y^2$
    (c) Both (a) and (b)
    (d) Neither (a) nor (b)

39. For what value of K is $16x^2 + 24xy + K$ a perfect square?
    (a) $9y^2$
    (b) $18y^2$
    (c) $3y^2$
    (d) $16y^2$

40. If $2y + \dfrac{1}{2y} = l$ and $2y - \dfrac{1}{2y} = m$, then $l^2 - m^2 =$ _____.
    (a) 2
    (b) 4
    (c) 6
    (d) 0

Level 3

Directions for questions 41 to 65: Select the correct alternative from the given choices.

41. If $x^2 - y^2 = 28$ and $x + y = 7$ then $(x - y)^2 =$ _____.
    (a) 8
    (b) 4
    (c) 16
    (d) 12

42. Factorise $a^4 + a^3 + a^2 + a$.
    (a) $(a + 1)(a^2 + 1)(a - 1)$
    (b) $a(a + 1)(a^2 + 1)$
    (c) $(a^2 + 1)(a^2 + 1)$
    (d) None of these

43. Factorise $(2a + 3b)^2 - (3a - 2b)^2$.
    (a) $(5a + b)(5a - b)$
    (b) $(a + 5b)(a - 5b)$
    (c) $(5a + b)(5b - a)$
    (d) $(5a + b)(5b + a)$

44. What is the remainder when $(2x^2 + 3x + 7)$ is divided by $(x + 2)$?
    (a) 3
    (b) 9
    (c) 7
    (d) 5

45. What is the quotient when $(x^3 + 8)$ is divided by $(x^2 - 2x + 4)$?
    (a) $x - 2$
    (b) $x + 2$
    (c) $x + 1$
    (d) $x - 1$

46. Which of the following is a perfect square?
    (a) $9x^2 + 24xy + 4y^2$
    (b) $4x^2 + 12xy + 3y^2$
    (c) $25x^2 - 10xy + 4y^2$
    (d) $25x^2 - 30xy + 9y^2$

47. If $x + y = 7$ and $xy = 2$, then $x^2 - y^2 =$ _____. $(x > y)$.
    (a) $7\sqrt{46}$
    (b) $7\sqrt{44}$
    (c) $7\sqrt{41}$
    (d) $7\sqrt{43}$

48. If $x + \dfrac{1}{x} = 2$, then $x^{100} - \dfrac{1}{x^{100}} =$ _____.
    (a) 0
    (b) 1
    (c) 2
    (d) 2100

49. Factorise $x^2 + a^2 + 2a + 2x + 2ax$.
    (a) $(x + a)(x + a + 2)$
    (b) $(x + a)(x - a - 2)$
    (c) $(x - a)(x + a + 2)$
    (d) $(x - a)(x - a + 2)$

50. For what value of p is $9x^2 + 18xy + p$ a perfect square?
    (a) $9y^2$
    (b) $3y$
    (c) $6y^2$
    (d) $4y^2$

51. If $2y + \dfrac{1}{2y} = 3$, then $16y^4 + \dfrac{1}{16y^4} =$ _____.
    (a) 81
    (b) 79
    (c) 49
    (d) 47

52. If $496 \times 492 = x^2 - 4$ $(x > 0)$, then $x =$ _____.
    (a) 495
    (b) 494
    (c) 493
    (d) 496

53. Factorise $x^4 + x^2 + 1$.
    (a) $(x^2 - x - 1)(x^2 + x - 1)$
    (b) $(x^2 + x + 1)(x^2 - x + 1)$

(c) $(x^2 - x + 1)(x^2 + x)$
(d) $(x^2 + x - 1)(x^2 - 1)$

54. If $x + \dfrac{1}{x} = a$ and $x - \dfrac{1}{x} = b$, then $a^2 - b^2 =$ _____.

   (a) 4 (b) 3
   (c) 2 (d) None of these

55. Factorise $y^2 + 2xy + 2xz - z^2$.

   (a) $(x - y + z)(y + z)$
   (b) $(x + y + z)(y - z)$
   (c) $(y - z)(y + z + 2x)$
   (d) $(y + z)(y - z + 2x)$

56. If $A = (x - a)(x - b)(x - c) \ldots (x - z)$, then the number of terms in the expansion of $(a + A)(b + A)(c + A) \ldots (z + A)$ is _____.

   (a) 1 (b) 27
   (c) 56 (d) 54

57. $(4x^4 + 19x^2 + 25) \div (2x^2 - x + 5) =$ _____.

   (a) $2x^2 - 9x + 5$
   (b) $2x^2 + 9x + 5$
   (c) $2x^2 + x + 5$
   (d) $2x^2 - x + 5$

58. If $P = 8x^4 + 6x^3 - 15x^2 + 27x - 20$ and $Q = 2x^2 + 3x - 4$, then find the remainder when P is divided by Q.

   (a) 0 (b) −1
   (c) −8 (d) −4

59. $x^3 + xy^2 - x^2y - y^3 =$ _____.

   (a) $(x^2 + y^2)(x + y)$
   (b) $(x^2 + y^2)(x - y)$
   (c) $(x - y)(x + y)^2$
   (d) $(x + y)(x - y)^2$

60. If $y - \dfrac{1}{y} = 3$, then find $y^4 + \dfrac{1}{y^4}$.

   (a) 119 (b) 117
   (c) 123 (d) 125

61. If $64x^2 + \dfrac{1}{64x^2} = 20$, then $8x - \dfrac{1}{4x}$ can be

   (a) 2 (2) 8
   (3) 1 (4) 4

62. If $p + q = 15$ and $pq = 54$, then $p - q$ can be

   (a) 3 (b) 5
   (c) 4 (d) 6

63. Divide $x^2 + 7x + 12$ by $x + 3$ and find the quotient.

   (a) $x + 4$ (b) $x - 4$
   (c) $2x + 4$ (4) $2x - 4$

64. What should be subtracted from $x^3 + 2x^2 - 3x + 10$, so that the difference is a multiple of $x - 2$?

   (a) 10 (b) 20
   (c) 15 (d) 30

65. If $C = (7x + 9)$ and $D = (4x^2 + 8x + 5)$, then the degree of the product CD is _____.

   (a) 3 (b) 4
   (c) 2 (d) 1

ASSESSMENT TESTS

Test 1

Directions for questions 1 to 11: Select the correct alternative from the given choices.

1. Factorise: $a^3 - 3b^2 + 3a^2 - ab^2$
   The following are the steps involved in solving the above problem. Arrange them in sequential order.

   (A) $(a + 3)(a^2 - b^2)$
   (B) Rearrange the terms as $a^3 + 3a^2 - 3b^2 - ab^2$
   (C) $a^2(a + 3) - b^2(3 + a)$
   (D) $(a + 3)(a + b)(a - b)$

   (a) BCDA (b) BCAD
   (c) BDCA (d) ABCD

2. The following are the steps involved in finding the value of $10\frac{1}{3} \times 9\frac{2}{3}$ by using an appropriate identity. Arrange them in sequential order.

(A) $(10)^2 - \left(\frac{1}{3}\right)^2 = 100 - \frac{1}{9}$

(B) $10\frac{1}{3} \times 9\frac{2}{3} = (10+\frac{1}{3})(10-\frac{1}{3})$

(C) $(10+\frac{1}{3})(10-\frac{1}{3}) = (10)^2 - \left(\frac{1}{3}\right)^2$
[∵ $(a+b)(a-b) = (a^2-b^2)$]

(D) $100 - \frac{1}{9} = 99 + 1 - \frac{1}{9} = 99\frac{8}{9}$

(a) BCDA (b) BACD
(c) ACBD (d) BCAD

3. If $a^2 + 16a + k$ is a perfect square, then find the value of k.

(a) 4 (b) 16
(c) 36 (d) 64

4. If $x + \frac{1}{x} = 2$, then $x^{2013} - \frac{1}{x^{2012}} =$ _____.

(a) 2 (b) 1
(c) 0 (d) −1

5. $(2x - 6y)^2 - (6y + 8x)^2 =$ _____.

(a) $(5x)(3x + 6y)$
(b) $-4(5x)(3x - 6y)$
(c) $-12(5x)(x + 2y)$
(d) $-6(5x)(x - 2y)$

6. The HCF of $64x^6y^4$, $48x^4y^8$ and $54x^5y^4$ is _____.

(a) $2x^4y^4$ (b) $6x^2y^2$
(c) $8x^4y^4$ (d) $8x^2y$

7. Factorise: $12a^3b^3 - 3ab$

(a) $3(a^2b^2 - 1)$
(b) $3ab(a - b)$
(c) $3ab(2ab + 1)(2ab - 1)$
(d) $3ab(a + 1)(b - 1)$

8. $(a^5 - a^3) \div (a^2 + a) =$ _____.

(a) $a^2(a + 1)$ (b) $a(a - 1)$
(c) $a^2(a - 1)$ (d) $a^3(a + 1)$

9. If $3x - \frac{1}{2x} = 3$, then find the value of $\frac{36x^4 + 1}{4x^2}$.

(a) 9 (b) 12
(c) 15 (d) 6

10. Find the square root of $3^{6n^2}(36)^{2a}(16)^b$.

(a) $3^{3n^2}6^{2a}2^{2b}$ (b) $3^{6n}6^a4^b$
(c) $3^{3n^2}6^a4^b$ (d) $3^{6n}6^a4^{2b}$

11. $\frac{x^4}{4} + \frac{y^3}{3} - \frac{3z^2}{5} + \frac{x^4}{3} - \frac{3y^3}{5} + \frac{z^2}{4} - \frac{3x^4}{5} + \frac{y^3}{4} + \frac{z^2}{3} =$ _____.

(a) $\frac{-1}{60}(x^4 + y^3 + z^2)$

(b) $\frac{1}{60}(x^4 - y^3 + z^2)$

(c) $\frac{x^4 + y^3 + z^2}{60}$

(d) $\frac{x^4 + y^3 - z^2}{60}$

Directions for questions 12 to 15: Match Column A with Column B.

Column A	Column B
12. $225x^2 - 625y^2 =$ ( )	(a) $25(x-2)(x-2)$
13. $x^2 - x - y - y^2 =$ ( )	(b) $25(3x-5y)(3x+5y)$
14. $x^2 - x - y^2 + y =$ ( )	(c) $(x+y)(x-y-1)$
15. $25x^2 - 100x + 100 =$ ( )	(d) $(x-y)(x+y-1)$
	(e) $(x+y)(x+y-1)$

2.22 Chapter 2

Test 2

Directions for questions 1 to 26: Select the correct alternative from the given choices.

16. Factorise $25x^2 - 30xy + 9y^2$.

    The following are the steps involved in solving the above problem. Arrange them in sequential order.

    (A) $(5x - 3y)^2$ $[\because a^2 - 2ab + b^2 = (a - b)^2]$
    (B) $(5x)^2 - 30xy + (3y)^2 = (5x)^2 - 2.(5x)(3y) + (3y)^2$
    (C) $(5x - 3y)(5x - 3y)$

    (a) ABC   (b) BCA
    (c) ACB   (d) BAC

17. The following are the steps involved in finding the value of $100\frac{3}{4} \times 99\frac{1}{4}$ by using an appropriate identity. Arrange them in sequential order.

    (A) $(100 + \frac{3}{4})(100 - \frac{3}{4}) = (100)^2 - (\frac{3}{4})^2$
        $[\because (a + b)(a - b) = a^2 - b^2]$
    (B) $10000 - \frac{9}{16} = 9999 + 1 - \frac{9}{16} = 9999\frac{7}{16}$
    (C) $(100)^2 - (\frac{3}{4})^2 = 10000 - \frac{9}{16}$
    (D) $100\frac{3}{4} \times 99\frac{1}{4} = (100 + \frac{3}{4})(100 - \frac{3}{4})$

    (a) DABC   (b) DACB
    (c) DBAC   (d) DCAB

18. If $x^2 + 8x + k$ is a perfect square, then find the value of k.

    (a) 1   (b) 4
    (c) 16   (d) 64

19. If $x + \frac{1}{x} = 2$, then $x^{2010} + x^{2009} = $ _____.

    (a) 4019   (b) 2
    (c) 0   (d) 1

20. $(3x - 4y)^2 - (4x + 3y)^2 = $ _____.

    (a) $(7x - y)(7y - x)$
    (b) $(x - 7y)(7x + y)$
    (c) $(x + 7y)(7x - y)$
    (d) $(y - 7x)(x + 7y)$

21. The HCF of $44 x^5y^4$, $88x^6y^5$ and $66x^7y^6$ is _____.

    (a) $22x^4y^4$ (2) $22x^5y^4$
    (3) $11x^7y^6$ (4) $44x^5y^4$

22. Factorise $x^4 + x^3 - x^2 - x$.

    (a) $(x^2 + 1)(x - 1)x$
    (b) $x(x + 1)^2(x - 1)$
    (c) $(x^2 + 1)(x - 1)^2$
    (d) $x(x - 1)^2(x + 1)$

23. $(a^4 - a^2) \div (a^3 + a^2) = $ _____.

    (a) $a^2 + 1$ (b) $a - 1$
    (c) $a + 1$ (d) $a^2 - 1$

24. If $x - \frac{1}{2x} = 2$, then find the value of $\frac{4x^4 + 1}{4x^2}$.

    (a) 10   (b) 14
    (c) 5   (d) 7

25. Find the square root of $4^{6n^2} (25)^{\frac{a}{2}} 9^{b^4}$.

    (a) $4^{3n^2} 5^{\frac{a}{4}} 3^{b^2}$ (b) $2^{6n^2} 5^{\frac{a}{2}} 3^{b^2}$
    (c) $2^{6n^2} 5^{\frac{a}{2}} 3^{b^4}$ (d) $4^{3n^2} 5^a 3^{b^2}$

26. $\frac{a^2}{2} + \frac{b^3}{3} - \frac{3c^3}{4} + \frac{a^2}{3} - \frac{3b^3}{4} + \frac{c^3}{2} - \frac{3a^2}{4} + \frac{b^3}{2} + \frac{c^3}{3} = $ _____.

    (a) $\frac{a^2 + b^3 + c^3}{12}$ (b) $\frac{a^2 + b^3 - c^3}{6}$
    (c) $\frac{6a^2 + 9b^3 - 12c^3}{24}$ (d) $\frac{6a^2 + 9b^3 - 12c^3}{12}$

Directions for questions 27 to 30: Match Column A with Column B.

Column A		Column B
27. $(a + b)^2 =$	( )	(a) $(x + 1)^2 (x - 1)$
28. $x^3 - x^2 - x + 1 =$	( )	(b) $(a - b) + 4ab$
29. $(a + b)(a^2 - ab + b^2) =$	( )	(c) $(a^3 + b^3)$
30. $(x + 1)(x^2 - 1) =$	( )	(d) $(x - 1)^2 (x + 1)$
		(e) $(a^3 - b^3)$

Expressions and Special Products

TEST YOUR CONCEPTS

Very Short Answer Type Questions

1. True
2. False
3. True
4. False
5. True
6. True
7. True
8. True
9. True
10. False
11. $5x^3 - 5x^2y$
12. $(2x - 1)$
13. $x^2 + (a + b)x + ab$
14. $(x + y)(x - y)$
15. $a^2 + 6ab + 9b^2$

16. (a)
17. (b)
18. (b)
19. (d)
20. (d)
21. (b)
22. (c)
23. (c)
24. (a)
25. (c)
26. (b)
27. (c)
28. (c)
29. (b)
30. (b)

Short Answer Type Questions

31. $7x^3 + 3x^2 - x - 3$
32. $-7x^3 + 3x^2 + 5x - 11$
33. $21x^3 + 6x^2 - 5x - 2$
34. $-21x^3 + 6x^2 + 13x - 26$
35. $3x^2 - 7x - 6$
36. $x^2 + x - 30$
37. $15x^3 - 21x^2 + 18x$
38. $12x^2 - 10x - 42$
39. $9x^2 + 24xy + 16y^2$
40. $25x^2 + 2 + \dfrac{1}{25x^2}$
41. $x^2 - 4xy + 4y^2$
42. $9x^2 - 2 + \dfrac{1}{9x^2}$
43. $x^2 - 4y^2$
44. $x^4 - y^4$

45. 10404
46. 2916
47. $105\dfrac{1}{6}$
48. 9025
49. 784
50. $90\dfrac{1}{4}$
51. 9879
52. 9936
53. $13x^2y^2z$
54. $10b$
55. $(x + y)(x^2 + y^2)$
56. $6x(a - b)$
57. 999964
58. (a) Perfect square (b) Not perfect square

Chapter 2

59. $x + 4$
60. $(a - b)(a + b)^2$
61. $12abc$
62. $(x - y)^2(x + y)^2$
63. 15
64. $\dfrac{13p}{15q}$
65. $x^2 - y^2$

Essay Type Questions

66. $9a^2 - 16b^2$
67. 15
68. 14
69. 1154
70. $a^3 + a^2b + ab^2 + b^3$, 0
73. $4(x + 2y)(x + y)$
74. $(8x - y)(-2x + 5y)$
75. $(a + 1)[a^3 - a^2 - 1]$
76. 34
77. ± 5
78. ± 2

CONCEPT APPLICATION

Level 1

| 1. (c) | 2. (a) | 3. (c) | 4. (c) | 5. (d) | 6. (c) | 7. (c) | 8. (b) | 9. (a) | 10. (c) |
| 11. (d) | 12. (a) | 13. (b) | 14. (c) | 15. (d) | 16. (f) | 17. (b) | 18. (d) | 19. (f) | 20. (a) |
| 21. (b) |

Level 2

| 22. (c) | 23. (d) | 24. (b) | 25. (c) | 26. (a) | 27. (b) | 28. (d) | 29. (b) | 30. (d) | 31. (c) |
| 32. (b) | 33. (b) | 34. (c) | 35. (b) | 36. (d) | 37. (c) | 38. (c) | 39. (a) | 40. (b) |

Level 3

41. (c)	42. (b)	43. (c)	44. (b)	45. (b)	46. (d)	47. (c)	48. (a)	49. (a)	50. (a)
51. (d)	52. (b)	53. (b)	54. (a)	55. (d)	56. (a)	57. (c)	58. (a)	59. (b)	60. (a)
61. (d)	62. (a)	63. (a)	64. (b)	65. (a)

ASSESSMENT TESTS

Test 1

| 1. (b) | 2. (d) | 3. (d) | 4. (c) | 5. (c) | 6. (a) | 7. (c) | 8. (c) | 9. (b) | 10. (a) |
| 11. (a) | 12. (b) | 13. (c) | 14. (d) | 15. (a) |

Test 2

| 16. (d) | 17. (b) | 18. (c) | 19. (b) | 20. (d) | 21. (b) | 22. (b) | 23. (b) | 24. (c) | 25. (c) |
| 26. (a) | 27. (b) | 28. (d) | 29. (c) | 30. (a) |

CONCEPT APPLICATION

Level 1

1. We know that terms which have same literal factors are called like terms. Only c and d are like terms.

   Hence, the correct option is (c)

2. Degree of $5x^2y + 3xy$ is same as degree of $5x^2y$ which is $2 + 1 = 3$.

   Hence, the correct option is (a)

3. $3x - \dfrac{3}{2} = 0$

   $\Rightarrow 3x = \dfrac{3}{2}$

   $\Rightarrow x = \dfrac{1}{2}$.

   Hence, the correct option is (c)

4. $x + y + z$ is a trinomial.

   Hence, the correct option is (c)

5. $x + 2 = 3$ is not an identity.

   Hence, the correct option is (d)

6. Let $2x + 3 = 0$.

   $\Rightarrow 2x = -3 \Rightarrow x = -\dfrac{3}{2}$

   $\therefore$ Zero of $2x + 3$ is $-\dfrac{3}{2}$.

   Hence, the correct option is (c)

7. $(x + y)^2 = x^2 + 2xy + y^2$

   It has three terms.

   $\therefore$ It is a trinomial.

   Hence, the correct option is (c)

8. $\sqrt{\dfrac{256a^4b^4}{625a^6b^2}} = \sqrt{\dfrac{(16)^2 b^2}{(25)^2 a^2}}$

   $= \dfrac{16b}{25a}$

   Hence, the correct option is (b)

9. $\dfrac{a^2 - b^2}{a - b} = \dfrac{(a+b)(a-b)}{a-b} = a + b$

   Hence, the correct option is (a)

10. (C), (B) and (A) is the sequential order from the first to the last.

    Hence, the correct option is (c)

11. (C), (B), (D) and (A) is the sequential order from the first to the last.

    Hence, the correct option is (d)

12. (C), (A), (D), (B) and (E) is the sequential order from the first to the last.

    Hence, the correct option is (a)

13. (B), (A), (D) and (C) is the sequential order from the first to the last.

    Hence, the correct option is (b)

14. $\to c$: $a + b = \dfrac{a^2 - b^2}{a - b} = \dfrac{16}{2} = 8$.

15. $\to d$: The degree of $(x - a)(x - b)(x - c)(x - d)$ is 4.

16. $\to f$: $4x^2 + 20xy + 25y^2$

    $= (2x)^2 + 2(2x)(5y) + (5y)^2$

    $= (2x + 5y)^2$.

17. $\to b$: $(5x^2 + 7x - 3) - (2x^2 + 8x - 8)$

    $= 5x^2 - 2x^2 + 7x - 8x - 3 + 8$

    $= 3x^2 - x + 5$.

18. $\to d$: $(3x^2 - 5) - (2x^2 - 5 + y^2)$

    $= 3x^2 - 2x^2 - 5 + 5 - y^2$

    $= x^2 - y^2$.

19. $\to f$: $9x^2 - 16y^2 = (3x)^2 - (4y)^2$

    $= (3x - 4y)(3x + 4y)$

20. $\to a$: $\dfrac{x^3 - y^3}{x - y} = \dfrac{(x - y)(x^2 + xy + y^2)}{(x - y)}$

    $= x^2 + xy + y^2$.

    (by division)

21. $\to b$: $(x + 2)(x + 3) = x^2 + 3x + 2x + 6$

    $= x^2 + 5x + 6$

    $\therefore$ The degree of $(x + 2)(x + 3)$ is 2.

2.26 Chapter 2

Level 2

22. $a + b = 4$ (1)

$a^2 - b^2 = (a + b)(a - b) = 36$ (2)

(2) ÷ (1) we get

$\dfrac{(a-b)(a+b)}{(a+b)} = \dfrac{36}{4} = 9$

$a - b = 9$

$\Rightarrow (a - b)^2 = 9^2 = 81.$

Hence, the correct option is (c)

23. $X = 3x^3 + 3x^2 + 3x + 3$

$Y = + 3x^2 - 3x + 3$

$\; - + -$

——————————————

$X - Y = 3x^3 + 6x$

Hence, the correct option is (d)

24. When $x = -1$, $2x^2 - 2x + 2$

$= 2(-1)^2 - 2(-1) + 2 = 6.$

When $x = 1$, $2x^2 - 2x + 2$

$= 2(1)^2 - 2(1) + 2 = 2.$

∴ The required sum $= 6 + 2 = 8.$

Hence, the correct option is (b)

25. Given $X = 2x^2$,

$Y = 4x^6 - 6x^4$ and $x = -1$.

$\dfrac{Y}{X} = \dfrac{4x^6 - 6x^4}{2x^2}$

$= 2x^4 - 3x^2$

$= 2(-1)^4 - 3(-1)^2 = -1.$

Hence, the correct option is (c)

26. Given, $x + \dfrac{1}{x} = 6$

Taking squares on both the sides, we get

$\left(x + \dfrac{1}{x}\right)^2 = 6^2$

$\Rightarrow x^2 + \dfrac{1}{x^2} + 2 = 36$

$\Rightarrow x^2 + \dfrac{1}{x^2} = 34.$

Hence, the correct option is (a)

27. Value of $x^2 + y^2 + 3xy = -5$ and $x = -2$

$\Rightarrow (-2)^2 + y^2 - 6y = -5$

$\Rightarrow y^2 - 6y + 9 = 0$

$\Rightarrow (y - 3)^2 = 0$

$\Rightarrow y - 3 = 0$

$\Rightarrow y = 3.$

Hence, the correct option is (b)

28. $(a + b)(c + d) = ac + ad + bc + bd$

∴ There are 4 terms.

Hence, the correct option is (d)

29. The recurring part of the decimal is called period.

Period of $10.23\overline{46}$ is 46.

Hence, the correct option is (b)

30. $= \dfrac{(79 + 9)(79 - 9)}{(89 + 9)(89 - 9)}$

$= \dfrac{88 \times 70}{98 \times 80} = \dfrac{11}{14}$

Hence, the correct option is (d)

31. $\sqrt{\dfrac{[9(x+y)^2]}{[12(x-y)^2]}}$

$= \dfrac{9(x+y)}{12(x-y)}$

$= \dfrac{3(x+y)}{4(x-y)}$

Hence, the correct option is (c)

32. Given, $A = (3x + 6)$ and $B = 2x^2 + 3x + 4$.

The degree of A is 1 and the degree of B is 2.

∴ Degree of AB is $= 1 + 2 = 3.$

Hence, the correct option is (b)

33. HCF$(18p^2qr, 24pq^2r$ and $27pqr^2)$

HCF$((pqr)(18p), (pqr)(24q), (pqr)(27r))$

$= (pqr)$HCF$(18p, 24q, 27r)$

$= (pqr)$HCF$(3(6p), 3(8q), 3(9r))$

$= (3pqr)$HCF$(6p, 8q, 9r)$

$= 3pqr(1) = 3pqr$

Hence, the correct option is (b)

Expressions and Special Products | **2.27**

34. $(x^3 - x^2)^2$

    $= (x^3)^2 - 2(x^3)(x^2) + (x^2)^2$

    $= x^6 - 2x^5 + x^4$.

    Degree = 6

    Hence, the correct option is (c)

35. $(998)^2 = (1000 - 2)^2$

    $= (1000)^2 - 2(1000)(2) + 2^2$

    $= 996004$

    Hence, the correct option is (b)

36. $x^4 + y^4 + 2x^2y^2 = (x^2)^2 + (y^2)^2 + 2x^2y^2 = (x^2 + y^2)^2$

    Hence, the correct option is (d)

37. $3x + \dfrac{1}{x} = 6$

    Squaring on both the sides, we get

    $\Rightarrow 9x^2 + \dfrac{1}{x^2} + 2(3x)\left(\dfrac{1}{x}\right) = 36$

    $(\because (a+b)^2 = a^2 + b^2 + 2ab)$

    $\Rightarrow 9x^2 + \dfrac{1}{x^2} + 6 = 36$

    $\Rightarrow 9x^2 + \dfrac{1}{x^2} = 30$.

    Hence, the correct option is (c)

38. Option (1) : $16a^2 + 36b^2 - 48ab$

    $(4a)^2 + (6b)^2 - 2(4a)(6b) = (4a - 6b)^2$ is a perfect square.

    Option (2) : $9x^2 + 18xy + 9y^2 = (3x + 3y)^2$ is a perfect square.

    Hence, the correct option is (c)

39. $16x^2 + 24xy + K$

    $= (4x)^2 + 2(4x)(3y) + K$ is a perfect square.

    $= (4x)^2 + 2(4x) \times (3y) + (3y)^2 = (4x + 3y)^2$

    $K = 9y^2$.

    Hence, the correct option is (a)

40. $2y + \dfrac{1}{2y} = l$

    $l^2 = (2y)^2 + 2 \times 2y \times \dfrac{1}{2y} + \left(\dfrac{1}{2y}\right)^2$

    $= 4y^2 + \dfrac{1}{4y^2} + 2$ (1)

    $2y - \dfrac{1}{2y} = m$

    $\Rightarrow \left(2y - \dfrac{1}{2y}\right)^2 = m^2$

    $= 4y^2 + \dfrac{1}{4y^2} - 2 \times \dfrac{1}{2y} \times 2y$

    $m^2 = 4y^2 + \dfrac{1}{4y^2} - 2$ (2)

    $l^2 - m^2 = (1) - (2)$

    $= \left(4y^2 + \dfrac{1}{4y^2} + 2\right) - \left(4y^2 + \dfrac{1}{4y^2} - 2\right) = 4$

    Hence, the correct option is (b)

Level 3

41. Given, $x^2 - y^2 = 28$ and $x + y = 7$

    $x^2 - y^2 = 28$

    $(x + y)(x - y) = 28$

    $\Rightarrow x - y = 4$

    $\Rightarrow (x - y)^2 = 4^2 = 16$.

    Hence, the correct option is (c)

42. $a^4 + a^3 + a^2 + a = a(a^3 + a^2 + a + 1)$

    $= a[a^2(a + 1) + a + 1]$

    $= a(a + 1)(a^2 + 1)$

    Hence, the correct option is (b)

43. $(2a + 3b)^2 - (3a - 2b)^2$

    $= [2a + 3b + 3a - 2b][2a + 3b - 3a + 2b]$

    $= (5a + b)(5b - a)$

    Hence, the correct option is (c)

44. $(x + 2)\ \overline{\smash{\big)}\ 2x^2 + 3x + 7}\ (2x - 1)$
    $\phantom{(x+2)\ \overline{\smash{\big)}\ }}2x^2 + 4x$
    $\phantom{(x+2)\ \overline{\smash{\big)}\ }}(-)(-)$
    $\phantom{(x+2)\ \overline{\smash{\big)}\ 2x^2}}\overline{-x + 7}$
    $\phantom{(x+2)\ \overline{\smash{\big)}\ 2x^2}}-x - 2$
    $\phantom{(x+2)\ \overline{\smash{\big)}\ 2x^2}}(+)(+)$
    $\phantom{(x+2)\ \overline{\smash{\big)}\ 2x^2}}\overline{9}$

    $\therefore$ The remainder = 9.

    Hence, the correct option is (b)

2.28 Chapter 2

45. $x^2 - 2x + 4 \overline{\smash{\big)}\, x^3 + 0x^2 + 0x + 8} \, (x + 2$
$\phantom{x^2 - 2x + 4 \overline{\smash{\big)}\,}} x^3 - 2x^2 + 4x$
$(-) (+) (-)$
$\overline{}$
$ + 2x^2 - 4x + 8$
$ + 2x^2 - 4x + 8$
$(-) (+) (-)$
$\overline{}$
$ 0$

∴ The quotient = $x + 2$.

Hence, the correct option is (b)

46. From the options,

Option(4): $25x^2 - 30xy + 9y^2$
$= (5x)^2 - 2 \times 5x \times 3y + (3y)^2$
$= (5x - 3y)^2$

∴ It is a perfect square.

Hence, the correct option is (d)

47. Given $x + y = 7$, and $xy = 2$

$x - y = \sqrt{(x+y)^2 - 4xy}$
$ = \sqrt{7^2 - 4(2)} = \sqrt{41}$

Now, $x^2 - y^2 = (x + y)(x - y)$
$ = 7\sqrt{41}$.

Hence, the correct option is (c)

48. Given $x + \dfrac{1}{x} = 2$

$x^2 + 1 = 2x \Rightarrow x^2 - 2x + 1 = 0$
$\Rightarrow (x - 1)^2 = 0$
$\Rightarrow x - 1 = 0$
$\Rightarrow x = 1$
$\Rightarrow x^{100} - \dfrac{1}{x^{100}} = 1 - 1 = 0$.

Hence, the correct option is (a)

49. $x^2 + a^2 + 2a + 2x + 2ax$
$= x^2 + a^2 + 2ax + 2a + 2x$
$= (x + a)^2 + 2(x + a)$
$= (x + a)(x + a + 2)$.

Hence, the correct option is (a)

50. Given, $9x^2 + 18xy + p$ is a perfect square.

If a trinomial is perfect square, then it must be in the form of $a^2 \pm 2ab + b^2$.

$9x^2 + 18xy + p = (3x)^2 + 2(3x)(3y) + p$
$\Rightarrow P = (3y)^2 = 9y^2$

Hence, the correct option is (a)

51. Given, $2y + \dfrac{1}{2y} = 3$.

Taking squares on both the sides,

$(2y + \dfrac{1}{2y})^2 = 3^2$

$4y^2 + \dfrac{1}{4y^2} + 2 = 9$

$4y^2 + \dfrac{1}{4y^2} = 7$

Again taking squares on both the sides,

$(4y^2 + \dfrac{1}{4y^2})^2 = 7^2$

$16y^4 + \dfrac{1}{16y^4} + 2 = 49$

$16y^4 + \dfrac{1}{16y^4} = 47$.

Hence, the correct option is (d)

52. Given, $496 \times 492 = x^2 - 4$
$\Rightarrow (494 + 2)(494 - 2) = x^2 - 2^2$
$\Rightarrow (494)^2 - 2^2 = x^2 - 2^2$
$x^2 = (494)^2$
$x = 494 \, (x > 0)$.

Hence, the correct option is (b)

53. $x^4 + x^2 + 1$

$x^4 + 2x^2 + 1 - x^2$
$= (x^2 + 1)^2 - x^2$
$= (x^2 + x + 1)(x^2 - x + 1)$.

Hence, the correct option is (b)

54. Given $x + \dfrac{1}{x} = a$ and $x - \dfrac{1}{x} = b$

Now $a^2 - b^2 = (a + b)(a - b)$

$= (x + \dfrac{1}{x} + x - \dfrac{1}{x})(x + \dfrac{1}{x} - x + \dfrac{1}{x})$

$= (2x)\left(\dfrac{2}{x}\right) = 4$.

Hence, the correct option is (a)

55. $y^2 + 2xy + 2xz - z^2$
$= y^2 - z^2 + 2xy + 2xz$
$= (y + z)(y - z) + 2x(y + z)$
$= (y + z)(y - z + 2x)$.
Hence, the correct option is (d)

56. $A = (x - a)(x - b)(x - c) \ldots (x - x) \ldots (x - z) = 0$
$(a + A)(b + A)(c + A) \ldots (z + A) = (a + 0)$
$(b + 0)(c + 0) \text{------} (z + 0) = a.b.c. \ldots z$
$\therefore$ There is only one term.
Hence, the correct option is (a)

57. $4x^4 + 19x^2 + 25$
$= (2x^2)^2 + (5)^2 + 20x^2 - x^2$
$= (2x^2 + 5)^2 - x^2$
$= (2x^2 + x + 5)(2x^2 - x + 5)$
$\Rightarrow \dfrac{4x^4 + 19x^2 + 25}{2x^2 - x + 5} = 2x^2 + x + 5$.
Hence, the correct option is (c)

58.
$$2x^2 + 3x - 4 \overline{) 8x^4 + 6x^3 - 15x^2 + 27x - 20} \ (4x^2 - 3x + 5$$
$ 8x^4 + 12x^3 - 16x^2$
$ (-) \ \ \ (-) \ \ \ (+)$
$ \overline{}$
$ -6x^3 + x^2 + 27x$
$ -6x^3 - 9x^2 + 12x$
$ + \ \ \ + \ \ \ -$
$ \overline{}$
$ 10x^2 + 15x - 20$
$ 10x^2 + 15x - 20$
$ (-) \ \ \ (-) \ \ \ (+)$
$ \overline{}$
$ 0$

$\therefore$ The remainder $= 0$.
Hence, the correct option is (a)

59. $x^3 + xy^2 - x^2y - y^3$
$= x(x^2 + y^2) - y(x^2 + y^2)$
$= (x^2 + y^2)(x - y)$.
Hence, the correct option is (b)

60. $y - \dfrac{1}{y} = 3$
Squaring on both sides, we get
$y^2 + \dfrac{1}{y^2} - 2 = 9$ $\quad (\because (a - b)^2 = a^2 + b^2 - 2ab)$

$y^2 + \dfrac{1}{y^2} = 11$
Again squaring on both the sides, we get
$y^4 + \dfrac{1}{y^4} + 2 = 121$
$y^4 + \dfrac{1}{y^4} = 119$.
Hence, the correct option is (a)

61. $64x^2 + \dfrac{1}{16x^2} = 20$

$(8x)^2 + \left(\dfrac{1}{4x}\right)^2 = 20$

$(8x)^2 + \left(\dfrac{1}{4x}\right)^2 - 2(8x)\left(\dfrac{1}{4x}\right) + 2(8x)\left(\dfrac{1}{4x}\right) = 20$

$\left(8x - \dfrac{1}{4x}\right)^2 + 4 = 20$

$ (\because (p - q)^2 = p^2 + q^2 - 2pq)$

$\left(8x - \dfrac{1}{4x}\right)^2 = 16$

$8x - \dfrac{1}{4x} = 4$.
Hence, the correct option is (d)

62. $(p - q)^2 = (p + q)^2 - 4pq$
$ (\because (p - q)^2 = p^2 + q^2 - 2pq$ and $(p + q)^2 = p^2 + q^2 + 2pq)$
$\Rightarrow (p - q)^2 = (15)^2 - 4(54)$
$= 225 - 216 = 9$
$\Rightarrow p - q = 3$.
Hence, the correct option is (a)

63.
$$x + 3 \overline{) x^2 + 7x + 12} \ (x + 4$$
$ x^2 + 3x$
$ \overline{}$
$ 4x + 12$
$ 4x + 12$
$ \overline{}$
$ 0$

Quotient $= x + 4$
Hence, the correct option is (a)

2.30 Chapter 2

64.
$$
\begin{array}{r}
x^2 + 4x + 5 \\
x-2 \overline{\smash{\big)}\ x^3 + 2x^2 - 3x + 10} \\
\underline{x^3 - 2x^2} \\
4x^2 - 3x + 10 \\
\underline{+4x^2 - 8x} \\
+5x + 10 \\
\underline{+5x - 10} \\
20
\end{array}
$$

$\Rightarrow$ 20 is the remainder.

$\Rightarrow$ 20 should be subtracted to make $x^3 + 2x^2 - 3x + 10$ a multiple of $x - 2$.

Hence, the correct option is (b)

65. $C = 7x + 9$

$D = (4x^2 + 8x + 5)$

$C \times D = (7x + 9)(4x^2 + 8x + 5)$

$ = 28x^3 + x^2(56 + 36) + x(72 + 35) + 45$

$ = 28x^3 + 92x^2 + 107x + 45.$

Degree = 3.

Hence, the correct option is (a)

ASSESSMENT TESTS

Test 1

Solutions for questions 1 to 11:

1. (B), (C), (A) and (D) is the required sequential order.

   Hence, the correct option is (b)

2. (B), (C), (A), and (D) is the required sequential order.

   Hence, the correct option is (d)

3. $a^2 + 16a + k = (a)^2 + 2(a)(8) + k$ is a perfect square.

   $\Rightarrow k = 8^2 = 64.$

   Hence, the correct option is (d)

4. $x + \dfrac{1}{x} = 2 \Rightarrow \dfrac{x^2 + 1}{x} = 2 \Rightarrow x^2 + 1 = 2x$

   $\Rightarrow x^2 - 2x + 1 = 0 \Rightarrow (x-1)^2 = 0$

   $\Rightarrow (x-1) = 0 \Rightarrow x = 1$

   $\Rightarrow x^{2013} - \dfrac{1}{x^{2012}} = 1 - \dfrac{1}{1} = 1 - 1 = 0$

   Hence, the correct option is (c)

5. $(2x - 6y)^2 - (6y + 8x)^2$

   $= [2(x - 3y)]^2 - [2(3y + 4x)]^2$

   $= 4[(5x)(-3)(x + 2y)]$

   $= 2^2(x - 3y)^2 - 2^2(3y + 4x)^2$

   $= 4[(x - 3y)^2 - (4x + 3y)^2]$

   $= 4[(x - 3y + 4x + 3y)(x - 3y - 4x - 3y)]$

   $= 4[(5x)(-6y - 3x)] = -12(5x)(x + 2y)$

   Hence, the correct option is (c)

6. $64x^6y^4 = 2^6 \cdot x^6 \cdot y^4$

   $48x^4y^8 = 2^4 \times 3 \cdot x^4 \cdot y^8$

   $54x^5y^4 = 2 \times 3^3 \cdot x^5 \cdot y^4$

   HCF = $2\,x^4y^4$.

   Hence, the correct option is (a)

7. $12a^3b^3 - 3ab = 3ab(4a^2b^2 - 1)$

   $ = (3ab)((2ab)^2 - 1^2)$

   $ = 3ab(2ab + 1)(2ab - 1).$

   Hence, the correct option is (c)

8. $(a^5 - a^3) \div (a^2 + a)$

   $= a^3(a^2 - 1) \div a(a + 1)$

   $= \dfrac{a^3(a+1)(a-1)}{a(a+1)} = a^2(a - 1).$

   Hence, the correct option is (c)

9. $3x - \dfrac{1}{2x} = 3 \Rightarrow \left(3x - \dfrac{1}{2x}\right)^2 = 3^2$

   $\Rightarrow 9x^2 - 2 \cdot 3x \times \dfrac{1}{2x} + \dfrac{1}{4x^2} = 9$

   $\Rightarrow 9x^2 - 3 + \dfrac{1}{4x^2} = 9 \Rightarrow 9x^2 + \dfrac{1}{4x^2} = 12$

   $\Rightarrow \dfrac{36x^4 + 1}{4x^2} = 12.$

   Hence, the correct option is (b)

Expressions and Special Products

10. $\sqrt{3^{6n^2} \cdot (36)^{2a} \cdot (16)^b}$

    $= \left(3^{6n^2} \cdot 6^{4a} \cdot 2^{4b}\right)^{\frac{1}{2}}$

    $= 3^{3n^2} \cdot 6^{2a} \cdot 2^{2b}$

    Hence, the correct option is ()

11. $\dfrac{x^4}{4} + \dfrac{y^3}{3} - \dfrac{3z^2}{5} + \dfrac{x^4}{3}$

    $- \dfrac{3y^3}{5} + \dfrac{z^2}{4} - \dfrac{3x^4}{5} + \dfrac{y^3}{4} + \dfrac{z^2}{3}$

    $= \dfrac{x^4}{4} + \dfrac{x^4}{3} - \dfrac{3x^4}{5} + \dfrac{y^3}{3}$

    $- \dfrac{3y^3}{5} + \dfrac{y^3}{4} \dfrac{-3z^2}{5} + \dfrac{z^2}{4} + \dfrac{z^2}{3}$

    $= \dfrac{15x^4 + 20x^4 - 36x^4}{60} + \dfrac{20y^3 - 36y^3 + 15y^3}{60}$

    $+ \dfrac{-36z^2 + 15z^2 + 20z^2}{60}$

    $= \dfrac{-x^4}{60} + \dfrac{-y^3}{60} + \dfrac{-z^2}{60}$

    $= -\left(\dfrac{x^4 + y^3 + z^2}{60}\right)$

    Hence, the correct option is (a)

Solutions for questions 12 to 15:

12. → (b): $225x^2 - 625y^2 = (15x)^2 - (25y)^2$

    $= (15x + 25y)(15x - 25y)$

    $= 25(3x + 5y)(3x - 5y)$

13. → (c): $x^2 - x - y - y^2 = (x^2 - y^2) - x - y$

    $= (x + y)(x - y) - 1(x + y)$

    $= (x + y)(x - y - 1)$

14. → (d): $x^2 - x - y^2 + y = x^2 - y^2 - x + y =$

    $= (x + y)(x - y) - (x - y)$

    $= (x - y)(x + y - 1)$

15. → (a): $(5x)^2 - 2(5x)(10) + (10)^2 = (5x - 10)^2$

    $= 25(x - 2)(x - 2)$

Test 2

Solutions for questions 16 to 26:

16. (B), (A) and (C) is the required sequential order.

    Hence, the correct option is (d)

17. (D), (A), (C) and (B) is the required sequential order.

    Hence, the correct option is (b)

18. $x^2 + 8x + k = (x)^2 + 2(x)(4) + k$ is a perfect square.

    $\Rightarrow k = 4^2 = 16$.

    Hence, the correct option is (c)

19. $x + \dfrac{1}{x} = 2$

    $\Rightarrow \dfrac{x^2 + 1}{x} = 2$

    $\Rightarrow x^2 - 2x + 1 = 0$

    $(x - 1)^2 = 0$

    $x - 1 = 0 \Rightarrow x = 1$

    $\therefore x^{2010} + x^{2009} = 1 + 1 = 2.$

    Hence, the correct option is (b)

20. $(3x - 4y)^2 - (4x + 3y)^2$

    $= (3x - 4y + 4x + 3y)(3x - 4y - 4x - 3y)$

    $(\because (a^2 - b^2) = (a + b)(a - b))$

    $= (7x - y)(-7y - x)$

    $= (y - 7x)(x + 7y)$

    Hence, the correct option is (d)

21. $44x^5y^4 = 2^2 \times 11 \, x^5y^4$

    $88x^6y^5 = 2^3 \times 11 \, x^6 \cdot y^5$

    $66x^7y^6 = 2 \times 3 \times 11 \, x^7y^6$

    $\therefore$ The HCF of the given polynomials is $2 \times 11 x^5 y^4$
    $= 22x^5y^4$.

    Hence, the correct option is (b)

22. $x^4 + x^3 - x^2 - x$

    $= x^3(x + 1) - x(x + 1)$

    $= (x + 1)(x^3 - x)$

    $= (x + 1)(x)(x^2 - 1)$

$= x(x + 1)(x + 1)(x - 1)$

$= x(x - 1)(x + 1)^2.$

Hence, the correct option is (b)

23. $(a^4 - a^2) \div (a^3 + a^2)$

$= a^2(a^2 - 1) \div a^2(a + 1)$

$= \dfrac{a^2(a+1)(a-1)}{a^2(a+1)} = a - 1.$

Hence, the correct option is (b)

24. $x - \dfrac{1}{2x} = 2$

$x^2 + \dfrac{1}{4x^2} - 2 \cdot x \cdot \dfrac{1}{2x} = 4$

$x^2 + \dfrac{1}{4x^2} = 4 + 1$

$\dfrac{4x^4 + 1}{4x^2} = 5.$

Hence, the correct option is (c)

25. $\sqrt{4^{6n^2} \cdot (25)^{\frac{a}{2}} \cdot 9^{b^4}}$

$= \left[4^{6n^2} (25)^{\frac{a}{2}} 9^{b^4} \right]^{\frac{1}{2}}$

$= \left[2^{2 \times 6n^2} 5^{2 \times \frac{a}{2}} 3^{2 \times b^4} \right]^{\frac{1}{2}}$

$= 2^{6n^2} \cdot 5^{\frac{a}{2}} \cdot 3^{b^4}$

Hence, the correct option is (c)

26. $\dfrac{a^2}{2} + \dfrac{b^3}{3} - \dfrac{3c^3}{4} + \dfrac{a^2}{3} - \dfrac{3b^3}{4} + \dfrac{c^3}{2} - \dfrac{3a^2}{4} + \dfrac{b^3}{2} + \dfrac{c^3}{3}$

$= \dfrac{a^2}{2} + \dfrac{a^2}{3} - \dfrac{3a^2}{4} + \dfrac{b^3}{3}$

$- \dfrac{3b^3}{4} + \dfrac{b^3}{2} - \dfrac{3c^3}{4} + \dfrac{c^3}{2} + \dfrac{c^3}{3}$

$= \dfrac{6a^2 + 4a^2 - 9a^2}{12} + \dfrac{4b^3 - 9b^3 + 6b^3}{12}$

$+ \dfrac{-9c^3 + 6c^3 + 4c^3}{12}$

$= \dfrac{a^2 + b^3 + c^3}{12}.$

Hence, the correct option is (a)

Solutions for questions 27 to 30:

27. → (b): $(a - b)^2 + 4ab = (a + b)^2.$

28. → (d): $x^3 - x^2 - x + 1 = x^2(x - 1) - 1(x - 1)$

$= (x - 1)(x^2 - 1)$

$= (x - 1)(x - 1)(x + 1)$

$= (x - 1)^2 (x + 1).$

29. → (c): $(a + b)(a^2 - ab + b^2) = a^3 + b^3.$

30. → (a): $(x + 1)(x^2 - 1) = (x + 1)(x + 1)(x - 1) = (x + 1)^2 (x - 1).$

Chapter 3
Ratio and Its Applications

$$a) \; 5(3x) - 3(x-4) = 0$$
$$15x - 3x + 12 = 0$$
$$12x + 12 = 0$$
$$12x = -12$$
$$x = -12/12$$
$$x = -1 \qquad x^2 = 1$$

REMEMBER
Before beginning this chapter, you should be able to:
- Use quantities in your daily life
- Understand the concept of fractions

KEY IDEAS
After completing this chapter, you should be able to:
- Understand the meaning of ratio in simplest form, comparison of ratios and its properties
- Learn the proportion and variation and their types
- Know percentage as a relative term, and conversions, comparison, and points of percentage
- Learn and understand about profit and loss and simple interest
- Apply unitary method in solving the problems
- Understand concept of mandays; pipes and cisterns
- Solve problems related to boats and streams and races

INTRODUCTION

In our daily life, we come across different situations where we have to compare quantities such as heights, weights, distance or time. We can compare two quantities of the same kind in two ways—by subtracting the smaller from the bigger or by dividing one by the other.

For example, let us take the two numbers 2 and 3. We may say that 2 is 1 less than 3. This is comparison by subtraction.

We may also say that 2 is two thirds of 3. This is comparison by division. Comparison by division is called a **ratio**.

When we compare by subtraction units are involved, i.e., 2 seconds is 1 second less than 3 seconds. Similarly, ₹2 is ₹1 less than ₹3.

But when we compare by division (i.e., use ratios) units are not involved. 2 seconds is $\frac{2}{3}$ of 3 seconds. A ratio is simply a number. It has no units.

Thus a ratio is the quotient obtained when one quantity is divided by another of the same kind. It expresses how many times, or how many parts one quantity is of another of the same kind. The ratio of a and b is $\frac{a}{b}$. It can also be written as $a:b$, which is read as a is to b.

For example, the ratio of 5 litres and 9 litres is written as 5:9.

> **Note** The ratio of two quantities can be found, only when both the quantities are of the same kind. The ratio between 1 metre and 5 seconds cannot be found, as the quantities given are not of the same kind.

Terms of a Ratio

For a given ratio $a:b$, we say that a is the **first term** or **antecedent** and b is the **second term** or **consequent**. In the ratio 3:4, 3 is the antecedent while 4 is the consequent.

Properties of a Ratio

The value of a ratio remains the same, if both the terms of the ratio are multiplied or divided by the same non-zero quantity, i.e., if a is a real number and b, m are non-zero real numbers, the following conclusions hold true.

1. $\dfrac{a}{b} = \dfrac{ma}{mb} \Rightarrow a:b = am:bm,$

2. $\dfrac{a}{b} = \dfrac{\left(\dfrac{a}{m}\right)}{\left(\dfrac{b}{m}\right)} \Rightarrow a:b = \dfrac{a}{m} : \dfrac{b}{m}$

Simplest Form of a Ratio

The ratio of two or more quantities is said to be in the simplest form, if the highest common factor (HCF) of the quantities is 1. If the HCF of the quantities is not 1, then each quantity of the ratio is divided by the HCF to convert the ratio into its simplest form.

For example, suppose there are three numbers 6, 9 and 12. The HCF of the numbers is 3. Dividing each of 6, 9 and 12 by 3, the results obtained are 2, 3 and 4. Ratio of 6, 9 and 12 in the simplest form is 2:3:4.

Ratio and Its Applications   3.3

Suppose $a:b = 3:4$, then $a:b$ is equal to 3:4 or 6:8 or 9:12 and so on. The ratio of two quantities, in the simplest form is $a:b$. Now a and b may be equal to ($3x$) and ($4x$) respectively, when x must have been the common factor, that got cancelled.

EXAMPLE 3.1

Express 24:36 in its simplest form.

SOLUTION

The HCF of 24 and 36 is 12. We divide each term by 12.

Then, $24:36 = \dfrac{24}{12} : \dfrac{36}{12} = 2:3$.

∴ The ratio 24:36 in its simplest form is 2:3.

Comparison of Ratios

Two or more ratios can be compared by expressing the ratios as fractions or decimals.

EXAMPLE 3.2

Compare the ratios 2:3 and 4:5.

SOLUTION

(a) By converting to decimal:

$2:3 = \dfrac{2}{3} = 0.\overline{6}$

$4:5 = \dfrac{4}{5} = 0.8$

As $0.\overline{6} < 0.8$, $2:3 < 4:5$.

(b) By converting to fractions:

Ratios can be compared by converting them to like fractions.

$2:3 = \dfrac{2}{3} = \dfrac{2(5)}{3(5)} = \dfrac{10}{15}$

$4:5 = \dfrac{4}{5} = \dfrac{4(3)}{5(3)} = \dfrac{12}{15}$

As $\dfrac{10}{15} < \dfrac{12}{15}$, $2:3 < 4:5$.

Proportion

The equality of two ratios is called proportion.

> **Note** If $a:b = c:d$, i.e., $\dfrac{a}{b} = \dfrac{c}{d}$, then a, b, c and d are said to be in proportion.

1. a, b, c and d are respectively known as the first, second, third and the fourth terms.

2. The first and the fourth terms are called extremes while the second and the third terms are called means.

3. The product of extremes = The product of means, i.e., $ad = bc$.
4. If $\dfrac{a}{b} = \dfrac{c}{d}$, then the given proportion can be written as $\dfrac{b}{a} = \dfrac{d}{c}$, by taking reciprocals of terms on both sides. This relationship is known as **invertendo**.
5. If $\dfrac{a}{b} = \dfrac{c}{d}$, then multiplying both sides of the proportion by $\dfrac{b}{c}$, we get $\dfrac{a}{c} = \dfrac{b}{d}$. This relationship is known as **alternendo**.

Note If $\dfrac{a}{b} = \dfrac{c}{d} = \dfrac{e}{f}$ and l, m and n are any three non-zero numbers, then $\dfrac{a}{b} = \dfrac{c}{d} = \dfrac{e}{f} = \dfrac{la \pm mc \pm ne}{lb \pm md \pm nf}$.

Continued Proportion

Three quantities a, b and c are said to be in continued proportion if $\dfrac{a}{b} = \dfrac{b}{c}$.

Here, c is called the third proportional of a and b and b is called the mean proportional of a and c.

We have already learnt that, the product of means = the product of extremes.

$\Rightarrow b \times b = a \times c \Rightarrow b^2 = ac \Rightarrow b = \pm\sqrt{ac}$

$\therefore$ The mean proportional of a and c is $\sqrt{ac}$.

Variation

Different quantities are related to each other in different ways. We shall first study the relationship between two quantities. If one quantity varies, another quantity that is related to the first can vary in a number of ways. We shall study two such ways of variation.

Types of Variation

1. **Direct variation:** If two quantities are related to each other such that an increase (or decrease) in the first quantity results in a corresponding proportionate increase (or decrease) in the second quantity, then the two quantities are said to vary directly with each other.

    For example, at constant speed, distance covered varies directly as time, i.e., the longer the time for which the body travels the greater will be the distance covered.

    This is expressed as, distance travelled $\propto$ time.

2. **Inverse variation:** If two quantities are related to each other such that an increase (or decrease) in the first quantity results in a corresponding proportionate decrease (or increase) in the second quantity, then the two quantities are said to vary inversely with each other.

    For example, number of men working together to complete a job is inversely proportional to the time taken by them to finish that job. When the number of men increases, the time taken to finish the same job decreases.

    $\therefore$ The number of men (n) working together to complete a job is inversely proportional to the time taken (t) by them to finish it.

    This is expressed as, $n \propto \dfrac{1}{t}$.

In this chapter, we shall learn about the concept of percentages and its wider applications in day-to-day life situations. In order to solve problems in chapters like profit and loss, simple interest, etc., it is essential to have a thorough understanding of this topic.

Percentage

Per cent means 'for every hundred'.

A fraction, in which the denominator is 100, is a percentage. The denominator, i.e., 100 is denoted by a special symbol %, read as per cent.

For example, $\dfrac{10}{100} = 10\%$

$\dfrac{25}{100} = 25\%$

$\dfrac{x}{100} = x\%$

Since any ratio is a fraction, each ratio can also be expressed as a percentage.

For example, a ratio of $\dfrac{1}{2}$ can be converted to a percentage figure as $\dfrac{1}{2} = \dfrac{1(50)}{2(50)} = \dfrac{50}{100} = 50\%$.

Expressing $x\%$ as a fraction: Any percentage can be expressed as a decimal fraction by dividing the percentage figure by 100.

As $x\% = x$ out of $100 = \dfrac{x}{100}$.

So, $75\% = 75$ out of $100 = \dfrac{75}{100} = \dfrac{3}{4}$ or 0.75.

Expressing the fraction $\dfrac{a}{b}$ as a decimal and as a percentage: Any fraction can be expressed as a decimal (terminating or non-terminating but recurring) and any decimal fraction can be converted into percentage by multiplying it with 100.

$\dfrac{1}{2} = 0.5 = 50\%$

$\dfrac{1}{4} = 0.25 = 25\%$

$\dfrac{1}{5} = 0.2 = 20\%$

$\dfrac{1}{3} = 0.33\ldots = 33.33\ldots\%$

Problems Based on Basic Concepts

EXAMPLE 3.3

Express 40% as a fraction.

SOLUTION

$40\% = \dfrac{40}{100} = \dfrac{2}{5}$

$\therefore$ 40% as a fraction is $\dfrac{2}{5}$.

EXAMPLE 3.4
Express 27% as a decimal.

SOLUTION

$27\% = \dfrac{27}{100} = 0.27.$

EXAMPLE 3.5
Express $\dfrac{1}{4}$ as percent.

SOLUTION

$\dfrac{1}{4} = (\dfrac{1}{4} \times 100)\% \% = 25\%.$

EXAMPLE 3.6
Find 30% of 90.

SOLUTION

$30\% \text{ of } 90 = \dfrac{30}{100} \times 90 = 27.$

Percentage, A Relative Value

When you obtain 18 marks out of 20 marks in your maths unit test, this would be an absolute value. Let us say that you got 90% in the maths unit test, it is understood that you got 90 marks out of 100 marks. However, if the maximum marks for the unit test is 50, then the marks you got are 90% of 50 or $\dfrac{90}{100} \times 50 = 45.$

Hence, the actual score depends upon the maximum marks of the unit test and varies with the maximum marks. For example, if the maximum marks are 60, 90% of 60 = 54. If the maximum marks are 70, 90% of 70 = 63. As the maximum marks vary, your marks also vary. Hence percentage is a relative or comparative value. If the maximum marks are 100, your score is 90.

Comparison of Percentages

Let us say, in your class 30% of the students are girls and in IX class, 40% of the students are girls. Can you say that the number of girls of your class is less than the number of girls in the IX class? The answer depends on the total number of students in each class. If there are 50 students in your class, the number of girls in your class = $\dfrac{30}{100} \times 50 = 15$. If there are 30 students in IX class, then the number of girls in IX class = $\dfrac{40}{100} \times 30 = 12$. Though the percentage of girls in your class is less than the percentage of girls in IX class, the number of girls in your class may be more than that in IX class.

However, you can say that the percentage of girls in your class is less than the percentage of girls in IX class. But the number of girls is the two classes cannot be compared, if the total number of students in each class is not known.

However, when you say that the percentage of girls in your class is less than the percentage of girls in IX class, you can specify that it is less by 10 percentage points. (i.e., 40% − 30% = 10% points)

Percentage Points

It is the difference between two percentage values. It is not equal to either percentage increase or percentage decrease.

When two absolute values are given, four different percentage values can be calculated involving the two values.

1. The smaller as a percentage of the greater.
2. The smaller as a certain percentage less than the greater.
3. The greater as a percentage of the smaller.
4. The greater as a certain percentage greater than the smaller.

For example, consider two rods of lengths a and b. Let $a = 4$ m and $b = 6$ m.

We can express a as a percentage of b.

$\frac{a}{b} = \frac{4}{6} = \frac{2}{3} = \frac{200}{3}\% = 66\frac{2}{3}\%$, i.e., a is $66\frac{2}{3}\%$ of b.

We can also perform the comparison by both subtraction and division, i.e., compute $\frac{(b-a)}{b}$ and express that as a percentage $\frac{(b-a)}{b} = \frac{(6-4)}{6} = \frac{2}{6} = \frac{1}{3} = \frac{100}{3}\% = 33\frac{1}{3}\%$, i.e., a is $33\frac{1}{3}\%$ less than b.

Similarly, $\frac{b}{a} = \frac{6}{4} = \frac{3}{2} = (300/2)\% = 150\%$, i.e., b is 150% of a.

Finally, $\frac{(b-a)}{b} = \frac{(6-4)}{4} = \frac{1}{2} = (100/2)\% = 50\%$, i.e., b is 50% more than a.

The application of the concepts of ratio, proportion and variation are universal. We shall consider this application in four areas.

EXAMPLE 3.7

In a village, 40% of the population comprises of children below 14 years of age. If the ratio of number of men to the number of women is 2:1, then find the number of women in that village, given that the total population of the village is 5000.

SOLUTION

Total population = 5000.

Population of children = $\frac{40}{100} \times 5000 = 2000$.

Rest of the population = 3000.

Let the number of men and women be $2k$ and k respectively.

∴ $3k = 3000 \Rightarrow k = 1000$.

∴ Number of women = 1000.

1. Profit and Loss
2. Simple and Interest
3. Time and Work
4. Time and Distance

Profit and Loss

When a person buys an article for a certain price and then sells it for a different price, he makes a profit or incurs a loss.

Cost Price (C): The price at which an article is purchased is called its cost price.

Selling Price (S): The price at which an article is sold is called its selling price.

Profit and loss: If the selling price of an article is greater than its cost price, we say that there is a profit (or) gain and if it is less, then we say that there is a loss. If $S > C$, i.e., if $p > 0$, there is a profit. If $p < 0$, there is a loss. If $S = C$, there is no profit, no loss. Profit or loss is generally expressed as a percentage of the cost price.

EXAMPLE 3.8

If cost price of an article is ₹80 and its selling price is ₹100, then find the profit and profit percentage.

SOLUTION

Cost price (C) = ₹80.
Selling price (S) = ₹100.
As $S > C$, there is profit.
∴ Profit = $S - C$ = ₹100 − ₹80 = ₹20.
Profit percentage = $\left[\dfrac{\text{profit}}{C} \times 100\right]\%$ = $[(20/80) \times 100]\%$ = 25%.

EXAMPLE 3.9

Krishna sold a motor bike for ₹15000, losing 25%. Find the cost price of motor bike.

SOLUTION

S = ₹15,000, loss = 25%
$S = (100 - \text{loss}\%)\%$ of C.
$\Rightarrow 15,000 = \left[\dfrac{(100-25)}{100}\right](C) \Rightarrow C = 15,000 \times \dfrac{100}{75} \Rightarrow C$ = ₹20,000.

EXAMPLE 3.10

A trader allows 20% discount on the marked price of his articles. If the marked price is ₹150 and gain per cent is 20%, then find the cost price of each article.

SOLUTION

Marked price = ₹150.

Percentage of discount = 20%.

$S = 80\%$ of $150 = \left(\dfrac{80}{100}\right) \times 150 = ₹120.$

$S = ₹120$ and gain% = 20%.

$S = 120\%$ of C.

$120 = \dfrac{120}{100} \times C \Rightarrow C = 120 \times \dfrac{100}{120} \Rightarrow C = ₹100$

∴ Cost price = ₹100.

Overheads: All the expenditure incurred on transportation, repairs, etc., (if any) are categorised as overheads. These overheads are always included in the C.P. of the article.

Simple Interest

Interest is the money paid by a borrower to the lender for using the money for a specified period of time. For example, if person A borrows ₹100 from person B, for a period of one year on the condition that he would repay ₹110 at the end of a year, the additional money of ₹10 is the interest. A is the borrower and B is the lender and ₹100 is the principal. The definitions of certain terms which are used frequently are given below.

Principal or sum: The money borrowed from an agency or an individual for a certain period of time is called the principal or sum.

Amount: The principal together with the interest is called the amount, i.e., Amount (A) = Principal (P) + Interest (I).

Rate of interest: The interest on ₹100 per annum is called the rate of interest per annum.

Simple Interest: If the principal remains the same for the entire loan period, then the interest paid is called the simple interest.

Formula for the computation of simple interest: Let P be the principal in rupees, R be the rate of interest and T denote the number of years.

Then, Simple Interest (S.I.) = $\dfrac{PTR}{100}$

Also, Amount (A) = Principal (P) + Simple Interest (S.I.) = $P + \dfrac{PTR}{100} = P\left(1 + \dfrac{RT}{100}\right)$.

EXAMPLE 3.11

Find the simple interest on ₹2560 for 3 years at 15% per annum.

SOLUTION

Principal (P) = ₹2560

Time period (T) = 3 years

Rate (R) = 15% p.a.

Simple interest $(I) = \dfrac{PTR}{100}$

$= 2560 \times 3 \times \dfrac{15}{100} = 128 \times 9 = ₹1152.$

EXAMPLE 3.12

At what rate of per cent ₹1500 amounts to ₹2100 in 4 years?

SOLUTION

Time period $(T) = 4$ years
Amount $(A) = ₹2100$
Principal $(P) = ₹1500$
Interest $= A - P$
$\qquad = ₹(2100 - 1500) = ₹600.$

We have, $I = \dfrac{PTR}{100}$

$600 = 1500 \times 4 \times \dfrac{R}{100}$

$\Rightarrow 600 = 60 \times R \Rightarrow R = \dfrac{600}{60} \Rightarrow R = 10.$

∴ Rate of interest $= 10\%$.

EXAMPLE 3.13

In how many years will a certain sum amounts to 6 times itself at 25% per annum simple interest?

SOLUTION

Let the sum be P.

$5P = \text{Interest} = \dfrac{P \times 25 \times t}{100}$

$t = 20$ years.

Time and Work

Work to be done is usually considered as one unit. It may be constructing a wall or a road, filling up or emptying a tank or cistern or eating certain amount of food. Time is measured in days, hours, etc.

There are some basic assumptions that are made while solving the problems on time and work. These are taken for granted and are not specified in every problem.

1. If a person (or one member of the workforce) does some work in a certain number of days, then we assume (unless otherwise explicitly stated in the problem) that he does the work uniformly, i.e., he does the same amount of work every day.

   For example, if a person can complete a work in 15 days, we assume that he completes $\dfrac{1}{15}$ th of the work in one day.

   If a person completes a piece of work in 4 days, we assume that he completes $\dfrac{1}{4}$ th of the work on each day and conversely, if a person can complete $\dfrac{1}{4}$ th of work in one day, we assume that he can complete the total work in 4 days.

2. If there is more than one person (or members of the "workforce") carrying out the work, it is assumed that each person (or members of the workforce), unless otherwise specified, does the same amount of work each day. This means that they share the work equally.

If two people together can do the work in 8 days, it means that each person can complete it in 16 days. This, in turn means, each person can do $\frac{1}{16}$ th of the work per day.

If a man works three times as fast as a boy does, the man takes one-third of the time the boy takes to complete the work. If the boy takes 12 days to complete the work, then the man takes 4 days to complete the work.

This method is known as **'UNITARY METHOD'**, i.e., the time taken per **'Unit Work'** or number of persons required to complete 'Unit Work' or work completed by 'Unit Person' in 'Unit Time', etc., is what is first calculated.

Note If two persons, A and B can individually do some work in p days and q days respectively, we can find out how much work can be done by them together in one day. Since A can do $\frac{1}{p}$ part of the work in one day and B can do $\frac{1}{q}$ part of the work in one day and both of them together can do $\left[\frac{1}{p} + \frac{1}{q}\right]$ part of the work in one day.

From this, we can find out the number of days that they take to complete the work.

We should recollect the fundamentals of variation (direct and inverse) here.

1. When the number of days is constant, work and men are directly proportional to each other, i.e., if the work to be done increases, more number of men are required to complete the work in the same number of days.

2. When the number of men is constant, work and days are directly proportional, i.e., if the work increases, more days are required, provided the work is to be completed by the same number of men.

3. When the work is constant, number of men and days are inversely proportional, i.e., if the number of men increases, less days are required to complete the same work and vice-versa.

   The concept of MANDAYS is very important and useful here. The number of men multiplied by the number of days required to complete the work will give the number of *mandays*. Here work is measured in terms of *mandays*. The total number of *mandays* representing a specific task will remain a constant. So, if we change one of the variables - men or days, then the other will change accordingly, so that their product will remain constant (remember from our knowledge of VARIATION, two variables whose product is a constant, are said to be inversely proportional to each other). The two variables 'men' and 'days' are inversely proportional to each other.

EXAMPLE 3.14

If 10 persons can make 20 articles in a day, then how many persons are required to make 10 articles in a day?

SOLUTION

Let the number of persons required be x.

The number of persons varies directly with the number of articles in a constant period

$$\therefore \frac{10}{20} = \frac{x}{10} \Rightarrow x = \frac{100}{20} = 5.$$

Hence, in general we can say that:

If M_1 men can do W_1 units of work in D_1 days working H_1 hours per day and M_2 men can do W_2 units of work in D_2 days working H_2 hours per day, then $\dfrac{M_1 D_1 H_1}{W_1} = \dfrac{M_2 D_2 H_2}{W_2}$.

EXAMPLE 3.15

A pipe can fill an empty tank in 12 minutes and another pipe can fill it in 24 minutes. If both the pipes are kept open simultaneously, then in how many minutes will the tank get filled?

SOLUTION

The part of the tank filled in 1 minute $= \dfrac{1}{12} + \dfrac{1}{24}$

$$= \dfrac{2+1}{24} = \dfrac{3}{4} = \dfrac{1}{8}$$

∴ They can fill the tank in 8 minutes.

Time and Distance

Let us consider an object moving uniformly. It covers equal distances in equal intervals of time. For example, if in 1 second it covers 5 metres, in the next second it covers another 5 metres, i.e., in 2 seconds it covers 10 metres, in 3 seconds it covers 15 metres and so on. The distance which the object covers in unit time is called its speed. Hence, the speed of the object whose motion is described above is 5 metres per second.

Speed: The distance covered per unit time is called **speed**, i.e., Speed $= \dfrac{\text{Distance}}{\text{Time}}$.

The above relationship between the three quantities; distance, speed and time can also be expressed as follows:

Distance = Speed × Time (or) Time $= \dfrac{\text{Distance}}{\text{Speed}}$

If two bodies travel with the same speed, the distance covered varies directly as time and it is written as Distance ∝ Time. Further, if two bodies travel for the same period of time, the distance covered varies directly as the speed, i.e., Distance ∝ Speed. If two bodies travel the same distance, time varies inversely as speed, i.e., Time $\propto \dfrac{1}{\text{Speed}}$.

Distance is usually measured in kilometres, metres or miles; time in hours or seconds and speed in km/h (also denoted by kmph) or miles/h (also denoted by mph) or metres/second (denoted by m/s).

1 km per hour $= \dfrac{1 \times 1000 \text{ m}}{3600 \text{ sec}} = \dfrac{5}{18}$ m/sec.

Note To convert speed in kmph to m/sec, multiply it with $\dfrac{5}{18}$.

To convert speed in m/sec to kmph, multiply it with $\dfrac{18}{5}$.

Average Speed: The average speed of a body travelling at different speeds for different time periods is defined as follows.

$$\text{Average speed} = \frac{\text{Total distance travelled}}{\text{Total time taken}}$$

EXAMPLE 3.16

(a) Express 72 km/h in m/s.

(b) Express 25 m/s in km/h.

SOLUTION

(a) 72 km/h = 72 × $\frac{5}{18}$ m/s = 20 m/s.

(b) 25 m/s = 25 × $\frac{18}{5}$ km/h = 90 km/h.

EXAMPLE 3.17

A man travels 100 km in 5 hrs. If he increases his speed by 5 km/h, then find the time taken by him to travel the same distance.

SOLUTION

Speed = $\frac{\text{Distance}}{\text{Time}}$ = $\frac{100}{5}$ = 20 km/h.

Increased speed = 20 + 5 = 25 km/h.

Time taken to travel at 25 km/h = $\frac{100}{25}$ = 4 h.

Relative speed: The speed of one moving body in relation to another moving body is called the relative speed, i.e., it is the speed of one moving body as observed from the second moving body.

If two bodies are moving in the same direction, the relative speed is equal to the difference of the speeds of the two bodies.

If two bodies are moving in the opposite direction, the relative speed is equal to the sum of the speeds of the two bodies.

Trains: In time and distance topic, we come across many problems on trains. While passing a stationary point or a telegraph/telephone pole completely, a train has to cover its entire length. Hence, the distance travelled by the train to pass a stationary point/a telegraph/a telephone pole is equal to its own length.

While passing a platform, a train has to cover its own length as well as the length of the platform. Hence the distance travelled by the train to pass a platform/a bridge/the other train is equal to the total length of the train and the length of the platform/the bridge/the other train.

While overtaking another train (when the trains move in the same direction) or while crossing another moving train (when the trains move in the same or opposite directions) a train has to cover its own length as well as the length of the other train. Hence in this case, the distance travelled by the train, is equal to the total length of the two trains, but the speed at which this distance is covered is the relative speed of the train with respect to the other train.

Chapter 3

EXAMPLE 3.18

If a train x, 200 m long, crosses a bridge of length 300 m in 30 sec, then find the speed of the train (in km/h).

SOLUTION

To cross the bridge, the distance travelled by the train = 200 m + 300 m = 500 m.

Time taken = 30 sec.

$\therefore$ Speed of the train $= \dfrac{\text{Distance}}{\text{Time}} = \dfrac{500}{30}$

$= \dfrac{50}{30}$ m/sec $= \dfrac{50}{30} \times \dfrac{18}{5}$ km/h = 60 km/h.

EXAMPLE 3.19

A train of length 240 m crosses a platform in 20 seconds. If the speed of the train is 72 km/h, then find the length of the platform.

SOLUTION

Speed $= \dfrac{\text{Distance}}{\text{Time}}$

$72 \times \dfrac{5}{18} = \dfrac{240 + \text{Length of the platform}}{20 \text{ seconds}}$

$20 = \dfrac{240 + l}{20}$ (Assume the length of the platform as l)

$400 + 240 + l$

$l = 400 - 240 = 160$ m.

$\therefore$ The length of the platform = 160 m.

Boats and streams: Problems related to boats and streams are different in the computation of relative speed from those of trains/cars.

When a boat is moving in the same direction as the stream or water current, the boat is said to be moving **WITH THE STREAM OR DOWNSTREAM**.

When a boat is moving in a direction opposite to that of the stream or water current, it is said to be moving **AGAINST THE STREAM OR UPSTREAM**.

If the boat is moving with a certain speed in water that is not moving, then the speed of the boat is called the **SPEED OF THE BOAT IN STILL WATER**.

When the boat is moving upstream, the speed of the water opposes (and hence reduces) the speed of the boat.

When the boat is moving downstream, the speed of the water aids (and thus increases) the speed of the boat. Thus, we have

Speed of the boat against the stream = Speed of the boat in still water − Speed of the stream.

Speed of the boat with the stream = Speed of the boat in still water + Speed of the stream.

These two speeds, the speed of the boat against the stream and the speed of the boat with the stream are speeds with respect to the bank.

If u is the speed of the boat downstream and v is the speed of the boat upstream, then we have the following two relationships.

Speed of the boat in still water = $\dfrac{(u+v)}{2}$

Speed of the water current = $\dfrac{(u-v)}{2}$

In some problems, instead of a boat, it may be a swimmer. But the approach is exactly the same.

EXAMPLE 3.20

Speed of a boat in still water is 10 km/h and the speed of the stream is 5 km/h. Find the total time taken to travel a distance of 20 km upstream and 30 km downstream

SOLUTION

Speed of boat upstream = 10 − 5 = 5 km/h.

Speed of boat downstream = 10 + 5 = 15 km/h.

Total time taken = $\dfrac{\text{Distance}}{\text{Speed}} = \dfrac{20}{5} + \dfrac{30}{15} = 4 + 2 = 6$ h.

EXAMPLE 3.21

A boat travels a distance of 50 km downstream in 5 h and 80 km upstream in 10 h. Find the speed of the boat and speed of the current.

SOLUTION

Let speed of the boat in still water be x kmph and speed of current be y kmph.

Speed of the boat downstream = $(x + y)$ kmph.

Speed of the boat upstream = $(x - y)$ kmph.

$\therefore x + y = \dfrac{50}{5} = 10$ (1)

and $x - y = \dfrac{80}{10} = 8$ (2)

On adding (1) and (2), we get

$2x = 18$

$x = 9 \Rightarrow y = 1$

$\therefore$ Speed of the boat in still water = 9 kmph.

Speed of current = 1 kmph.

Points to Remember

- $a:b$ is same as $\dfrac{a}{b}$.
- Comparison of ratios can be done by making the consequents equal.
- If $a:b = c:d$, then a, b, c and d are in proportion.
- If a, b, c and d are in proportion, then $ad = bc$.
- If b is the mean proportional of a and c (i.e., $a:b = b:c$), then $b^2 = ac$.
- If $x \propto y$, then $\dfrac{x}{y}$ is constant.
- If $x \propto \dfrac{1}{y}$, then xy is constant.
- Per cent means per hundred.
- Profit or loss percentage is always calculated on cost price.
- Discount is always offered on marked price.
- If the time taken to complete a piece of work in n days, then the part of the work done in 1 day is $\dfrac{1}{n}$.
- Relative speed = The sum of the speeds (when bodies are moving in opposite direction).
- Relative speed = The difference of the speeds (when bodies are moving in same direction).
- 1 km/h = $\dfrac{5}{18}$ m/s and 1 m/s = $\dfrac{18}{5}$ km/h.
- Distance = Speed × Time.

Formulae

- Average = $\dfrac{\text{The sum of the quantities}}{\text{Number of quantities}}$.
- Percentage increase = $\left(\dfrac{\text{Greater value} - \text{Smaller value}}{\text{Smaller value (or) Initial value}} \times 100\right)\%$
- Percentage decrease = $\left(\dfrac{\text{Greater value} - \text{Smaller value}}{\text{Greater value (or) Initial value}} \times 100\right)\%$
- When $S > C$, Profit = $S - C$.
- When $S < C$, Loss = $C - S$.
- Distance = Speed × Time.
- Average speed = $\dfrac{\text{Total distance travelled}}{\text{Total time taken}}$.
- $I = \dfrac{PTR}{100}$
- Amount = Principal + Interest
- $\dfrac{M_1 D_1}{W_1} = \dfrac{M_2 D_2}{W_2}$

Ratio and Its Applications

TEST YOUR CONCEPTS

Very Short Answer Type Questions

Directions for questions 1 to 10: State whether the following statements are true or false.

1. If $a:b = 3:4$, then $4a:3b = 1:1$.
2. Inverse ratio of 5:2 is 2:5.
3. Simplest form of 24:36 is 4:9.
4. If $\frac{1}{2} : \frac{1}{3} : \frac{1}{6} = 3:x$, then $x = 1:2$.
5. If $\frac{x}{6} = \frac{7}{3}$, then $x = 14$.
6. If 1, 2, 3 and x are in proportion, then $x = 1$.
7. The cost of 3 pens is ₹30, then the cost of 2 pens is ₹20.
8. If 3 persons can construct x chairs in 6 days, then 6 persons can construct x chairs in 4 days.
9. If SP = ₹22 and CP = ₹18, then profit = ₹4.
10. If CP = ₹100 and loss = ₹20, then SP = ₹80.

Directions for questions 11 to 20: Fill in the blanks.

11. If interest on ₹x for 2 years at $R\%$ p.a. is ₹80, then interest on ₹$2x$ for 1 year at $R\%$ p.a. is ____.
12. Per cent means _____.
13. Compound ratio of 2:3 and 3:4 is _____.
14. Ratio of 2 rupees 50 paise and 3 rupees 75 paise is _____.
15. The average of 23, 24, 28 and 29 is _____.
16. The ratio of the number of boys and the number of girls in a class is 9:11. If the number of boys is 18, then the number of girls is _____.
17. Mean proportion of 2 and 8 is _____.
18. 50% of 50 = _____.
19. If 25% of x is 100, then x = _____.
20. If $a:b = 2:3$ and $b:c = 3:4$, then $a:c$ = _____.

Directions for questions 21 to 30: Select the correct alternative from the given choices.

21. 25:30 = _____.
    (a) 4:5 (b) 5:6
    (c) 6:5 (d) 6:7

22. $\frac{1}{12} : \frac{1}{60}$ = _____.
    (a) 4:1 (b) 1:4
    (c) 5:1 (d) 1:5

23. $\frac{2}{5}$ as a per cent is _____.
    (a) 30% (b) 35%
    (c) 40% (d) 45%

24. $\frac{5}{4}$ as a per cent is _____.
    (a) 80% (b) 125%
    (c) 120% (d) 130%

25. 12% as a fraction is _____.
    (a) $\frac{3}{25}$ (b) $\frac{4}{25}$
    (c) $\frac{3}{20}$ (d) $\frac{6}{25}$

26. 1.25% as a fraction is _____.
    (a) $\frac{1}{40}$ (b) $\frac{1}{60}$
    (c) $\frac{1}{80}$ (d) $\frac{1}{70}$

27. If CP = ₹200 and SP = ₹250, then find the profit or loss.
    (a) ₹50 loss (b) ₹50 profit
    (c) ₹25 profit (d) ₹25 loss

28. CP = ₹120 and SP = ₹80, then find the profit or loss.
    (a) ₹40 loss (b) ₹60 loss
    (c) ₹40 profit (d) ₹60 profit

29. Which of the following fractions is equivalent to 25%?
    (a) $\frac{1}{4}$ (b) $\frac{1}{5}$
    (c) $\frac{1}{3}$ (d) $\frac{1}{2}$

30. If $\frac{5}{6}$ of 29% of $y = 29$, then find the value of y.
    (a) 290 (b) 58
    (c) 120 (d) 100

Short Answer Type Questions

31. Divide ₹175 in the ratio 1:2:4.
32. Find the fourth proportional of 2, 3 and 6.
33. Find the average of 22 kg, 24 kg, 33 kg, 42 kg and 39 kg.

Directions for questions 34 and 35: Calculate simple interest and amount in the following cases.

34. ₹2500 for 3 years at 12% per annum.
35. ₹3650 for 2 years 6 months at 16% per annum.
36. If $a:b = 5:6$ and $b:c = 3:4$, then find $a:b:c$.
37. The cost of 12 books is ₹144. Find the cost of 18 books.
38. Twelve men are required to complete a work in 24 days. How many men are required to complete twice the work in 16 days?
39. In a class, there are 20 boys and 15 girls. What percentage of the number of boys is to the number of girls?
40. By selling a television for ₹6600, a trader gains 10%. Find the cost price of the television for the trader.
41. The average height of 20 boys is 120 cm and the average height of 10 girls is 117 cm. Find the average height of these 30 students.
42. A piece of cloth which is 10 m in length is cut into two pieces such that the length of the one piece is equal to one-fourth the length of the other. Find the lengths of the two pieces.
43. The ratio of the costs of two vehicles is 3:2 and the total cost of the two vehicles is ₹50,000. Find the individual costs of the vehicles.
44. If a number is divided into three parts which are in the ratio 1:1:2, then what percentage of the total value is the sum of the equal parts?
45. In a right triangle, the length of the sides is in the ratio 5:4:3. Find the length of individual sides, if the difference between the longest side and the smallest side is 2 cm.
46. The price of a commodity is reduced to its $\dfrac{5}{6}$ th. What is the percentage reduction in the price?
47. Find a number such that the ratio of $\dfrac{5}{2}$ to the number is the same as the ratio of the number to $\dfrac{5}{8}$.
48. Town A is 200 km far from town B. A person can travel in a car from town A to town B in 4 hours. If he travels at a constant speed, then what is the speed of the car (in kmph)?
49. Ajay's home is 150 m away from a bus stop. At what speed should he travel to reach his house from the bus stop in two minutes?
50. Find the time taken by a car running at 25 m/s to cover a distance of 360 km (in hours)?
51. A 200 m long train passes a pole in 20 sec. What is the speed of the train (in m/s)?
52. Kiran alone can do a piece of work in 15 days and Aman can do the same in 10 days. If they start working together, in how many days can the work be completed?
53. If 256 is the third proportional of a and b, where $a:b = 3:16$, then find the value of $a + b$.
54. If $\dfrac{15x + 16y}{25x + 4y} = \dfrac{7}{6}$, then find $x:y$.
55. If $m:n = 3:5$ and $m:r = 2:3$, then find $n:r$.
56. The present ages of Pavan and Kalyan are in the ratio of 4:5. The product of their ages numerically is equal to 980. Find the difference between their ages (in years).
57. Aashritha scored 85, 90, 98, 78, 89 and 70 marks in six different subjects. The maximum mark in each subject is 100. Find her average score.
58. A shop owner sold a sofa set for ₹10,000, losing ₹2500. Find his loss percentage.
59. P can do a piece of work in 25 days and Q can do the same work in 15 days. In how many days they can complete the work by working together?
60. A certain sum becomes doubled in 5 years. Find the rate of interest.
61. A train crosses a pole in 12 seconds; when it is traveling at 54 km/h. Find the length of the train.
62. Calculate the amount on ₹2500 at 12% per annum for 8 years under simple interest.

Essay Type Questions

63. Deepti received ₹500 as pocket money. She bought a new dress using 40% of her pocket money. She bought crackers worth ₹100 and also a gift for her parents with the remaining amount. What was the cost of the gift she bought for her parents?

64. In a board exam the pass percentages of a certain school for the past three years are 87%, 90% and 92%. If 400 students appear for the exam every year, then how many students have passed the exam in the past three years?

65. A certain concrete mix needs cement, fine aggregate and coarse aggregate in the ratio 1:2:4. If the contractor has 25 kg of fine aggregate, how much quantity of the cement and that of the coarse aggregate is required to prepare the mix?

66. The ratio of Ram's savings to his expenditure is 5:2 and that of Manu is 4:3. If Ram's expenditure is $\frac{1}{3}$ rd of Manu's expenditure and the sum of their expenditures is ₹3000, find the salaries of Ram and Manu respectively.

67. The ratio of the number of boys to that of girls in a school is 3:4. Two-third of the number of boys are cricket players and the rest are football players. One-fourth of the number of girls are badminton players and the rest are handball players. Which game has the greatest number of participants if the total strength of the school is 1400?

68. In an exam, there were 50 questions and each question carries 4 marks. Sahil scored 80% marks in the test. Later 5 questions were proved logically incorrect and the score of each student was calculated out of the rest of the questions. What is Sahil's new score, if he hadn't answered any of the deleted questions?

69. Ramchand took a loan of ₹12,000 from his neighbour and agreed to repay it after 2 years at the rate of 8% per annum, simple interest. At the end of 2 years, his neighbour demanded an interest at the rate of 12%, what is the extra amount paid by the borrower?

70. How much time will a 200 m long train running at 15 m/s take to cross a bridge of length 355 m?

71. A farmer can plough a farm in 10 days by working 5 hours a day. In how many days can 5 farmers plough 10 such farms working at 5 hours a day?

72. By working together, A and B can finish a work in 15 days. If B alone can finish the work in 20 days, in how many days can A alone finish the work?

73. A pipe can fill an empty tank in 5 hours and there is an emptying pipe which can empty the full tank in 7 hours. If both the pipes are kept open simultaneously, then in how many hours will the tank be filled?

74. A trader marked his articles 20% above cost price. The cost price per article is ₹250. If he allows 10% discount on the marked price, then find his profit percentage.

75. A train of length 120 m, travelling at 30 m/s takes 30 seconds to overtake another train of length 240 m. If both the trains are travelling in the same direction, then find the speed of the longer train.

76. Two pipes A and B can fill a tank in 5 h and 8 h respectively. Both the pipes are opened simultaneously and pipe B is closed after one hour. Find the time taken by A to fill the remaining part of the tank.

77. The present ages of Ram and Sita are in ratio of 5:6. The product of their ages (in years) is 120. Find the difference in their ages (in years).

78. Ram can do a work in 20 days. Raj can do $\left(\frac{1}{30}\right)$ th of a piece of work in a day. In how many days can they do the same work, working together?

79. Find the value of y, if $8:5 = y^2:40$ $(y > 0)$.

80. A train crosses a stationary pole in 20 seconds when it is travelling at 54 km/h. Find the length of the train (in m).

CONCEPT APPLICATION
Level 1

Directions for questions 1 to 31: Select the correct alternative from the given choices.

1. If $x:y = 1:1$, then $\dfrac{3x + 4y}{5x + 6y} =$ _____.
   (a) $\dfrac{7}{11}$ (b) $\dfrac{17}{11}$
   (c) $\dfrac{17}{23}$ (d) $\dfrac{4}{5}$

2. If $x = 100$, then 120% of $x =$ _____.
   (a) 125 (b) 115
   (c) 110 (d) 120

3. 40% of 150 + 60% of 150 = _____.
   (a) 200 (b) 150
   (c) 250 (d) 180

4. Interest = Amount − _____.
   (a) Principal (b) Profit
   (c) Loss (d) None of these

5. If $a = 2b$, then $a:b =$ _____.
   (a) 2:1 (b) 1:2
   (c) 3:4 (d) 4:3

6. If a person travels 60 km in 4 hours in a car, then the speed of the car is _____.
   (a) 12 km/h (b) 15 km/h
   (c) 20 km/h (d) 24 km/h

7. 1 km/h = _____ m/sec.
   (a) $\dfrac{18}{5}$ (b) $\dfrac{10}{17}$
   (c) $\dfrac{5}{18}$ (d) $\dfrac{17}{10}$

8. If a person travels 150 km in 5 hours, then he travels 180 km in _____ hours.
   (a) 3 (b) 4
   (c) 8 (d) 6

9. Distance covered by a car with 45 km/h in 2 h is _____.
   (a) 75 km (b) 72 km
   (c) 45 km (d) 90 km

10. If a train 100 m long takes 10 seconds to cross a telegraph post, then how much time does it take to cross a bridge of length 100 m (in seconds).
    (a) 12 (b) 15
    (c) 20 (d) 24

11. If six men can do a piece of work in 6 days, then 3 men can do the same work in _____ days.
    (a) 10 (b) 12
    (c) 15 (d) 18

12. If A can do a work in 10 days, then A's one day work is _____.
    (a) $\dfrac{1}{10}$ (b) $\dfrac{2}{5}$
    (c) $\dfrac{3}{7}$ (d) Cannot be determined

13. If B can do three-fourth of a work in 3 days, then the time taken to complete the whole work is _____.
    (a) 6 days (b) 5 days
    (c) 4 days (d) 3 days

14. If A can type 60 letters in a minute and B can type 120 letters in a minute, then both can type _____ letters in a minute.
    (a) 180 (b) 200
    (c) 220 (d) 240

15. If X can do a piece of work in n days, then X's p days' work is _____.
    (a) $\dfrac{n}{p}$ (b) $\dfrac{p}{n}$
    (c) $\dfrac{n}{2p}$ (d) None of these.

16. Find the value of x, if $80:60 = x:12$.
    (a) 16 (b) 7
    (c) 24 (d) 50

17. If $x:y = 2:3$ and $y:z = 2:3$, then $x:z = $ _____.
    (a) 2:3
    (b) 3:4
    (c) 5:7
    (d) 4:9

18. The cost of 30 mangoes is ₹150. What is the cost of 60 mangoes (in ₹)?
    (a) 400
    (b) 300
    (c) 350
    (d) 500

19. 50% of 150 + 70% of 300 = _____.
    (a) 295
    (b) 285
    (c) 265
    (d) 275

20. 55% of 1000 ÷ 60% of 2000 = ?
    (a) $\dfrac{11}{24}$
    (b) $\dfrac{12}{25}$
    (c) $\dfrac{13}{24}$
    (d) $\dfrac{14}{25}$

21. If $x\%$ of 2000 = 600, then find the value of x.
    (a) 60
    (b) 30
    (c) 40
    (d) 50

22. Which of the following fractions is equivalent to $16.\overline{6}\%$?
    (a) $\dfrac{1}{6}$
    (b) $\dfrac{3}{6}$
    (c) $\dfrac{7}{40}$
    (d) $\dfrac{8}{60}$

23. Find the value of x, if $4:3 = x^2:12$. $(x > 0)$
    (a) 16
    (b) 4
    (c) 9
    (d) 3

24. If $\dfrac{2}{5}$ of 50% of $x = 10$, then find x.
    (a) 100
    (b) 50
    (c) 25
    (d) 80

25. If $x\%$ of $\dfrac{12}{5}$ is $\dfrac{18}{25}$, then find x.
    (a) 30
    (b) 40
    (c) 72
    (d) 50

26. Which of the following fractions is equivalent to 12.5%?
    (a) $\dfrac{1}{8}$
    (b) $\dfrac{1}{6}$
    (c) $\dfrac{1}{5}$
    (d) $\dfrac{1}{12}$

27. If $\dfrac{4}{7}$ of 49% of $x = 21$, then the value of x is _____.
    (a) 125
    (b) 98
    (c) 84
    (d) 75

28. The following are the steps involved in finding the total time taken by both A and B together to complete a work, when A takes 4 days and B takes 12 days to complete the work individually. Arrange them in sequential order.
    (A) A and B's 1 day work = $\dfrac{1}{4} + \dfrac{1}{12} = \dfrac{4}{12} = \dfrac{1}{3}$.
    (B) ∴ Both A and B together can complete the entire work in 3 days.
    (C) A's 1 day work = $\dfrac{1}{4}$. B's 1 day work = $\dfrac{1}{12}$.
    (a) ABC
    (b) CAB
    (c) CBA
    (d) ACB

29. The following are the steps involved in solving the problem given below. Arrange them in sequential order. "If Arun purchased an article at ₹500 and sold at ₹750. Find the profit/loss percentage".
    (A) As $SP > CP$, there is a profit.
    (B) Profit % = $\dfrac{\text{Profit}}{CP} \times 100\% = \dfrac{250}{750} \times 100\%$
    (C) Profit = $SP - CP$ = ₹750 − ₹500 = ₹250
    (D) Given CP = ₹500 and SP = ₹750
    (E) ∴ Percentage of profit = $33\dfrac{1}{3}\%$
    (a) DCABE
    (b) DABCE
    (c) DCAEB
    (d) DACBE

30. The following are the steps involved in finding the value of $x:y$, if $\dfrac{3x + 4y}{2x + 5y} = \dfrac{4}{3}$. Arrange them in sequential order from the first to the last.
    (A) $9x + 12y = 8x + 20y$
    (B) Given, $\dfrac{3x + 4y}{2x + 5y} = \dfrac{4}{3}$
    (C) $\dfrac{x}{y} = 8 \Rightarrow x:y = 8:1$
    (D) $(3x + 4y)3 = (2x + 5y)4$
    (E) $x = 8y$
    (a) BDAEC
    (b) BDCEA
    (c) BADEC
    (d) BADCE

Chapter 3

31. The following are the steps involved in finding the simple interest on ₹1000 for 3 years at the rate of 15%. Arrange them in sequential order from the first to the last.

    (A) ∴ Simple interest (in rupees) = $\frac{100 \times 3 \times 15}{100}$.

    (B) We have $I = \frac{PTR}{100}$.

    (C) $10 \times 3 \times 15 = 450$

    (a) BCA (b) BAC
    (c) ABC (d) CBA

Directions for questions 32 to 35: Match the Column A with Column B.

Column A		Column B
32. If SP = ₹120 and CP = ₹100 then	( )	(a) 2
33. If SP = ₹75 and CP = ₹100, then	( )	(b) 5
34. If a:b = 4:3 and b:c = 3:2, then a/c is	( )	(c) Profit is ₹20

35. 18 km/h = ____ m/s ( ) (d) Loss is ₹20

(e) Profit % is 25%

(f) Loss % is 25%

Directions for questions 36 to 39: Match the Column A with Column B.

Column A		Column B
36. Six men can do a piece of work in 6 days. Two men can do the work in	( )	(a) 5:8
37. If A's 1 day work is 1/6, then A can do the entire work in	( )	(b) 8 hours
38. The simplest form of 60:96 is	( )	(c) 6 days
39. Time taken to travel 360 km at 45 kmp/h is	( )	(d) 10:16

(e) 18 days

(f) 6 hours

Level 2

Directions for questions 40 to 95: Select the correct alternative from the given choices.

40. The ages of Ram and Shyam are in the ratio 7:6 and the sum of their ages is 78 years. Find the age of Ram and Shyam respectively (in years).

    (a) 36, 43 (b) 42, 36
    (c) 30, 48 (d) 40, 38

41. If a:b = 2:5, then find the value of $\frac{(3a+2b)}{(4a+b)}$.

    (a) $\frac{16}{13}$ (b) $\frac{13}{16}$
    (c) $\frac{25}{22}$ (d) $\frac{20}{21}$

42. Find the third proportional of 3 and 27.

    (a) 243 (b) 256
    (c) 289 (d) 225

43. Anand saves 55% of his income. If his income is ₹11000, then find his expenditure (in ₹).

    (a) 6050 (b) 7450
    (c) 4950 (d) 3550

44. Rahul scored 180 marks in the first test and 150 marks in the second test. The maximum marks in each test is 200. What is the decrease in his performance in percentage points?

    (a) 20% (b) 15%
    (c) 25% (d) 23%

45. In the above problem, what is the percentage decrease in his performance?

    (a) 15% (b) $13\frac{1}{3}\%$
    (c) 20% (d) $16\frac{2}{3}\%$

46. Find the simple interest for ₹1500 at 8% per annum for 3 years.

    (a) ₹400 (b) ₹360
    (c) ₹450 (d) ₹500

47. Calculate the amount on ₹20000 at 5% per annum for 5 years under simple interest (in ₹).

    (a) 5000 (b) 25000
    (c) 24000 (d) 20000

48. What distance will a body cover in 20 sec travelling at 50 m/s?
 (a) 700 m (b) 1000 m
 (c) 1500 m (d) 600 m

49. Twelve persons can do a piece of work in 20 days. How many persons are required to do the same in 24 days?
 (a) 20 (b) 10
 (c) 15 (d) 18

50. Sahil takes 1h to travel from his house to a market. If his speed is 1.5 m/s, then how far is the market from his house?
 (a) 6 km (b) 5 km
 (c) 5.4 km (d) 3 km

51. The cost of 30 mangoes is less than the cost of 40 mangoes by ₹50. Find the cost of 50 mangoes (in ₹).
 (a) 400 (b) 300
 (c) 250 (d) 350

52. If the simple interest on a certain sum for 2 years at the rate of 5% per annum is ₹4000, then find the sum (in ₹).
 (a) 46000 (b) 44000
 (c) 40000 (d) 48000

53. Nikhil is twice as efficient as Anil. Nikhil takes 30 days to complete the work alone; in how many days can Anil finish the work?
 (a) 60 days (b) 30 days
 (c) 20 days (d) 40 days

54. A gold smith can make three necklaces in 5 days. How many necklaces can he make in 30 days?
 (a) 19 (b) 18
 (c) 20 (d) 25

55. A train of length 200 m crosses a platform of length 100 m in 12 seconds. Find the speed of the train (in m/s).
 (a) 40 (b) 25
 (c) 60 (d) 18

56. A train travelling at 90 km/h crosses a pole in 8 seconds. Find the length of the train (in m).
 (a) 250 (b) 200
 (c) 150 (d) 178

57. A can do a piece of work in 16 days and B can do the same work in 20 days. In how many days can they do the same work, working together?
 (a) $8\frac{1}{9}$ (b) $9\frac{8}{9}$
 (c) $8\frac{8}{9}$ (d) $9\frac{1}{8}$

58. In how many years will a certain sum become 3 times itself at 25% per annum under simple interest?
 (a) 5 (b) 8
 (c) 12 (d) 6

59. Calculate the amount on ₹25000 at 8% per annum for 6 years under simple interest.
 (a) ₹35000 (b) ₹37000
 (c) ₹45000 (d) ₹47000

60. A trader purchased a bicycle for ₹2500 and sold at ₹2700. Find his profit percentage.
 (a) 8% (b) 10%
 (c) 6% (d) 4%

61. Akhil scored 62, 70, 56, 85, 77 and 70 marks in six different subjects. The maximum mark in each subject is 100. Find his average score.
 (a) 68 (b) 65
 (c) 78 (d) 70

62. If $\dfrac{25p+14q}{5p+7q} = \dfrac{8}{3}$, then find $p:q$.
 (a) 3:5 (b) 2:3
 (c) 2:5 (d) 5:4

63. The present ages of Renu and Seenu are in the ratio 5:6. The sum of their present ages is 44 in years. Find the difference of their ages (in years).
 (a) 4 (b) 5
 (c) 8 (d) 2

64. If 343 is the third proportional of a and b, where $a:b = 1:7$, then find the value of $a + b$.
 (a) 14 (b) 24
 (c) 56 (d) 63

65. If $a:b = 4:5$ and $b:c = 2:3$, then $a:c = $ _____.
 (a) 4:3 (b) 8:15
 (c) 8:9 (d) 5:3

Level 3

66. A varies directly with B when C is constant and inversely with C when B is constant. $A = 6$, when $B = 2$ and $C = 3$. What is the value of A, when $B = 3$ and $C = 18$?
    (a) $\dfrac{1}{2}$ (b) $\dfrac{5}{2}$
    (c) $\dfrac{3}{2}$ (d) $\dfrac{7}{2}$

67. The mean proportional of a and b is 10 and the value of a is four times the value of b. Find the value of $a + b$ ($a > 0$ and $b > 0$).
    (a) 20 (b) 25
    (c) 101 (d) 29

68. If $\dfrac{30x + 2y}{4x + y} = \dfrac{16}{3}$, then find $x : y$.
    (a) 7 : 13 (b) 6 : 13
    (c) 5 : 13 (d) 4 : 13

69. If 243 is the third proportional of a and b, where $b : a = 9 : 1$, then find the values of a and b.
    (a) 3, 27 (b) 2, 18
    (c) 4, 36 (d) 1, 9

70. Find the difference between interest obtained for ₹1000 at 12% per annum for 3 years and that for ₹1500 at 8% per annum for $1\dfrac{1}{2}$ years (in ₹).
    (a) 360 (b) 300
    (c) 180 (d) 200

71. Which of the following yields maximum interest for 2 years?
    (a) ₹1500 at 8% per annum
    (b) ₹1000 at 11% per annum
    (c) ₹2000 at 5% per annum
    (d) ₹900 at 20% per annum

72. The present ages of Seeta and Geeta are in the ratio 3:4. The product of their ages (in years) is 48. Find the sum of their ages (in years).
    (a) 7 (b) 21
    (c) 14 (d) 28

73. A shop keeper bought 10 kg of rice for ₹200, 5 kg of rice at ₹15 per kg and 5 kg at ₹25 per kg. Find the average cost of the rice per kg.
    (a) ₹25 (b) ₹15
    (c) ₹10 (d) ₹20

74. Amit scored 50, 60, 45, 80 and 75 marks in five different subjects. The maximum mark in each subject is 100. Find his average percentage of marks.
    (a) 60% (b) 63%
    (c) 62% (d) 65%

75. In an office, one day, 20% of the staff members were on leave in the morning session. On that same day, 10% of the staff, who were on leave, came for the afternoon session. If total strength of the office is 400, then how many were present for the evening session? (The persons who were present in the morning session were also present in the afternoon session)
    (a) 360 (b) 380
    (c) 364 (d) 328

76. If $\dfrac{a}{b} = \dfrac{x}{y} = \dfrac{p}{q}$, then $\dfrac{6a + 9x + 2p}{6b + 9y + 2q} = $ ____.
    (a) $\dfrac{a}{b}$ (b) $\dfrac{x}{y}$
    (c) $\dfrac{p}{q}$ (d) All of these

77. The income of a person A increases from ₹12,000 to ₹14,000 and the expenditure increases from 40% to 50%. Find the percentage increase or decrease in the monthly savings of A.
    (a) $2\dfrac{7}{9}\%$ increase (b) $2\dfrac{7}{9}\%$ decrease
    (c) 3% increase (d) 3% decrease

78. Twelve persons can make 360 toys in 8 days. If three more persons join them, then how many toys can be made in 8 days?
    (a) 370 (b) 369
    (c) 450 (d) 359

79. In how many years will a certain sum becomes five times itself at 20% per annum simple interest?

(a) 20 (b) 18
(c) 16 (d) 12

80. A can do $\frac{1}{10}$ th of a piece of work in a day. B can do the same work in 15 days. In how many days can they do the same work, working together?
    (a) 5 (b) 6
    (c) 8 (d) 9

81. X can construct a wall in 2 days. Y can construct it in 3 days and Z can construct it in 4 days. If they construct the wall, working together and receive an amount of ₹11700, then find the share of Y.
    (a) ₹3900 (b) ₹3800
    (c) ₹3700 (d) ₹3600

82. Two pipes P_1 and P_2 can fill a tank in 8 hours and 5 hours respectively. They are opened together, after an hour, pipe P_2 is closed, in how many more hours will the tank be filled?
    (a) $\frac{27}{5}$ h (b) $\frac{27}{2}$ h
    (c) $\frac{23}{5}$ h (d) $\frac{28}{5}$ h

83. A pipe can fill an empty tank in 8 hours. After it is kept open for 2 hours another pipe which can fill the tank in 6 hours is opened. In how many more hours will the tank be filled?
    (a) $\frac{18}{7}$ h (b) $\frac{19}{7}$ h
    (c) $\frac{19}{9}$ h (d) $\frac{18}{8}$ h

84. Raj travels a distance of one kilometre at a speed of 1 m/s and four kilometres at a speed of 30 km/h. Find the total time taken to cover the entire distance (in seconds).
    (a) 1480 (b) 900
    (c) 950 (d) 1050

85. Teena takes 12 hours to go from her city to another city by train. If she goes by car, whose speed is 1.5 times that of the train, then how much time will she take to cover the same distance?
    (a) 8 h (b) 6 h
    (c) 9 h (d) 10 h

86. Mohan takes 15 min to go to a market and 200 sec to go to a bus stop travelling at same speed. What is the ratio of distances in two cases?

(a) 4.5:1 (b) 5:1
(c) 1:4.5 (d) 4.5:6

87. A train crosses a post in 10 sec when it is travelling at 72 km/h. Find the length of the train (in m).
    (a) 100 m (b) 200 m
    (c) 300 m (d) 150 m

88. A 150 m long train crosses a platform in 15 seconds. What is the length of the platform if the speed of the train is 90 km/h?
    (a) 225 m (b) 300 m
    (c) 200 m (d) 250 m

89. A train overtakes a person walking at 1 m/s in 5 seconds. If the speed of the train is 65 m/s, then what is the length of the train?
    (a) 300 m (b) 320 m
    (c) 400 m (d) 400 m

90. Two trains travelling in opposite direction at 60 kmph and 50 kmph crossed each other in 9 seconds. If the train which is travelling at 60 kmph crossed a pole in 6 seconds, then find the length of the train which is travelling at 50 kmph.
    (a) 130 m (b) 160 m
    (c) 175 m (d) 200 m

91. Two trains are travelling in opposite direction at 72 km/h and 90 km/h. They cross each other in 16 seconds. If the length of the slower train is 320 m, then find the length of the faster train (in m).
    (a) 380 (b) 420
    (c) 400 (d) 425

92. Two pipes P_1 and P_2 can fill a tank in 6 hours and 3 hours respectively. Both the pipes are opened simultaneously and pipe P_1 is closed after 1 hour. Find the time taken by P_2 to fill the remaining part of the tank. (in hours)
    (a) 1 (b) $1\frac{1}{2}$
    (c) 2 (d) $2\frac{1}{2}$

93. In a class, there are 30 boys and 20 girls. The average weight of 30 boys is 60 kg and average weight of 20 girls is 35 kg. Find the average weight of the class.

3.26 Chapter 3

(a) 40 kg (b) 50 kg
(c) 60 kg (d) 45 kg

94. Radha takes 4 hours to travel from her village to her friend's city by bus. If she goes by car, whose speed is $\frac{2}{3}$ times that of the bus, then how much time will she take to cover the same distance?

(a) 3 hours (b) 8 hours
(c) 12 hours (d) 6 hours

95. A pipe can fill an empty tank in 12 hours. After it is kept open for 3 hours, another pipe which can fill the tank in 8 hours is opened. In how many more hours will the tank be filled?

(a) $\frac{17}{7}$ h (b) $\frac{18}{5}$ h
(c) $\frac{20}{9}$ h (d) $\frac{16}{5}$ h

ASSESSMENT TESTS

Test 1

Directions for questions 1 to 11: Select the correct alternative from the given choices.

1. The ages of A and B are in the ratio 5:4 and the difference of their ages is 10 years. Find the ratio of their ages 8 years before.

   The following are the steps involved in solving the above problem. Arrange them in sequential order.

   (A) Let the age of A be $5x$ years and the age of B be $4x$ years.

   (B) 8 years ago, A's age was 42 years and B's age was 32 years. The required ratio = 42:32 = 21:16.

   (C) $5x - 4x = 10 \Rightarrow x = 10$.

   (D) A's age = $5x = 5 \times 10 = 50$ years and B's age = $4x = 4 \times 10 = 40$ years.

   (a) ACDB (b) ABCD
   (c) BCDA (d) None of these

2. The following are the steps involved in finding the value of x from $4:(x-2) = (x+2):8$.

   Arrange them in sequential order.

   (A) $4:(x-2) = (x+2):8 \Rightarrow (x-2)(x+2) = 32$
   (B) $x^2 = 36$
   (C) $x^2 - 4 = 32$
   (D) $x = \pm 6$

   (a) ABCD (b) ACBD
   (c) ACDB (d) ADBC

3. Find the value of $x:y$, if $\frac{7x-15y}{4x+6y} = \frac{5}{6}$.

   (a) 60:11 (b) 30:11
   (c) 27:13 (d) 17:13

4. The value of 30% of 60% of 200 is _____.

   (a) 23 (b) 24
   (c) 18 (d) 36

5. The fourth proportional of 9, 8 and 18 is _____.

   (a) 16 (b) 9
   (c) 72 (d) None of these

6. The mean proportional of 27 and 3 is _____.

   (a) 6 (b) 3
   (c) 9 (d) 27

7. Which of the following is equivalent to $3.\overline{6}$%?

   (a) $\frac{11}{100}$ (b) $\frac{11}{300}$
   (c) $\frac{22}{100}$ (4) $\frac{36}{100}$

8. What distance (in m) will a train cover in 2 minutes travelling at 108 km/h?

   (a) 3600
   (b) 360
   (c) 306
   (d) None of these

9. ₹3000 is lent out at 3% per annum for 20 years under simple interest. Calculate the amount (in ₹).

    (a) 1800   (b) 1080
    (c) 3600   (d) 4800

10. If the cost of 25 pencils is ₹75, then what is the cost of 75 pencils (in ₹)?

    (a) 225   (b) 150
    (c) 180   (d) 100

11. Sixteen men, working at 6 hours a day, can do a piece of work in 15 days. In how many days can 18 men working at 10 hours a day do the same work?

    (a) 18   (b) 10
    (c) 16   (d) 8

Directions for questions 12 to 15: Match Column A with Column B.

Column A		Column B
12. The average of the sum of the squares of the first five natural numbers is	( )	(a) 5:1
13. If $x = 5y$, then $x:y$ is	( )	(b) 20%
14. The cost price of an article is ₹200 and its selling price is ₹180. Find the loss percentage.	( )	(c) 10%
15. At what rate of interest will ₹250 amount to triple its present value in 10 years at simple interest (p.a.)%.	( )	(d) 11
		(e) 1:5
		(f) 15%
		(g) 10

Test 2

Directions for questions 16 to 26: Select the correct alternative from the given choices.

16. The following are the steps involved in finding the value of x, if $32x:24 = x^3:12$. Arrange them in sequential order.

    (A) $2x^2 = 32 \Rightarrow x^2 = 16$
    (B) $24x^2 = 12 \times 32$
    (C) $32x:24 = x^3:12 \Rightarrow 24x^3 = 12 \times 32x$
    (D) $x = \pm 4$

    (a) CDAB   (b) CBAD
    (c) ABCD   (d) BCDA

17. The ages of Ram and Laxman are in the ratio of 4:3 and the sum of their ages is 49 years. Find the ratio of their ages after 7 years. The following are the steps involved in solving the above problem. Arrange them in sequential order.

    (A) Given, the ratio of ages is 4:3.
    (B) Let Ram's age be $4x$ years and Laxman's age be $3x$ years. $4x + 3x = 49 \Rightarrow 7x = 49 \Rightarrow x = 7$.
    (C) Ratio of ages after 7 years = 35:28 = 5:4.
    (D) Ram's age = $4 \times 7 = 28$ years. Laxman's age = $3 \times 7 = 21$ years.

    Ram's age after 7 years = 35 years and Laxman's age after 7 years = 28 years.

    (a) ABCD   (b) BCDA
    (c) CDAB   (d) ABDC

18. Find the value of $x:y$, if $\dfrac{8x - 10y}{3x + 4y} = \dfrac{4}{5}$.

    (a) 55:29   (b) 36:17
    (c) 33:14   (d) 16:11

19. The value of 20% of 40% of 100 is ____.

    (a) 12   (b) 16
    (c) 8   (d) 25

20. The fourth proportional of 3, 7, and 15 is ____.

    (a) 35   (b) 15
    (c) 7   (d) 21

21. The mean proportional of 64 and 4 is ____.

    (a) 8   (b) 32
    (c) 4   (d) 16

22. Which of the following fractions is equivalent to $2.\overline{9}\%$?

3.28 Chapter 3

(a) $\dfrac{3}{100}$ (b) 3

(c) 2.9   (d) $\dfrac{2.9}{100}$

23. What distance (in m) will a train cover in 1 minute travelling at 72 km/h?
    (a) 120   (b) 1200
    (c) 1400   (d) 720

24. ₹2000 is lent out at 2% per annum for 10 years under simple interest. Calculate the amount (in ₹).
    (a) 1400   (b) 2400
    (c) 200   (d) 1500

25. If the cost of 15 mangoes is ₹180, then what is the cost of 25 mangoes (in ₹)?
    (a) 220   (b) 360
    (c) 200   (d) 300

26. Twelve men, working at 8 hours a day, can do a piece of work in 18 days. In how many days can 24 men working at 6 hours a day can do the same work?
    (a) 12   (b) 16
    (c) 20   (d) 30

Directions for questions 27 to 30: Match Column A with Column B.

Column A		Column B
27. The average of the first 10 natural numbers is	( )	(a) 20 years
28. If $a:b = 2:3$ and $b:c = 9:7$, then $a:b:c$ is	( )	(b) 5.5
29. If the cost price of an article is ₹12 and the selling price of the article is ₹15. Find the profit percentage.	( )	(c) 6:9:7
30. In how many years will the amount of ₹500 double at the rate of 5% p.a. under simple interest?	( )	(d) 5%
		(e) 25%
		(f) 2:3:7
		(g) 25 years
		(h) 6

Ratio and Its Applications   3.29

TEST YOUR CONCEPTS

Very Short Answer Type Questions

1. True
2. True
3. False
4. False
5. True
6. False
7. True
8. False
9. True
10. False
11. ₹80
12. Per hundred
13. 1:2
14. 2:3
15. 26
16. 22
17. 4
18. 25
19. 400
20. 1:2
21. (b)
22. (c)
23. (c)
24. (b)
25. (a)
26. (c)
27. (b)
28. (a)
29. (a)
30. (c)

Short Answer Type Questions

31. ₹25, ₹50, ₹100
32. 9
33. 32 kg
34. ₹3400
35. ₹5110
36. 5:6:8
37. ₹216
38. 36
39. 75%
40. ₹6000
41. 119 cm
42. 2 m, 8 m
43. ₹30000, ₹20000
44. 50%
45. 5 cm, 4 cm, 3 cm.
46. $16.\overline{6}\%$
47. $\pm \dfrac{5}{4}$
48. 50 kmph
49. 1.25 m/s
50. 4 h
51. 10 m/sec
52. 6 days
53. 57
54. 4:5
55. 10:9
56. 7 years
57. 85
58. 20%
59. $9\dfrac{3}{8}$ days
60. 20%
61. 180 m
62. ₹4900

ANSWER KEYS

Chapter 3

Essay Type Questions

63. ₹200
64. 1076
65. 50 kg
66. ₹2625, ₹5250
67. handball team
68. $88\frac{8}{9}\%$
69. ₹960
70. 37 sec
71. 20 days
72. 60 days
73. 17.5 hrs.
74. 8%
75. 18 m/s
76. $3\frac{3}{8}$ h
77. 2
78. 12
79. 8
80. 300 m

CONCEPT APPLICATION

Level 1

1. (a)	2. (d)	3. (b)	4. (a)	5. (a)	6. (b)	7. (c)	8. (d)	9. (d)	10. (c)
11. (b)	12. (a)	13. (c)	14. (a)	15. (b)	16. (a)	17. (d)	18. (2)	19. (b)	20. (a)
21. (b)	22. (a)	23. (b)	24. (b)	25. (a)	26. (a)	27. (d)	28. (b)	29. (d)	30. (a)
31. (b)	32. (c)	33. (f)	34. (a)	35. (b)	36. (e)	37. (c)	38. (a)	39. (b)	

Level 2

40. (b)	41. (a)	42. (a)	43. (c)	44. (b)	45. (d)	46. (b)	47. (b)	48. (b)	49. (b)
50. (c)	51. (c)	52. (c)	53. (a)	54. (b)	55. (b)	56. (b)	57. (c)	58. (b)	59. (b)
60. (a)	61. (d)	62. (c)	63. (a)	64. (c)	65. (b)				

Level 3

66. (c)	67. (b)	68. (c)	69. (a)	70. (c)	71. (d)	72. (c)	73. (d)	74. (c)	75. (d)
76. (d)	77. (b)	78. (c)	79. (a)	80. (b)	81. (d)	82. (a)	83. (a)	84. (a)	85. (a)
86. (a)	87. (b)	88. (a)	89. (b)	90. (c)	91. (c)	92. (b)	93. (b)	94. (d)	95. (b)

ASSESSMENT TESTS

Test 1

| 1. (a) | 2. (b) | 3. (a) | 4. (d) | 5. (a) | 6. (c) | 7. (b) | 8. (a) | 9. (d) | 10. (a) |
| 11. (d) | 12. (d) | 13. (a) | 14. (c) | 15. (b) | | | | | |

Test 2

| 16. (b) | 17. (d) | 18. (c) | 19. (c) | 20. (a) | 21. (d) | 22. (a) | 23. (b) | 24. (b) | 25. (d) |
| 26. (a) | 27. (b) | 28. (c) | 29. (e) | 30. (a) | | | | | |

Ratio and Its Applications 3.31

CONCEPT APPLICATION
Level 1

1. $\dfrac{7}{11}$
2. 120
3. 150
4. Principal
5. 2:1
6. 15 km/h
7. $\dfrac{5}{18}$
8. 6
9. 90 km
10. 20 seconds
11. 12 days
12. $\dfrac{1}{10}$
13. 4
14. 180
15. $\dfrac{p}{n}$
16. $80:60 = x:12$
    $\Rightarrow \dfrac{80}{60} = \dfrac{x}{12}$
    $\Rightarrow x = \dfrac{80 \times 12}{60} = 16.$
    Hence, the correct option is (a)
17. $x:y = 2:3$
    $y:z = 2:3$
    HCF of terms represented by y is HCF of 3 and 2 is 6.
    $\therefore x:y = 2 \times 2:3 \times 2 = 4:6$
    $y:z = 2 \times 3:3 \times 3 = 6:9$
    $\Rightarrow x:y:z = 4:6:9$
    $\Rightarrow x:z = 4:9$
    Hence, the correct option is (d)

18. Cost of 30 mangoes = ₹150
    Cost of one mango = $\dfrac{150}{30}$ = ₹5
    $\therefore$ Cost of 60 mangoes = 60×5 = ₹300.
    Hence, the correct option is (b)
19. 50% of 150 + 70% of 300
    $= \left(\dfrac{50}{100} \times 150\right) + \left(\dfrac{70}{100} \times 300\right)$
    $= 75 + 210 = 285.$
    Hence, the correct option is (b)
20. 55% of 1000 ÷ 60% of 2000
    $= \left(\dfrac{55}{100} \times 1000\right) + \left(\dfrac{60}{100} \times 2000\right)$
    $= 550 \div 1200 = \dfrac{11}{24}$
    Hence, the correct option is (a)
21. $\dfrac{x}{100} \times 2000 = 600$
    $\Rightarrow 20x = 600$
    $\Rightarrow x = 30.$
    Hence, the correct option is (b)
22. $16.\overline{6}\% = \dfrac{16.\overline{6}}{100} = \dfrac{1}{6}$
    Hence, the correct option is (a)
23. Given, $\dfrac{4}{3} = \dfrac{x^2}{12}$
    $\Rightarrow \dfrac{48}{3} = x^2$
    $\Rightarrow x^2 = 16.$
    $\Rightarrow x = 4\ (x > 0).$
    Hence, the correct option is (b)
24. $\dfrac{2}{5} \times \dfrac{50}{100} \times x = 10$
    $\Rightarrow x = 50.$
    Hence, the correct option is (b)

3.32 Chapter 3

25. $\dfrac{x}{100} \times \dfrac{12}{5} = \dfrac{18}{25} =$

   $\Rightarrow x = 30.$

   Hence, the correct option is (a)

26. $12.5\% = \dfrac{12.5}{100} = \dfrac{125}{10 \times 100} = \dfrac{1}{8}$

   Hence, the correct option is (a)

27. $\dfrac{4}{7} \times \dfrac{49}{100} \times x = 21$ (given)

   $\dfrac{7}{25} \times x = 21, x = 75.$

   Hence, the correct option is (d)

28. (C), (A) and (B) is the required sequential order.

   Hence, the correct option is (b)

29. (D), (A), (C), (B) and (E) is the required sequential order.

   Hence, the correct option is (d)

30. (B), (D), (A), (E) and (C) is the required sequential order.

   Hence, the correct option is (a)

31. (B), (A) and (C) is the required sequential order.

   Hence, the correct option is (b)

32. → c: Profit = SP − CP = ₹120 − ₹100 = ₹20

33. → f: Loss = CP − SP = ₹100 − ₹75 = ₹25.

   Loss % = $\dfrac{25}{100} \times 100\% = 25\%.$

34. → a:

   $a:b = 4:3, b:c = 3:2$

   $\Rightarrow \left(\dfrac{a}{b}\right)\left(\dfrac{b}{c}\right) = \left(\dfrac{4}{3}\right)\left(\dfrac{3}{2}\right) \Rightarrow \dfrac{a}{c} = 2.$

35. → b:

   18 km/h = $18\left(\dfrac{5}{18}\right)$ m/sec = 5 m/s.

36. → e:

   We have $M_1 D_1 = M_2 D_2$

   $\Rightarrow 6 \times 6 = 2 \times D_2 \Rightarrow D_2 = 18$ days.

37. → c:

   A's 1 day work is $\dfrac{1}{6}$.

   ∴ A can do the entire work in 6 days.

38. → a:

   HCF of 60 and 96 is 12.

   ∴ $60:96 = \dfrac{60}{12} : \dfrac{96}{12} = 5:8$

39. → b:

   We have, $T = \dfrac{D}{S}$

   $\Rightarrow T = \dfrac{360}{45} = 8$ hours.

Level 2

40. The ratio of ages of Ram and Shyam = 7:6.

    Let the ages of Ram and Shyam be $7x$ years and $6x$ years respectively.

    Sum of their ages = 78.

    $\Rightarrow 7x + 6x = 78 \Rightarrow 13x = 78$

    $\Rightarrow x = 6.$

    ∴ Age of Ram = 7 × 6 = 42 years.

    Age of Shyam = 6 × 6 = 36 years.

    Hence, the correct option is (b)

41. Let $a = 2k$ and $b = 5k.$

    $\Rightarrow \dfrac{3a + 2b}{4a + b} = \dfrac{3(2k) + 2(5k)}{4(2k) + 5k} = \dfrac{16k}{13k} = \dfrac{16}{13}$

    Hence, the correct option is (a)

42. Let the third proportional of 3 and 27 be x.

    $\Rightarrow \dfrac{3}{27} = \dfrac{27}{x}$

    $\Rightarrow 3x = 27^2 \Rightarrow x = 243.$

    Hence, the correct option is (a)

43. Income of Anand = ₹11000.

Percentage of income saved = 55%.

∴ Percentage of income spend = 45%.

∴ Expenditure of Anand = 45% of ₹11000.

$= \dfrac{45}{100} \times ₹11000 = ₹4950$

Hence, the correct option is (c)

44. Rahul's score in first test = 180.

Rahul's score in second test = 150.

Decrease in percentage points

$= \dfrac{\text{Score in first test} - \text{Score in second test}}{\text{Total score}} \times 100.$

$= \dfrac{180 - 150}{200} \times 100 = 15\%.$

Hence, the correct option is (b)

45. Marks scored by Rahul in the first test = 180.

Marks scored by Rahul in the second test = 150.

Percentage decrease in his performance

$= ((180 - 150)/180) \times 100 = \dfrac{30}{180} \times 100.$

$= \dfrac{100}{6} = 16\dfrac{2}{3}\%$

Hence, the correct option is (d)

46. $S.I. = \dfrac{P \times N \times R}{100} = \dfrac{1500 \times 8 \times 3}{100} = ₹360$

Hence, the correct option is (b)

47. $SI = \dfrac{P \times N \times R}{100} = \dfrac{20000 \times 5 \times 5}{100} = 5000.$

Amount $(A) = I + P$

where I is the interest and P is the principal.

$A = 5000 + 20000 = ₹25000.$

Hence, the correct option is (b)

48. Distance = Speed × Time.

$= 50 \times 20 = 1000$ m.

Hence, the correct option is (b)

49. When amount of work is constant, number of persons vary inversely with number of days required to complete the work.

Let x number of persons can do the work in 24 days.

∴ $x \times 24 = 20 \times 12 \Rightarrow x = 10.$

Hence, the correct option is (b)

50. Distance = Speed × Time

$\Rightarrow 1.5 \times 3600 = 5400$ m $= 5.4$ km.

Hence, the correct option is (c)

51. Let the cost of each mango be ₹x.

$\Rightarrow 40x - 30x = 50$

$\Rightarrow 10x = 50 \Rightarrow x = 5.$

∴ Cost of 50 mangoes = $50 \times ₹5 = ₹250.$

Hence, the correct option is (c)

52. Let the sum be ₹x.

Given, $\dfrac{x \times 2 \times 5}{100} = 4000 \Rightarrow x = ₹40000.$

Hence, the correct option is (c)

53. Given, Anil is half as efficient as Nikhil.

∴ Anil takes double the number of days that Nikhil takes, i.e., 60 days.

Hence, the correct option is (a)

54. Number of necklaces made in 5 days = 3.

∴ Number of necklaces made in 30 days = $\left(\dfrac{3}{5}\right)(30) = 18.$

Hence, the correct option is (b)

55. Speed = $\dfrac{\text{Distance}}{\text{Time}}$

Distance = Length of the train + Length of the platform.

$= 200 + 100 = 300$ m.

∴ Speed = $\dfrac{300}{12} = 25$ m/s.

Hence, the correct option is (b)

3.34 Chapter 3

56. 90 km/h = $90 \times \dfrac{5}{18}$ m/s = 25 m/s.

    $\therefore$ Length of the train = Speed × Time = $25 \times 8 = 200$ m.

    Hence, the correct option is (b)

57. A can do $\dfrac{1}{16}$ th of the work in one day.

    B can do $\dfrac{1}{20}$ th of the work in one day.

    The part of the work done by A and B together

    $\dfrac{1}{16} + \dfrac{1}{20} = \dfrac{5+4}{80} = \dfrac{9}{80}$.

    $\therefore$ Both together can complete the work in $\dfrac{80}{9}$ days, i.e., $8\dfrac{8}{9}$ days.

    Hence, the correct option is (c)

58. Let the sum be ₹P

    $\therefore$ Amount = ₹3p

    $\Rightarrow$ Interest = ₹2p

    Given, R = 25%,

    $I = \dfrac{PTR}{100}$, $2P = \dfrac{P \times T \times 25}{100}$, $200 = 25T$,

    $T = \dfrac{200}{25} = 8$

    Time period = 8 years.

    Hence, the correct option is (b)

59. Given, p = ₹25000, R = 8% and T = 6 years.

    Amount (in ₹) = $P\left(1 + \dfrac{TR}{100}\right)$

    $= 25000\left(1 + \dfrac{6 \times 8}{100}\right)$

    $= 25000\left(1 + \dfrac{6 \times 2}{25}\right)$

    $= 25000 \times \dfrac{37}{25}$

    $= 1000 \times 37 = 37000$.

    Hence, the correct option is (b)

60. CP of bicycle = ₹2500, SP of bicycle = ₹2700.

    Profit = SP − CP

    $= ₹2700 − ₹2500 = ₹200$

    Profit% = $\dfrac{\text{Profit}}{\text{CP}} \times 100\%$

    $= \dfrac{200}{2500} \times 100\% = 8\%$

    Hence, the correct option is (a)

61. Average score = $\dfrac{62 + 70 + 56 + 85 + 77 + 70}{6}$

    $= \dfrac{420}{6} = 70$

    Hence, the correct option is (d)

62. $\dfrac{25p + 14q}{5p + 7q} = \dfrac{8}{3}$

    $(25p + 14q) \times 3 = 8 \times (5p + 7q)$

    $75p + 42q = 40p + 56q$

    $75p − 40p = 56q − 42q$, $35p$

    $= 14q$, $\dfrac{p}{q} = \dfrac{14}{35} = \dfrac{2}{5}$

    $\therefore p:q = 2:5$.

    Hence, the correct option is (c)

63. Let the present ages of Renu and Seenu be 5x and 6x years.

    Sum of their ages = $5x + 6x = 11x$.

    Given that $11x = 44$, $x = 4$ years.

    $\therefore$ Their present ages are 20, 24 years.

    $\therefore$ The difference between their ages = $24 − 20 = 4$ years.

    Hence, the correct option is (a)

64. Given, $a:b = b:343$

    $b^2 = 343a$ (1)

    and also given, $a:b = 1:7$.

    $a:b = 1:7$, $\dfrac{a}{b} = \dfrac{1}{7} \Rightarrow b = 7a$

    From (1), $(7a)^2 = 343a$

    $\Rightarrow a = 7$

    $\Rightarrow b = 7 \times 7 = 49$

    $\therefore a + b = 7 \times 49 = 56$.

    Hence, the correct option is (c)

Ratio and Its Applications 3.35

65. $a:b = 4:5$

$b:c = 2:3$

$\Rightarrow a:b = 4(2):5(2) = 8:10$

and $b:c = 2(5):3(5) = 10:15$

$\therefore a:b:c = 8:10:15$

$\Rightarrow a:c = 8:15.$

Hence, the correct option is (b)

Level 3

66. Given, $A \propto B$ \hfill (1) (C in constant)

$A \propto \dfrac{1}{C}$ \hfill (2) (B is constant)

From (1) and (2), we get

$A = \dfrac{KB}{C}$, where K is constant.

$\dfrac{AC}{B} = K$

(i) $A = 6, B = 2, C = 3$

$\therefore K = \dfrac{6 \times 3}{2} = 9.$

(ii) $A = ?, B = 3, C = 18$

$\dfrac{A \times 18}{3} = 9$

$\Rightarrow A = \dfrac{3}{2}$

Hence, the correct option is (c)

67. Given, $10^2 = a \times b$ \hfill (1)

$\Rightarrow 100 = ab$

and $a = 4b$ \hfill (2),

From (1) and (2), $100 = 4b^2.$

$\Rightarrow b = 5$ and $a = 20$

$\therefore a + b = 25.$

Hence, the correct option is (b)

68. $\dfrac{30x + 2y}{4x + y} = \dfrac{16}{3}$

$\Rightarrow 90x + 6y = 64x + 16y$

$\Rightarrow 26x = 10y$

$\Rightarrow x:y = 5:13.$

Hence, the correct option is (c)

69. Given, $\dfrac{b}{a} = 9 \Rightarrow b = 9a.$ \hfill (1)

Given, $\dfrac{a}{b} = \dfrac{b}{243}.$

$\Rightarrow b^2 = 243a.$ \hfill (2)

From (1) and (2),

$81a^2 = 243a$

$\Rightarrow a = 3 \Rightarrow b = 27.$

Hence, the correct option is (a)

70. Interest obtained for ₹1000 at 12% per annum for 3 years = $\dfrac{1000 \times 12 \times 3}{100}$ = ₹360.

Interest obtained for ₹1500 at 8% per annum for 1.5 years = $\dfrac{1500 \times 8 \times 1.5}{100}$ = ₹180.

$\therefore$ Required difference = ₹180.

Hence, the correct option is (c)

71. Here $N = 2y$, it is the same for all cases.

Interest for 1500 at 8% per annum = $\dfrac{1500 \times 8 \times 2}{100}$

= ₹240.

Interest for 1000 at 11% per annum = $\dfrac{1000 \times 11 \times 2}{100}$

= ₹220.

Interest for 2000 at 15% per annum = $\dfrac{2000 \times 5 \times 2}{100}$

= ₹200.

Interest for 900 at 20% per annum = $\dfrac{900 \times 20 \times 2}{100}$

= ₹360.

$\therefore$ Maximum interest is obtained for ₹900 at 20% per annum.

Hence, the correct option is (d)

Chapter 3

72. Let the age of Seeta be $3x$ and the age of Geeta be $4x$.

$4x \times 3x = 48 \Rightarrow 12x^2 = 48 \Rightarrow x = 2$

Seeta's age = $6y$ and Geeta's age = 8 years.

∴ The sum of their ages = 14 years.

Hence, the correct option is (c)

73. Total cost of 20 kg of rice = ₹200 + ₹15 × 5 + ₹25 × 5 = ₹400.

Average cost of rice per kg = ₹$\dfrac{400}{20}$ = ₹20.

Hence, the correct option is (d)

74. Average percentage of marks

$= \dfrac{\text{Sum of the marks scored by Amit}}{\text{Total marks}} \times 100$

$= \dfrac{50 + 60 + 45 + 80 + 75}{500} \times 100 = 62\%.$

Hence, the correct option is (c)

75. Total strength = 400.

20% of 400 = 80.

If 80 were absent, then the number of people present in morning = 400 − 80 = 320.

In the afternoon, 10% of 80, i.e., 8 turned up.

∴ Number of people present in the evening = 320 + 8 = 328.

Hence, the correct option is (d)

76. Let, $\dfrac{a}{b} = \dfrac{x}{y} = \dfrac{p}{q} = k$ (1)

∴ $a = bk$, $x = yk$ and $p = qk$.

$\dfrac{6a + 9x + 2p}{6b + 9y + 2q} = \dfrac{6(bk) + 9(yk) + 2(qk)}{6b + 9y + 2q}$

$= k\left(\dfrac{6b + 9y + 2q}{6b + 9y + 2q}\right) = k$

From (1), k is $\dfrac{a}{b}$ or $\dfrac{x}{y}$ or $\dfrac{p}{q}$.

Hence, the correct option is (d)

77. Savings = Income − Expenditure

In the first case, savings = 60% of 12000 = ₹7200.

In the second case, savings = 50% of 14000 = ₹7000.

∴ The percentage decrease in savings = $\left(\dfrac{(7200 - 7000)}{7200}\right) \times 100 = 2\dfrac{7}{9}\%$

Hence, the correct option is (b)

78. The number of toys is directly proportional to the number of persons working. (Number of days is constant). Let the required number of persons be x.

∴ $\dfrac{12}{15} = \dfrac{360}{x} \Rightarrow x = 450.$

∴ 15 persons can make 450 toys in 8 days.

Hence, the correct option is (c)

79. Let the sum of ₹x.

∴ Amount = ₹$5x$.

Simple interest = Amount − Principal = ₹$5x$ − ₹x = ₹$4x$.

Let the number of years be n.

We have simple interest = $\dfrac{PTR}{100}$

$\Rightarrow 4x = \dfrac{x \times n \times 20}{100}$

$\Rightarrow 4 = \dfrac{n}{5} \Rightarrow n = 20.$

∴ In 20 years, the sum becomes 5 times itself.

Hence, the correct option is (a)

80. A's 1 day work = $\dfrac{1}{10}$.

As B can do the work in 15 days, B's 1 day work = $\dfrac{1}{15}$

$(A + B)$'s 1 day work = $\dfrac{1}{10} + \dfrac{1}{15}$.

$= \dfrac{3 + 2}{30} = \dfrac{5}{30} = \dfrac{1}{6}$

A and B together can finish the entire work in 6 days.

Hence, the correct option is (b)

81. X's 1 day work = $\dfrac{1}{2}$

Y's 1 day work = $\dfrac{1}{3}$

Z's 1 day work = $\dfrac{1}{4}$

Ratio of shares of X, Y and Z is equal to ratio of their capacities, i.e., $\dfrac{1}{3}:\dfrac{1}{2}:\dfrac{1}{4}$, i.e., 6:4:3.

∴ Share of $Y = \dfrac{4}{13}(11700) = ₹3600$.

Hence, the correct option is (d)

82. In one hour, $1/8 + 1/5 = (13/40)$th of the tank gets filled.

∴ $1 - \dfrac{13}{40} = \dfrac{27}{40}$ part of the tank is still empty.

∴ $\dfrac{27}{40}$ part of the tank can be filled in $= \dfrac{8 \times 27}{40} = \dfrac{27}{5}$ h.

Hence, the correct option is (a)

83. Part of the tank filled in 2 h $= \dfrac{1}{8} \times 2 = \dfrac{1}{4}$.

The part of tank to be filled $= 1 - \dfrac{1}{4} = \dfrac{3}{4}$.

Part of tank filled in 1 h by both of the pipes $= \dfrac{1}{8} + \dfrac{1}{6} = \dfrac{7}{24}$.

∴ Time taken to fill $\dfrac{3}{4}$ th of the tank by both the pipes.

$= \dfrac{3}{4} \times \dfrac{24}{7} = \dfrac{18}{7}$ h.

Hence, the correct option is (a)

84. Time taken to cover 1 km = 1000 ÷ 1 = 1000 seconds.

Time taken to cover 4 km $= \dfrac{4000}{30 \times \dfrac{5}{18}}$ seconds

= 480 seconds.

Total time taken by him = 1000 + 480 = 1480 seconds.

Hence, the correct option is (a)

85. Let the distance between her city and another city be d km and the speed of the train be s kmph.

⇒ $\dfrac{d}{s} = 12$

Gives, Speed of the car = 1.5 s.

⇒ $\dfrac{d}{1.5s} = \dfrac{12}{1.5} = 8$ h.

Hence, the correct option is (a)

86. ∴ Required ratio of distances = Ratio of time taken in two cases = (15 × 6):200

= 9:2 = 4.5:1

Hence, the correct option is (a)

87. To cross the post, the train has to travel its own length.

Length of the train $= 72 \times \dfrac{5}{18} \times 10 = 200$ m.

(∴ Distance = Speed × Time)

Hence, the correct option is (b)

88. To cross the bridge, train has to travel the length of the train and the bridge.

Let the length of platform be l m.

Given, $\dfrac{l + 150}{90 \times \dfrac{5}{18}} = 15$.

(∴ Distance/Speed = Time)

⇒ $l + 150 = 25 \times 15$

⇒ $l = 375 - 150 = 225$ m.

Hence, the correct option is (a)

89. Relative speed = 65 − 1 = 64 m/s.

∴ Length of the train = Relative Speed × Time.

= 64 × 5 = 320 m.

Hence, the correct option is (b)

90. Let the length of the train, which is traveling at 60 kmph or $\dfrac{50}{3}$ m/sec is l_1 m and the length of the train traveling at 50 kmph or $\dfrac{125}{9}$ m/sec is l_2 m.

Given, $\dfrac{l_1}{\dfrac{50}{3}} = 6$

⇒ $l_1 = \dfrac{50}{3} \times 6 = 100$ m.

It is also given that, $\dfrac{l_1 + l_2}{\dfrac{50}{3} + \dfrac{125}{9}} = 9$

⇒ $l_1 + l_2 = 9\left(\dfrac{150 + 125}{9}\right)$

⇒ $100 + l_2 = 275$

⇒ $l_2 = 175$ m.

Hence, the correct option is (c)

3.38 Chapter 3

91. Relative speed = (72 + 90) km/h = 162 km/h.

$$= 162 \times \frac{5}{18} = 45 \text{ m/s.}$$

Time taken to cross each other

$$= \frac{\text{The sum of the lengths of two trains}}{\text{Relative speed}}.$$

$$16 = \frac{320 + T}{45} \quad \text{(where } T \text{ be the length of the faster train)}$$

$$720 = 320 + T$$

$$\therefore T_2 = 720 - 320 = 400 \text{ m.}$$

Hence, the correct option is (c)

92. Pipe P_1 can fill $\frac{1}{6}$ of the tank in one hour.

Pipe P_2 can fill $\frac{1}{3}$ of the tank in one hour.

Pipes P_1 and P_2 together can fill $\left(\frac{1}{6} + \frac{1}{3}\right)$ of the tank in one hour.

The part of the tank to be filled = $1 - \left(\frac{1}{6} + \frac{1}{3}\right)$.

$$= 1 - \left(\frac{1+2}{6}\right) = \frac{6-3}{6} = \frac{3}{6} = \frac{1}{2}.$$

Time taken by the pipe P_2 to fill $\frac{1}{2}$ of tank at the

rate of $\frac{1}{3}$ per hour = $\frac{\left(\frac{1}{2}\right)}{\left(\frac{1}{3}\right)} = \frac{1}{2} \times \frac{3}{1} = 1\frac{1}{2}$ h.

Hence, the correct option is (b)

93. The total weight of the boys = 30 × 60 = 1800 kg.

The total weight of the girls = 20 × 35 = 700 kg.

The total weight of the class = (1800 + 700) kg = 2500 kg.

The average weight of the class = $\frac{2500}{30+20}$ = 50 kg.

Hence, the correct option is (b)

94. Let the speed of bus be s kmph.

The speed of the car = $\frac{2}{3}$ s kmph.

Let the distance between the village and the city be d km.

Speed × Time = Distance covered.

For bus,

$$d = s \times 4 \quad (1)$$

For car,

$$d = \frac{2s}{3} \times t \quad (2)$$

⇒ from (1) and (2)

$$s \times 4 = \frac{2s}{3} \times t$$

$t = 6$ hours.

Hence, the correct option is (d)

95. The part of the tank filled by tap 1 is $\frac{1}{12}$ th part.

In 3 hours, it has filled $3 \times \frac{1}{12} = \frac{1}{4}$ th part of tank.

Remaining part of the tank, to be filled = $1 - \frac{1}{4}$ = $\frac{3}{4}$ th part of tank.

In one hour, both pipes can fill $\left(\frac{1}{12} + \frac{1}{8}\right)$ th part of tank.

$$= \frac{3+2}{24} = \frac{5}{24} \text{ th part of tank.}$$

$\frac{3}{4}$ th of tank can be filled in $\frac{\frac{3}{4}}{\frac{5}{24}} = \frac{18}{5}$ hours.

Hence, the correct option is (b)

Ratio and Its Applications   **3.39**

ASSESSMENT TESTS
Test 1

Solutions for questions 1 to 11:

1. (A), (C), (D) and (B) is the required sequential order.
   Hence, the correct option is (a)

2. (A), (C), (B) and (D) is the required sequential order.
   Hence, the correct option is (b)

3. $\dfrac{7x - 15y}{4x + 6y} = \dfrac{5}{6}$

   $42x - 90y = 20x + 30y$

   $22x = 120y$

   $\Rightarrow x:y = 120:22 = 60:11$

   Hence, the correct option is (a)

4. 30% of 60% of 200 $= \dfrac{30}{100} \times \dfrac{60}{100} \times 200$

   $= 18 \times 2 = 36$

   Hence, the correct option is (d)

5. Let the fourth proportional be x.

   $\Rightarrow 9:8 = 18:x \Rightarrow \dfrac{9}{8} = \dfrac{18}{x}$

   $x = \dfrac{18 \times 8}{9} = 16$.

   Hence, the correct option is (a)

6. We know if b is the mean proportional of a and c then $b^2 = ac$.

   $\therefore$ The mean proportional of 27 and 3 is $\sqrt{27 \times 3}$
   $= \sqrt{81} = 9$.

   Hence, the correct option is (c)

7. Let $x = 3.\overline{6}$.

   $\Rightarrow \quad x = 3.6666666 \ldots$

   $10x = 36.666666 \ldots$

   $\Rightarrow 10x = 36.66666666 \ldots$

   $\quad\quad x = 3.666666666 \ldots$

   $\underline{(-)\quad\quad\quad\quad\quad\quad\quad\quad\quad}$

   $9x = 33.000$

   $\Rightarrow x = \dfrac{33}{9}$

   $\therefore 3.\overline{6}\% = \dfrac{33}{9}\% = \dfrac{33}{900} = \dfrac{11}{300}$.

   Hence, the correct option is (b)

8. 108 km/h $= 108 \times \dfrac{5}{18}$ m/sec $= 30$ m/sec.

   The distance travelled in 2 minutes $= 30 \times 2 \times 60$ m $= 3600$ m.

   Hence, the correct option is (a)

9. S.I. $= \dfrac{3000 \times 3 \times 20}{100} = ₹1800$.

   $\therefore$ Amount = Principal + Interest = ₹3000 + ₹1800 = ₹4800.

   Hence, the correct option is (d)

10. The cost of 75 pencils $= \dfrac{75 \times 75}{25} = 75 \times 3 = ₹225$.

    Hence, the correct option is (a)

11. We have $\dfrac{M_1 D_1 H_1}{W_1} = \dfrac{M_2 D_2 H_2}{W_2}$

    $M_1 = 16, D_1 = 15, H_1 = 6$

    $M_2 = 18, D_2 = ?, H_2 = 10$ and $W_1 = W_2 = 1$

    $\dfrac{16 \times 15 \times 6}{1} = \dfrac{18 \times D_2 \times 10}{1}$

    $\Rightarrow D_2 = \dfrac{16 \times 15 \times 6}{18 \times 10} = 8$ days.

    Hence, the correct option is (d)

Solutions for questions 12 to 15:

12. → (d): Average $= \dfrac{1^2 + 2^2 + 3^2 + 4^2 + 5^2}{5}$

    $= \dfrac{1 + 4 + 9 + 16 + 25}{5} = \dfrac{55}{5} = 11$.

13. → (a): $x = 5y \Rightarrow \dfrac{x}{y} = \dfrac{5}{1} \Rightarrow x:y = 5:1$

14. → (c): $CP = ₹200$, $SP = ₹180$, Loss = ₹20

    The loss percentage $= \dfrac{20}{200} \times 100\% = 10\%$

3.40 Chapter 3

15. → (b): Principal = ₹250
Amount = ₹750
Simple Interest = ₹500
Rate = ?

Time = 10 years
$$\Rightarrow 500 = \frac{250 \times R \times 10}{100}$$
$R = 20\%$

Test 2

Solutions for questions 16 to 26:

16. (C), (B), (A) and (D) is the required sequential order.
Hence, the correct option is (b)

17. (A), (B), (D) and (C) is the required sequential order.
Hence, the correct option is (d)

18. $\dfrac{8x - 10y}{3x + 4y} = \dfrac{4}{5}$

$\Rightarrow 40x - 50y = 12x + 16y$

$\Rightarrow 28x = 66y$

$\Rightarrow x:y = 66:28 = 33:14$

Hence, the correct option is (c)

19. $\dfrac{20}{100} \times \dfrac{40}{100} \times 100 = 8$

Hence, the correct option is (c)

20. Let the fourth proportional be x.

$3:7 = 15:x \Rightarrow \dfrac{7 \times 15}{3} = x \Rightarrow x = 35.$

Hence, the correct option is (a)

21. The mean proportion = $\sqrt{64 \times 4} = \sqrt{256} = 16.$
Hence, the correct option is (d)

22. Let $x = 2.\overline{9} = 2.99999\ldots$

$10x = 29.99999\ldots$

$10x = 29.999999\ldots$

$\underline{-\ x = 2.999999\ldots}$

$9x = 27.0000\ldots$

$x = \dfrac{27}{9} = 3$

$\therefore 2.\overline{9}\% = 3\% = \dfrac{3}{100}.$

Hence, the correct option is (a)

23. 72 km/h = $72 \times \dfrac{5}{18}$ m/sec = 20 m/sec.

1 minute = 60 sec.

Distance travelled = $20 \times 60 = 1200$ m.

Hence, the correct option is (b)

24. S.I. = $\dfrac{2000 \times 2 \times 10}{100} = 400.$

Amount = 2000 + 400 = ₹2400.

Hence, the correct option is (b)

25. The cost of 25 mangoes = $\dfrac{180 \times 25}{15} = 300.$

Hence, the correct option is (d)

26.

Men	Hours	Days	Work
12	8	18	1
24	6	? (D_2)	1

We have

$\dfrac{M_1 D_1 H_1}{W_1} = \dfrac{M_2 D_2 H_2}{W_2}$

$\Rightarrow \dfrac{12 \times 8 \times 8}{1} = \dfrac{24 \times 6 \times D_2}{1}$

$\Rightarrow D_2 = 12.$

Hence, the correct option is (a)

Solutions for questions 27 to 30:

27. → (b): Average

$= \dfrac{1 + 2 + 3 + 4 + 5 + 6 + 7 + 8 + 9 + 10}{10}$

$= \dfrac{55}{10} = 5.5$

28. → (c): $a:b = 2:3$ and $b:c = 9:7$

$a:b = 2:3$

⇒ $a:b = 6:9$ (Multiplying it by 3)
⇒ $a:b = 6:9$ and $b:c = 9:7$
∴ $a:b:c = 6:9:7$

29. → (e): $C.P. = ₹12$, $S.P. = ₹15$
∴ Profit $= 15 - 12 = ₹3$
Profit% $= \dfrac{\text{Profit}}{\text{Cost Price}} \times 100\%$
$= \dfrac{3}{12} \times 100\% = 25\%$

30. → (a): Principal $= ₹500$
Amount $= ₹1000$
∴ Interest $= 1000 - 500 = ₹500$.
$S.I. = \dfrac{P \times T \times R}{100}$
⇒ $500 = \dfrac{500 \times T \times 5}{100}$
⇒ $T = 20$ years.

Chapter 4

Indices

a) $5(3x) - 3(x-4) = 0$
$15x - 3x + 12 = 0$
$12x + 12 = 0$
$12x = -12$
$x = -12/12$
$x = -1 \qquad x^2 = 1$

REMEMBER

Before beginning this chapter, you should be able to:
- Recognise terms like constant, variable, etc.
- Have thorough knowledge of mathematical operators

KEY IDEAS

After completing this chapter, you should be able to:
- Understand the terms like exponentiation, index, etc.
- Explain the laws of indices
- Solve exponential equations
- Use exponents on larger numbers

INTRODUCTION

The sum of n instances of a number x is n times x, denoted as nx, i.e., multiplication is repeated addition. Multiplying by n means adding n instances of a number.

For example, $n + n = 2 \times n$

$n + n + n = 3 \times n$

$n + n + n + \ldots\ x$ times $= x \times n$.

The product of n instances of a number x is denoted as x^n. This is an exponential expression, read as x to the power of n or x raised to n and the operation (of multiplying n instances of x) is called exponentiation or involution. Exponentiation is repeated multiplication. Raising x to the power of n (where, n is a natural number) is the product of n instances of x.

For example, $n \times n = n^2$

$n \times n \times n = n^3$

$n \times n \times n \times \ldots\ x$ times $= n^x$

In the expression n^x, n is called the base and x is the index (plural, indices) or exponent. The entire expression is called the xth power of n.

For example, $6 \times 6 \times 6 \times 6 \times 6 \times 6 \times 6$ can be written as 6^7. Here 6 is the base and 7 is the index (or exponent).

We shall look at the rules/properties pertaining to these exponential expressions in this chapter.

Initially, we shall consider only the positive integral values of the indices. We shall also see what meaning we can assign to exponential expressions in which the index is 0 or a negative integer or a rational number. In higher classes we shall consider all real values of the index. As for the base itself, it can be any real number. We shall first verify the following laws for positive integral values of the index.

Laws of Indices

1. $a^m \times a^n = a^{m+n}$ **(Product of powers)**

    For example, (a) $3^5 \times 3^7 = 3^{5+7} = 3^{12}$

    (b) $\left(\dfrac{3}{4}\right)^6 \times \left(\dfrac{3}{4}\right) \times \left(\dfrac{3}{4}\right)^3 = \left(\dfrac{3}{4}\right)^{6+1+3} = \left(\dfrac{3}{4}\right)^{10}$

    (c) $5^2 \times 5^4 \times 5^7 \times 5^{10} \times 5^3 = 5^{2+4+7+10+3} = 5^{26}$

    (d) $\left(\sqrt{3}\right)^4 \times \left(\sqrt{3}\right)^7 = \left(\sqrt{3}\right)^{4+7} = \left(\sqrt{3}\right)^{11}$

    Note $a^{m_1} \times a^{m_2} \times a^{m_3} \ldots \times a^{m_n} = a^{m_1+m_2+m_3+\ldots+m_n}$

2. $a^m \div a^n = a^{m-n}$, $a \neq 0$ **(Quotient of powers)**

    For example, (a) $9^6 \div 9^2 = 9^{6-2} = 9^4$

    (b) $\left(\dfrac{4}{5}\right)^{12} \div \left(\dfrac{4}{5}\right)^7 = \left(\dfrac{4}{5}\right)^{12-7} = \left(\dfrac{4}{5}\right)^5$

    Note We now consider what meaning we can assign to a^0.

Indices **4.3**

If we want these laws to be true for all values of m and n, i.e., even for $n = m$, from above (ii) we get

$$\frac{a^m}{a^m} = a^{m-m} \text{ or } 1 = a^0$$

We see that if we define a^0 as 1, this law will be true even for $n = m$. Therefore, we define a^0 as 1, provided $a \neq 0$. When $a = 0$, $a^{n-n} = a^n/a^n = 0/0$, which is not defined.

∴ 0^0 is not defined.

Note We shall now consider, what meaning we can assign to a^n, where n is a negative integer.

We have $a^m \times a^n = a^{m+n}$, consider $a^n \times a^{-n} = a^{n+(-n)}$ (if we want the law to be true) $= a^0 = 1$.

∴ $a^n \times a^{-n} = 1$.

$\Rightarrow a^{-n} = \dfrac{1}{a^n}$ and $\dfrac{1}{a^{-n}} = a^n$. If we define a^{-n} as $\dfrac{1}{a^n}$, this law is true even for negative value of n.

For example, $4^{-3} = \dfrac{1}{4^3}$, $3^{-1} = \dfrac{1}{3}$, $a^{-m} = \dfrac{1}{a^m}$ (provided $a \neq 0$).

Note $\left(\dfrac{a}{b}\right)^{-1} = \dfrac{1}{\frac{a}{b}} = \dfrac{b}{a}$

3. $(a^m)^n = a^{m \times n}$ *(power of a power)*

   For example, (a) $\left(7^3\right)^4 = 7^{3 \times 4} = 7^{12}$

   (b) $\left[\left(\dfrac{6}{7}\right)^3\right]^5 = \left(\dfrac{6}{7}\right)^{3 \times 5} = \left(\dfrac{6}{7}\right)^{15}$

   Note $[(a^m)^n]^p = a^{mnp}$ and so on.

4. $(ab)^n = a^n \times b^n$ *(power of a product)*

   For example, (a) $(21)^4 = (3 \times 7)^4 = 3^4 \times 7^4$

   (b) $(210)^9 = (2 \times 3 \times 5 \times 7)^9 = 2^9 \times 3^9 \times 5^9 \times 7^9$

   In problems, we may often want to write $a^n \times b^n$ as $(ab)^n$.

   For example, (a) $32 \times 243 = 2^5 \times 3^5 = (2 \times 3)^5 = 6^5$

   (b) $\dfrac{64}{125} \times \dfrac{27}{1331} = \dfrac{4^3}{5^3} \times \dfrac{3^3}{11^3} = \left(\dfrac{4}{5}\right)^3 \times \left(\dfrac{3}{11}\right)^3 = \left(\dfrac{4}{5} \times \dfrac{3}{11}\right)^3 = \left(\dfrac{12}{55}\right)^3$

   Note $(a\,b\,c\,d\,\ldots\,z)^n = a^n\,b^n\,c^n\,d^n\,\ldots\,z^n$

5. $\dfrac{a^n}{b^n}$ *(Power of a quotient)*

   For example, (a) $\left(\dfrac{3}{4}\right)^5 = \dfrac{3^5}{4^5}$

   (b) $\left(\dfrac{12}{7}\right)^6 = \left(\dfrac{4 \times 3}{7}\right)^6 = \dfrac{(4 \times 3)^6}{7^6} = \dfrac{4^6 \times 3^6}{7^6}$

   In problems, we may want to write down $\dfrac{a^n}{b^n}$ as $\left(\dfrac{a}{b}\right)^n$.

For example, (a) $\dfrac{5^4}{7^4} = \left(\dfrac{5}{7}\right)^4$

(b) $\dfrac{(6/7)^5}{(5/3)^5} = \left(\dfrac{6/7}{5/3}\right)^5 = \left(\dfrac{18}{35}\right)^5$

6. $\left(\dfrac{a}{b}\right)^{-n} = \left(\dfrac{b}{a}\right)^n$

For example, (a) $\left(\dfrac{7}{8}\right)^{-4} = \left[\dfrac{8}{7}\right]^4$

(b) $\left(\dfrac{1}{3}\right)^{-1} = \left(\dfrac{3}{1}\right)^1 = 3$

Note $\left(\dfrac{1}{a}\right)^{-1} = \left(\dfrac{a}{1}\right) = a$

EXAMPLE 4.1

Which is greater between 5^{1014} and 9^{676}?

SOLUTION

Let us take the G.C.D. of 1014 and 676, which is 338.
5^{1014} can be written as $(5^3)^{338}$, i.e., $(125)^{338}$.
Similarly, 9^{676} can be written as $(9^2)^{338} = (81)^{338}$.
Clearly, $5^{1014} > 9^{676}$.

EXAMPLE 4.2

What should be multiplied to 6^{-2} so that the product may be equal to 216?

SOLUTION

Let the required number be x.

$\therefore x \times 6^{-2} = 216$

$\Rightarrow x \times \dfrac{1}{6^2} = 6^3$ ($\because a^{-m} = 1/a^m$)

$\Rightarrow x = 6^3 \times 6^2$

$\Rightarrow x = 6^{3+2}$ ($\because a^m \times a^n = a^{m+n}$)

$\Rightarrow x = 6^5$

$\therefore$ The required number is 6^5.

Exponential Equation

1. If $a^m = a^n$, then $m = n$, if $a \neq 0$, $a \neq 1$ and $\neq -1$.

   For example, (a) If $4^n = 4^2$, then $n = 2$

   (b) If $3^q = 81$, i.e., $3^q = 3^4$, then $q = 4$.

2. If $a^n = b^n$, then $a = b$. (when n is odd)

For example, (a) if $3^5 = b^5$, then $b = 3$.

(b) If $8^{2p+1} = (5a)^{2p+1}$, then $8 = 5a$ and $a = \dfrac{8}{5}$.

3. If $a^n = b^n$, $n \neq 0$, then $a = \pm b$. (when n is even)

For example, (a) If $(p)^6 = (3)^6$, then $p = \pm 3$.

(b) $(-1)^2 = 1$, $(-1)^4 = 1$, $(-1)^8 = 1$

But $(-1)^1 = -1$, $(-1)^3 = -1$, $(-1)^5 = -1$

$\Rightarrow (-1)^n = 1$ when n is even and $(-1)^n = -1$ when n is odd.

EXAMPLE 4.3

If $\left(\dfrac{a}{b}\right)^{\frac{5}{14}} + \left(\dfrac{b}{a}\right)^{\frac{5}{14}} = 6$, then find the value of $\left(\dfrac{a}{b}\right)^{\frac{5}{7}} + \left(\dfrac{b}{a}\right)^{\frac{5}{7}}$.

SOLUTION

$\left(\dfrac{a}{b}\right)^{\frac{5}{14}} + \left(\dfrac{b}{a}\right)^{\frac{5}{14}} = 6$

Squaring on both the sides, we get

$\left(\dfrac{a}{b}\right)^{\frac{5}{7}} + \left(\dfrac{b}{a}\right)^{\frac{5}{7}} + 2\left(\dfrac{a}{b}\right)^{\frac{5}{14}}\left(\dfrac{b}{a}\right)^{\frac{5}{14}} = 36$

$\Rightarrow \left(\dfrac{a}{b}\right)^{\frac{5}{7}} + \left(\dfrac{b}{a}\right)^{\frac{5}{7}} + 2 = 36$

$\Rightarrow \left(\dfrac{a}{b}\right)^{\frac{5}{7}} + \left(\dfrac{b}{a}\right)^{\frac{5}{7}} = 34$

EXAMPLE 4.4

If $\left[\dfrac{7}{5}\right]^{\frac{2x+4}{4}} = \left(\dfrac{125}{343}\right)^{-\frac{2}{3}}$, then find the value of x.

SOLUTION

$\left[\dfrac{7}{5}\right]^{\frac{2x+4}{4}} = \left(\dfrac{125}{343}\right)^{-\frac{2}{3}}$

$\Rightarrow \left(\dfrac{7}{5}\right)^{\frac{2x+4}{4}} = \left(\dfrac{343}{125}\right)^{\frac{2}{3}}$

$$\Rightarrow \left(\frac{7}{5}\right)^{\frac{2x+4}{4}} = \left(\frac{7}{5}\right)^2$$

$$\Rightarrow \frac{2x+4}{4} = 2$$

$2x + 4 = 8 \Rightarrow x = 2.$

Standard Form of a Number

How do you read this number 49,000,000,000,000,000,000? It is difficult to read.

Such very large numbers are difficult to read, understand and even compare. By expressing such large numbers in power form it is easy to understand, read and to compare. Using exponents we can write such large numbers into shorter form for example 49,000,000,000,000,000,000 = 49×10^{18}

EXAMPLE 4.5

Express the following numbers in shorter form by expressing as product of a number and a power of 10.

(a) 27,000,000   (b) 30,000,000,000   (c) 0.000000043

SOLUTION

(a) 27,000,000 = 27×10^6

(b) 30,000,000,000 = 3×10^{10}

(c) $0.000000043 = \dfrac{43}{1000000000} = \dfrac{43}{10^9} = 43 \times 10^{-9}$

A number is said to be in its standard form when it is expressed as a product of a decimal number from 1 to 9 and a power of 10 for example 2,84,932 can be expressed as 2.84932×100000

∴ 2,84,932 = 2.84932×10^5

∴ 2.84932×10^5 is the standard form of 2,84,932.

EXAMPLE 4.6

Express the following numbers in the standard form.

(a) 7584300   (b) 493.8721   (c) 5.876   (d) 0.0000079

SOLUTION

(a) 7584300 = 7.5843×10^6

(b) 493.8721 = 4.938721×10^2

(c) 5.876 = 5.876×10^0

(d) 0.0000079 = 7.9×10^{-6}

Points to Remember

- $a + a + a + \ldots(n \text{ times}) = na$
- $a \times a \times a \times \ldots(n \text{ times}) = a^n$
- $a^m \times a^n = a^{m+n}$
- $a^m \times b^m = (ab)^m$
- $\dfrac{a^m}{b^m} = \left(\dfrac{a}{b}\right)^m$
- $\dfrac{a^m}{a^n} = a^{m-n}$ or $\dfrac{1}{a^{n-m}}$
- $(a^m)^n = a^{m \times n}$
- $a^0 = 1$ (where $a \neq 0$)
- $a^{-n} = \dfrac{1}{a^n}$
- $\dfrac{1}{a^{-m}} = a^m$
- If $x^m = x^n$, then $m = n$ (where $x \neq 1$, $x \neq 0$).
- If $x^m = y^m$, then $x = y$ (when m is odd).
- If $x^m = y^m$, then $x = \pm y$ (when m is even).
- $(-1)^x = 1$, when x is even.
- $(-1)^x = -1$, when x is odd.

TEST YOUR CONCEPTS

Very Short Answer Type Questions

Directions for questions 1 to 10: State whether the following statements are true or false.

1. $\dfrac{a^m}{a^n} = a^{m/n}\ (a \neq 0)$

2. $x^{2009} \times \dfrac{1}{x^{2008}} = x$

3. If $a^0 = 1$, then a is any real number.

4. If $2^x = 4^2$, then $x = 4$.

5. If $\dfrac{1}{3^x} = 3^2$, then $x = 2$.

6. $(-1)^{2008} = -1$.

7. If $x^y = x^{2^3}$, then $y = 6$. $(x \neq 0, x \neq 1)$

8. If $5^x = 7^x$, then $x = 2$.

9. $(6^0 - 7^0) = 0$.

10. The value of $3(x^2 - 3x + 2)$, when $x = 2$ is 0.

Directions for questions 11 to 18: Fill in the blanks.

11. Power notation of -1331 is _____.

12. $\dfrac{1}{x^{-4}} =$ _____.

13. $\left(\dfrac{2}{3}\right)^{-2} =$ _____.

14. If $x \times 2^{-5} = 2^5$, then $x =$ _____.

15. $\left(\dfrac{3}{5}\right)^3 \left(\dfrac{25}{27}\right) =$ _____.

16. $1350 =$ _____. (in power notation)

17. $\dfrac{625}{1296} =$ _____. (in power notation)

18. If $x = 2$ and $y = 4$, then $x^{\frac{x}{y}} + y^{\frac{x}{y}} =$ _____.

Directions for questions 19 to 25: Select the correct alternative from the given choices.

19. Radius of the first orbit of hydrogen is $\dfrac{0.529}{100000000}$ cm. Its value by using powers of 10 is _____.

    (a) 0.529×10^{-9} cm
    (b) 5.29×10^{-9} cm
    (c) 0.0529×10^{-9} cm
    (d) 5.29×10^{-10} cm

20. Velocity of light is 30000000000 cm/sec. Its value by using powers of 10 is _____.

    (a) 3×10^8 cm/sec
    (b) 3×10^9 cm/sec
    (c) 3×10^{10} cm/sec
    (d) 3×10^{11} cm/sec

21. If $a + b + c = 0$, then find the value of $\dfrac{(x^a)^3}{x^{-3b} \cdot x^{-3c}}$

    (a) 0
    (b) 1
    (c) -1
    (d) 3

22. If $a = 36$, then find the value of $a^{36^0} - a^{0^{36}}$

    (a) 36
    (b) 0
    (c) 1
    (d) 35

23. If $abc = 0$, then find the value of $\dfrac{\left[(x^a)^b\right]^{2c}}{x^{abc}}$

    (a) 3
    (b) 0
    (c) -1
    (d) 1

24. Simplify $\dfrac{\sqrt{144} + \sqrt{256}}{3^2 - 2}$

    (a) 8
    (b) 4
    (c) -4
    (d) -8

25. $(0.000729)^{\frac{1}{3}} =$ _____.

    (a) 0.030
    (b) $\sqrt{0.09}$
    (c) 0.18
    (d) 0.09

Short Answer Type Questions

26. Evaluate: $(1 + 3 + 5 + 7 + 9 + 11 + 13 + 15)^{\frac{5}{6}}$

27. Identify the greater number. $(2^2)^3$ and 2^{2^3}

28. Simplify: $\dfrac{(x^3 y^2 z)^3}{(xy^2 z)^2}$

29. If $2^x = 240$, then find 2^{x-4}.

30. If $(3x^4)^3 = 3^{3^3}$, then find x. (where $x > 0$)

31. Simplify $(x^{a-b})^{a+b} \cdot (x^{b-c})^{b+c} \cdot (x^{(c-a)})^{(c+a)}$ where $x \neq 0$ and $x \neq 1$.

32. Solve for x: $(25)^{x+2} = (125)^{2-x}$

33. Find the value of $32^{40^{78}} + 83^{32^{047}}$.

34. Simplify: $\left[\dfrac{2}{5}\right]^4 \left(\dfrac{25}{8}\right)^3 \left(\dfrac{125}{4}\right)^2$

35. If $x = \sqrt{18}$, then find the value of $\dfrac{x^5 + x^4}{x^3}$.

36. Find the value of $(-1)^0 + (-1)^1 + (-1)^2 + \ldots + (-1)^{100}$.

37. Arrange 6^{-3}, 5^{-3}, 4^{-3} and 3^{-3} in the ascending order.

38. A teacher wanted to distribute a total amount of ₹1296 equally among x number of students. If each student gets x rupees, then find x.

39. What should be multiplied to 3^{-4} so that the product is 6?

40. If $xyz = 0$, then find the value of $\left[(k^x)^y\right]^z - \left[(k^y)^z\right]^x - \left[(k^z)^x\right]^y$.

41. If $9^{2x-7} = (27)^{x-4}$, then find the value of 3^x.

42. Vicky scored x marks in Maths and Rakesh scored x^3 marks in Maths. The product of their marks was 256. Find their marks.

43. Simplify: $3x^{-2} y^{-3} z^2 \times 5x^2 \times y \times z^3 \div x^3 y^2 z^{-1}$

44. If $\dfrac{1}{(343)^{3y+2}} = 49$, then find the value of y.

45. Simplify: $\left(\dfrac{125}{216}\right)^{\frac{2}{3}} \times \left(\dfrac{36}{5}\right)^{+2} \div \left(\dfrac{25}{6}\right)^{-2}$

Essay Type Questions

46. If $x^{(5+a)^2} \times x^{(5-a)^2} = x^{40}$, then find a.

47. Simplify: $\left[2^{\frac{1}{ab}}\right]^c \left[2^{\frac{1}{bc}}\right]^a \left[2^{\frac{1}{ac}}\right]^b$ when $a^2 + b^2 + c^2 = 2abc$

48. If $3^x = 900$, then find 3^{x+2} and 3^{x-2}.

49. If $1800 = 2^a \times 3^b \times 5^c$, then find $a + b + c$.

50. By what number should we multiply 3^4 so that the product is equal to $\dfrac{1}{27}$?

51. By what number should $(8/27)^{-3}$ be divided so that the quotient is equal to $(27/8)^{-3}$?

52. Prove that $\dfrac{(x^{-1} + y^{-1})}{x^{-1}} + \dfrac{(x^{-1} + y^{-1})}{y^{-1}} = \dfrac{(x+y)^2}{xy}$.

53. Which is greater between 3^{48} and 2^{72}?

54. Find the value of $\dfrac{k}{1 - x^{a-b}} + \dfrac{k}{1 - x^{b-a}}$.

55. Find the least integer value which satisfies $x^4 > 1000$.

56. If $x^y \times y = 1215$ where $x, y \in N$, $x \neq 1$ and $y \neq 1$, then find xy.

57. If $8^{x^y} = 4096$ where x and y are positive integers and $x \neq 1$, then show that either $x = y$ or $x > y$.

58. If $432 > 2y^3$, then find the greatest possible integer value of y.

59. If $3^x = 9^y = 27^z = 729$, then show that $x + y + z = 11$.

60. Which is the least among $(3)^{25}$, $(9)^{12}$, $(27)^3$ and $(81)^2$?

4.10 Chapter 4

CONCEPT APPLICATION
Level 1

Directions for questions 1 to 20: Select the correct alternative from the given choices.

1. 2^{3^2} = _____.
   (a) 64 (b) 32
   (c) 256 (d) 512

2. $3^{2^{0^5}}$ = _____.
   (a) 0 (b) 1
   (c) 3 (d) 9

3. 200000000 = _____.
   (a) 2×10^8 (b) 2×10^7
   (c) 2×10^{10} (d) 2×10^9

4. In $2x^5$, base is _____.
   (a) 2 (b) $2x$
   (c) x (d) 5

5. $(2^3)^4$ = _____.
   (a) (2^{4^3}) (b) (2^{3^4})
   (c) $(2^4)^3$ (d) None of these

6. Find the value of $3^4\left[\left(\dfrac{2}{3}\right)^2 + \left(\dfrac{2}{3}\right) - \left(\dfrac{2}{3}\right)^3\right]$.
   (a) 86 (b) 66
   (c) 68 (d) 88

7. Value of $\left(\dfrac{1024}{243}\right)^{\frac{3}{5}}$ is _____.
   (a) $\dfrac{128}{27}$ (b) $\dfrac{32}{27}$
   (c) $\dfrac{64}{27}$ (d) $\dfrac{32}{9}$

8. Find the value of $(2^4 + 2^3)^{2/6}$.
   (a) $8(2)^{1/2}$ (b) $2(3)^{1/3}$
   (c) $64(2)^{1/2}$ (d) $4(3)^{1/3}$

9. $(25^2 - 15^2)^{\frac{3}{2}}$ = _____.
   (a) 4000 (b) 8000
   (c) 3125 (d) 1024

10. $(33^2 - 31^2)^{\frac{5}{7}}$ = _____.
    (a) 64 (b) 16
    (c) 32 (d) 4

11. If $abc = 0$, then find the value of $\left[\left(x^a\right)^b\right]^c$.
    (a) 1 (b) a
    (c) b (d) c

12. If $a + b + c = 0$, then find the value of $\sqrt{x^a \cdot x^b \cdot x^c}$.
    (a) 0 (b) 1
    (c) -1 (d) None of these

13. Simplify $\dfrac{\sqrt{36} + \sqrt{64}}{2^3 - 1}$.
    (a) 1 (b) 2
    (c) 3 (d) 4

14. If $a = 25$, then find the value of $a^{25^0} + a^{0^{25}}$.
    (a) 25 (b) 26
    (c) 24 (d) 0

15. Find the value of $(-1)^{301} + (-1)^{302} + (-1)^{303} + \ldots + (-1)^{400}$.
    (a) 1 (b) 101
    (c) 100 (d) 0

16. If $3^x = 6561$, then 3^{x-3} is _____.
    (a) 81 (b) 243
    (c) 729 (d) 27

17. The following are the steps involved in finding the value $(7 + x)^3$, when $(7x)^3 = 343$. Arrange them in sequential order.
    (A) $(7 + x)^3 = (7 + 1)^3 = 8^3 = 512$
    (B) $x^3 = \dfrac{343}{7^3} = \dfrac{7^3}{7^3} = 1$
    (C) $\Rightarrow x = 1$
    (D) $(7x)^3 = 343 \Rightarrow 7^3 \times x^3 = 343$
    (a) ABCD (b) DBCA
    (c) ACBD (d) BDCA

18. The following are the steps involved in finding the value of $\left(\dfrac{x^a}{x^b}\right)\left(\dfrac{x^b}{x^c}\right)\left(\dfrac{x^c}{x^a}\right)$. Arrange them in sequential order.

    (A) $x^0 = 1$
    (B) $x^{a-b+b-c+c-a}$
    (C) $\left(\dfrac{x^a}{x^b}\right)\left(\dfrac{x^b}{x^c}\right)\left(\dfrac{x^c}{x^a}\right) = x^{a-b} \cdot x^{b-c} \cdot x^{c-a}$

    (a) CBA (b) ACB
    (c) BCA (d) CAB

19. The following are the steps involved in finding the value of 3^{n-3}, when $3^n = 729$. Arrange them in sequential order.

    (A) $3^{n-3} = 3^{6-3} = 3^3 = 27$
    (B) $\Rightarrow n = 6$
    (C) $3^n = 729 \Rightarrow 3^n = 3^6$

    (a) ABC (b) BAC
    (c) CBA (d) CAB

20. The following are the steps involved in finding the value of $(a^{x+y})^{x-y} (a^{y+z})^{y-z} (a^{z+x})^{z-x}$. Arrange them in sequential order.

    (A) $a^{(x+y)(x-y)} \cdot a^{(y+z)(y-z)} \cdot a^{(z+x)(z-x)}$
    (B) $a^0 = 1$
    (C) $a^{x^2-y^2} \cdot a^{y^2-z^2} \cdot a^{z^2-x^2}$
    (D) $a^{x^2-y^2+y^2-z^2+z^2-x^2}$

    (a) ADCB (b) ACBD
    (c) ACDB (d) ADBC

Directions for questions 21 to 24: Match the Column A with Column B.

Column A		Column B
21. $\dfrac{2^{10}}{2^{-10}}$ ( )		(a) 1
22. 6^{18} ( )		(b) 0
23. If $2^x = 16$, then x ( )		(c) $(6^{10})^8$
24. $5^0 - 6^0$ ( )		(d) 2^{20}
		(e) 4
		(f) $2^{18} \times 3^{18}$

Directions for questions 25 to 28: Match the Column A with Column B.

Column A		Column B
25. 5^{-5} ( )		(a) 6^{12}
26. $6^{-6} \times 6^6$ ( )		(b) 1
27. If, $x^3 = 7^3$, then x ( )		(c) 5^5
28. $2^{1^{2^3}}$ ( )		(d) 7
		(e) $\dfrac{1}{5^5}$
		(f) 2

Level 2

Directions for questions 29 to 70: Select the correct alternative from the given choices.

29. Simplify $4a^{-1}b^{-2}c^3 \times 6a^3b^2c \div 12a^2c^4b$.

    (a) 2 (b) $2abc$
    (c) $2bc$ (d) $2b^{-1}$

30. If $3^{-x} = 3000$, then find $10^3 \times 3^{(2x+3)}$.

    (a) 3×10^{-3}
    (b) 27×10^{-3}
    (c) 9×10^{-3}
    (d) 3000

31. Find the value of $\dfrac{\left(-\dfrac{2}{3}\right)^2}{\left(-\dfrac{2}{3}\right)^3} + \dfrac{\left(-\dfrac{4}{9}\right)}{\left(\dfrac{2}{3}\right)^2}$.

    (a) $-\dfrac{1}{2}$ (b) $\dfrac{3}{2}$
    (c) $-\dfrac{5}{2}$ (d) $\dfrac{1}{2}$

32. If $(4)^{a+5} (16)^{2a} (32)^4 = (16)^{3a}$, then $a = $ ____.

    (a) 15 (b) 30
    (c) 24 (d) 20

33. If $2^n = 4096$, then 2^{n-5} is _____.
    (a) 128   (b) 64
    (c) 256   (d) 32

34. If $2^{3y-x} = 16$ and $2^{2y+x} = 2048$, then the value of y is _____.
    (a) 5   (b) 8
    (c) 6   (d) 3

35. If $x^y \times y^x = 72$, then find the sum of x and y where x and y are positive integers. ($x \neq 1$ and $y \neq 1$).
    (a) 12   (b) 10
    (c) 5   (d) 7

36. $\left(\dfrac{a^x}{a^y}\right)^z \times \left(\dfrac{a^y}{a^z}\right)^x \times \left(\dfrac{a^z}{a^x}\right)^y = $ _____. ($a \neq 0$ and $a \neq 1$)
    (a) 1   (b) 0
    (c) a^{xyz} (d) $a^{xy + yz + zx}$

37. Which is the greatest among $(5)^{23}$, $(25)^{11}$, $(625)^6$ and $(3125)^5$?
    (a) 5^{23} (b) $(25)^{11}$
    (c) $(625)^6$ (d) $(3125)^5$

38. $(0.01024)^{\frac{1}{5}} = $ _____.
    (a) $\sqrt{0.4}$ (b) 0.2
    (c) 0.4   (d) $\sqrt[3]{0.4}$

39. $\left(\dfrac{a}{b}\right)^{x+y+z} \div \left[\left(\sqrt{\dfrac{a}{b}}\right)^{-x} \times \left(\sqrt{\dfrac{a}{b}}\right)^{-y} \times \left(\sqrt{\dfrac{a}{b}}\right)^{-z}\right] = $ _____.
    (a) $[a^3/b^3]^{x+y+z}$ (b) $[a^2/b^2]^{x+y+z}$
    (c) $[a/b]^{(x+y+z)/2}$ (d) $[a/b]^{3(x+y+z)/2}$

40. If $\dfrac{1}{(512)^{2x+2}} = 2^{18}$, then find the value of x.
    (a) -2 (b) -3
    (c) -4 (d) -5

41. If $(2x)^5 = 100000$, then find the value of x.
    (a) 25   (b) 5
    (c) 10   (d) 2.5

42. If $(x + y)^3 = 1331$ and $(x - y)^5 = 243$, then find $x^2 - y^2$.
    (a) 33   (b) 22
    (c) 11   (d) 44

43. If $3^{2x-16} = 4^{3x-24}$, then $x = $ _____.
    (a) 9   (b) 6
    (c) 4   (d) 8

44. If $xyz = 0$, then find the value of $(a^x)^{zy} + (a^y)^{zx} + (a^z)^{xy}$.
    (a) 3   (b) 2
    (c) 1   (d) 0

45. Varun secured x marks in Maths and Rahul secured x^2 marks in Maths. The product of their marks was 729. Find their marks.
    (a) 3, 243   (b) 9, 81
    (3) 27, 27   (d) None of these

46. What should be multiplied to 2^{-6} so that the product is 1?
    (a) 16   (b) 32
    (c) 64   (d) 128

47. If $27^{x-5} = (32)^{x-7}$, then find the value of $\dfrac{1}{2^x}$.
    (a) 2^{17} (b) 2^{16}
    (c) 2^{15} (d) None of these

48. If $x = \sqrt{100}$, then find the value of $\dfrac{x^3 + x^2}{x}$.
    (a) 120   (b) 100
    (c) 110   (d) None of these

49. Which among 4^{-2}, 3^{-2}, 2^{-2} and 5^{-2} is the greatest?
    (a) 4^{-2} (b) 3^{-2}
    (c) 2^{-2} (d) 5^{-2}

50. A teacher wanted to distribute 2025 chocolates equally among x number of students. If each student gets x chocolates, find x.
    (a) 25   (b) 35
    (c) 45   (d) 55

Level 3

51. If $\sqrt{2} = 1.414$ and $\sqrt{7} = 2.646$, then find the value of $\sqrt{32} + \sqrt{252}$.
    (a) 20.839 (b) 21.532
    (c) 19.482 (d) 22.231

52. If $64 > x^3$, then the greatest possible integer value of x is _____.
    (a) 1 (b) 2
    (c) 3 (d) 4

53. Which of the following is the ascending order of 2^{1152}, 3^{768} and 5^{384}?
    (a) 5^{384}, 3^{768}, 2^{1152} (b) 5^{384}, 2^{1152}, 3^{768}
    (c) 3^{789}, 5^{314}, 2^{1152} (d) 2^{1152}, 3^{768}, 5^{384}

54. If $2^a = 4^b = 8^c = 64$, then which of the following relations hold true?
    (a) $a + b + c = 8$ (b) $a + b + c = 9$
    (c) $a + b + c = 10$ (d) $a + b + c = 11$

55. If $\sqrt{2.5} = 1.581$, then find the value of $\sqrt{0.625}$.
    (a) 0.7905 (b) 0.9426
    (c) 0.7632 (d) 0.9325

56. $\left(\dfrac{27}{343}\right)^{\frac{2}{3}} \times \left(\dfrac{343}{729}\right)^{\frac{2}{3}} \div \left(\dfrac{2401}{81}\right)^{-\frac{3}{4}} = $ _____.
    (a) $1/3 \times (7/3)^8$ (b) $1/9 \times (7/3)^7$
    (c) $1/9 \times (7/3)^6$ (d) $(7/3)^6$

57. If $(9)^{2x+4} = (243)^{2x-3.2}$, then $x = $ _____.
    (a) 12 (b) 10
    (c) 8 (d) 4

58. If $2^{2n-3} = 2048$, then $(4n + 3n^2) = $ _____.
    (a) 175 (b) 25
    (c) 125 (d) 75

59. If $7^{a^b} = 2401$ where a and b are positive integers and $a \neq 1$, then which of the following is (are) correct?
    I. $a = b$ II. $a > b$ III. $a < b$
    (a) Either I or III (b) Either II or III
    (c) Either I or II (d) I only

60. If $a^{b^c} = 729$, then find the minimum possible value of $a + b + c$ (where a, b and c are positive integers).
    (a) 29 (b) 13
    (c) 12 (d) 10

61. If $(2)^{2x-2} = (8)^{y-1} = (16)^{x-2.5}$, then find the sum of x and y.
    (a) 8 (b) 7
    (c) 9 (d) 6

62. $\left(\sqrt{1795} - \sqrt{1170}\right)^{3/4} \times \left(\sqrt{1795} + \sqrt{1170}\right)^{3/4} = $ _____.
    (a) 5 (b) 25
    (c) 125 (d) 625

63. If $a = (3^{-3} - 3^3)$, and $b = (3^3 - 3^{-3})$, then find the value of $\dfrac{a}{b} - \dfrac{b}{a}$.
    (a) 0 (b) 1
    (c) –1 (d) 2

64. Which of the following is the descending order of $(4)^{26}$, $(64)^9$ and $(256)^7$?
    (a) $(64)^9$, $(256)^7$, $(4)^{26}$ (b) $(4)^{26}$, $(64)^9$, $(256)^7$
    (c) $(256)^7$, $(64)^9$, $(4)^{26}$ (d) None of these

65. $\left(\dfrac{x^{5y-3} \times x^{3-2y}}{x^{4y-6} \times x^{2y-9}}\right)^{-\frac{4}{3}} = $ _____.
    (a) x^{3y+15} (b) x^{13-3y}
    (c) x^{4y-20} (d) x^{4y+18}

66. If $(xy)^a = z$, $(yz)^a = x$ and $(xz)^a = y$, then what is the value of a? (None of x, y and z is either 0 or 1)
    (a) 1 (b) $\dfrac{1}{2}$
    (c) $\dfrac{3}{2}$ (d) 0

67. $\dfrac{x^{-1} - y^{-1}}{z} + \dfrac{y^{-1} - z^{-1}}{x} + \dfrac{z^{-1} - x^{-1}}{y} = $ _____.
    (a) 1 (b) –1
    (c) 0 (d) 2

68. Find the value of $\dfrac{1}{1+x^{a-b}} + \dfrac{1}{1+x^{b-a}}$.
    (a) 0 (b) –1
    (c) 1 (d) x^{a+b}

4.14 Chapter 4

69. If $x^5 < 1000$, then find the greatest integer value of x.

(a) 1   (b) 2
(c) 3   (d) 4

70. If $x^y = y^x$, where x and y are distinct natural numbers, then find $x + y$.

(a) 6   (b) 8
(c) 4   (d) None of these

ASSESSMENT TESTS

Test 1

Directions for questions 1 to 11: Select the correct alternative from the given choices.

1. If $(x^2 - y^2)^4 = 256$ and $(x^2 + y^2)^5 = 243$, then find $x^4 - y^4$.

The following are the steps involved in solving the above problem. Arrange them in sequential order.

(A) $(x^2 - y^2)^4 = 256 = 4^4$ and $(x^2 + y^2)^5 = 3^5$
(B) $x^4 - y^4 = 12$
(C) $(x^2 - y^2)(x^2 + y^2) = 4 \times 3$
(D) $x^2 - y^2 = 4$ and $x^2 + y^2 = 3$

(a) ADCB   (b) ABCD
(c) ADBC   (d) ACDB

2. If $\sqrt{7} = 2.646$, then find the value of $(\sqrt{2} + \sqrt{14})^2$.

The following are the steps involved in solving the above problem. Arrange them in sequential order.

(A) $16 + 2\sqrt{2} \times \sqrt{2} \times \sqrt{7}$
(B) $16 + 2 \times 2 \times \sqrt{7}$
(C) $(\sqrt{2} + \sqrt{14})^2 = 2 + 14 + 2\sqrt{2} \times \sqrt{14}$
(D) $16 + 4 \times 2.646 = 16 + 10.584 = 26.584$

(a) ABCD   (b) ADBC
(c) CADB   (d) CABD

3. $3^4 \, x^3 \, y^6 \, z^4 \times 27x^{-2} \, y^{-3} \, z^{-6} \div 81x \, y^2 \, z^{-3} =$ _____.

(a) $27xy$ (b) $27yz$
(c) $27xz$ (d) $27xyz$

4. If $3^x = 243$, then $3^{\frac{x+1}{2}} =$ _____.

(a) 81   (b) 27
(c) 243   (d) 729

5. If $5^{-5y} = \dfrac{1}{3125}$ and $9^x = \dfrac{1}{81}$, then $x - y =$ _____.

(a) –5   (b) –2
(c) –3   (d) 2

6. Which is the greatest among the following?
$(\sqrt{3})^{30}$, $(\sqrt{27})^4$, $(\sqrt{81})^8$, and $(\sqrt[3]{9})^{12}$

(a) $(\sqrt{3})^{30}$ (b) $(\sqrt{27})^4$
(c) $(\sqrt[3]{9})^6$ (d) $(\sqrt{81})^8$

7. $\left(\dfrac{a^{x^4}}{a^{y^4}}\right)^{\frac{1}{x^2+y^2}} \times \left(\dfrac{a^{y^4}}{a^{z^4}}\right)^{\frac{1}{y^2+z^2}} \times \left(\dfrac{a^{z^4}}{a^{x^4}}\right)^{\frac{1}{z^2+x^2}} =$ _____.

(a) 1   (b) 2
(c) 3   (d) 4

8. $\left(\sqrt[6]{\dfrac{64}{125}}\right)^{-2} =$ _____.

(a) $\dfrac{5}{16}$ (b) $\dfrac{4}{5}$
(c) $\dfrac{5}{4}$ (d) $\dfrac{16}{5}$

9. If $x^y \times y^x = 256$, then find $y^2 - x^2$ (where x and y are positive integers and $x < y$).

(a) 15   (b) 12
(c) 14   (d) 16

10. $\left(\sqrt{1024} - \sqrt{24}\right)^{2/3} \left(\sqrt{1024} + \sqrt{24}\right)^{2/3} =$ _____.

(a) 100   (b) 1000
(c) 10   (d) None of these

11. Find the value of $\left(8^4 + 8^2\right)^{\frac{1}{2}}$.

(a) 84   (b) $8\sqrt{77}$
(c) 72   (d) $8\sqrt{65}$

Indices

Directions for questions 12 to 15: Match the Column A with Column B.

Column A		Column B
12. $100^{2^{3^{0^{10}}}}$ ( )		(a) 0
13. $(12^0 - 11^0)(3^1 - 1^3)$ ( )		(b) 12
14. $(128)^{5/7}$ ( )		(c) 32
15. If $5^x = 25^y = 625$, then $x^2 - y^2$ ( )		(d) 1
		(e) 100^2
		(f) 16

Test 2

Directions for questions 16 to 26: Select the correct alternative from the given choices.

16. If $(x - y)^3 = 216$ and $(x + y)^5 = 32$, then find $x^3 - y^3$.

   The following are the steps involved in solving the above problem. Arrange them in sequential order.

   (A) Therefore $x - y = 6$ and $x + y = 2$.
   (B) Solving $x - y = 6$ and $x + y = 2 \Rightarrow x = 4$, $y = -2$.
   (C) $x^3 - y^3 = 64 - (-2)^3 = 64 + 8 = 72$
   (D) $(x - y)^3 = 216 \Rightarrow (x - y)^3 = 6^3 \Rightarrow x - y = 6$ and $(x + y)^5 = 32 \Rightarrow (x + y)^5 = 2^5 \Rightarrow x + y = 2$.

   (a) DABC (b) DBAC
   (c) DCBA (d) BDCA

17. If $\sqrt{2} = 1.414$ and $\sqrt{7} = 2.646$, find the value of $(\sqrt{2} + 1)^2 + (\sqrt{7} + 1)^2$.

   The following are the steps involved in solving the above problem. Arrange them in sequential order.

   (A) 19.120
   (B) $11 + 2 \times 4.060 = 11 + 8.120$
   (C) $11 + 2(\sqrt{2} + \sqrt{7}) = 11 + 2(1.414 + 2.646)$
   (D) $(\sqrt{2} + 1)^2 + (\sqrt{7} + 1)^2 = 2 + 1 + 2\sqrt{2} + 7 + 1 + 2\sqrt{7}$

   (a) ABCD (b) DACB
   (c) DBCA (d) DCBA

18. $2^3 x^{+2} y^{+3} z^5 \times 5^2 x^{-5} y^{-5} z^{-5} \div 100 x^{-4} y^{-3} z^{-1} = $ _____.

   (a) $2xyz^2$ (b) $2xy^2z$
   (c) $2xyz$ (d) $2x^2yz$

19. If $2^n = 1024$, then $2^{\frac{n}{2}+2} = $ _____

   (a) 64 (b) 128
   (c) 256 (d) 512

20. If $2^{2y} = \frac{1}{4}$ and $3^{4x} = \frac{1}{81}$, then $x + y = $ _____.

   (a) 0 (b) -2
   (c) 1 (d) 2

21. Which is the greatest among $(9)^6$, $(27)^5$, $(81)^4$ and $(243)^2$?

   (a) $(9)^6$ (b) $(27)^5$
   (c) $(81)^4$ (d) $(243)^2$

22. $\left(\dfrac{a^{x^2}}{a^{y^2}}\right)^{\frac{1}{x+y}} \times \left(\dfrac{a^{y^2}}{a^{z^2}}\right)^{\frac{1}{y+z}} \times \left(\dfrac{a^{z^2}}{a^{x^2}}\right)^{\frac{1}{z+x}} = $ _____.

   (a) 1 (b) 2
   (c) 3 (d) 4

23. $\left(\sqrt[5]{\dfrac{3125}{243}}\right)^{-1} = $ _____.

   (a) $-\dfrac{3}{5}$ (b) $\dfrac{3}{5}$
   (c) $\dfrac{5}{3}$ (4) None of these

24. If $x^y \times y^x = 800$, then find $x - y$ (where x and y are positive integers and $x > y$).

   (a) 2 (2) 5
   (3) 4 (4) 3

Chapter 4

25. $\left(\sqrt{2013}-\sqrt{2012}\right)^{\frac{3}{5}}\left(\sqrt{2013}-\sqrt{2012}\right)^{\frac{3}{5}} = \underline{\quad}$.

   (a) 1   (b) 2

   (c) 3   (d) 6

26. Find the value of $(3^4 + 3^5)^{3/4}$.

   (a) 27

   (b) $27(2)^{3/2}$

   (c) $2^{3/2}$

   (d) None of these

Directions for questions 27 to 30: Match the Column A with Column B.

Column A		Column B
27. $2^{2^{2^{3^{0^{19}}}}}$ ( )		(a) 2
28. $\left(8^0 + 10^0\right)\left(2^1 - 2^2\right)$ ( )		(b) 16
29. $(512)^{4/9}$ ( )		(c) 256
30. If $3^{x-1} = 3^{6-y} = 27$, then $x^2 + y^2$ ( )		(d) 1
		(e) 25
		(f) 1024

Indices 4.17

TEST YOUR CONCEPTS

Very Short Answer Type Questions

1. False
2. True
3. False
4. True
5. False
6. False
7. False
8. False
9. True
10. True
11. $(-11)^3$
12. x^4
13. $\left(\dfrac{3}{2}\right)^2$
14. 2^{10}
15. $\dfrac{1}{5}$
16. $2 \times 3^3 \times 5^2$
17. $\dfrac{5^4}{2^4 \times 3^4}$
18. 6
19. (b)
20. (c)
21. (b)
22. (d)
23. (d)
24. (b)
25. (d)

Short Answer Type Questions

26. 32
27. 2^{2^3}
28. $x^7 \cdot y^2 \cdot z$
29. 15
30. ± 9
31. 1
32. $\dfrac{2}{5}$
33. 521
34. $\dfrac{5^8}{2^9}$
35. 90
36. 1
37. $6^{-3}, 5^{-3}, 4^{-3}$ and 3^{-3}
38. 36
39. 486
40. -1
41. 9
42. 4, 64
43. $15x^{-3}y^{-4}z^6$
44. $-\dfrac{8}{9}$
45. 5^4

Essay Type Questions

46. 1
47. 4
48. 8100; 100
49. 7
50. 3^{-7}
51. $\left(\dfrac{3}{2}\right)^{18}$
53. 3^{48}
54. k
55. 6
56. 15
58. 5
60. $(81)^2$

ANSWER KEYS

Chapter 4

CONCEPT APPLICATION

Level 1

1. (d)   2. (c)   3. (a)   4. (c)   5. (c)   6. (b)   7. (c)   8. (b)   9. (b)   10. (c)
11. (a) 12. (b) 13. (b) 14. (b) 15. (d) 16. (b) 17. (b) 18. (a) 19. (c) 20. (c)
21. (d) 22. (f) 23. (e) 24. (b) 25. (e) 26. (b) 27. (d) 28. (f)

Level 2

29. (d) 30. (a) 31. (c) 32. (a) 33. (a) 34. (d) 35. (c) 36. (a) 37. (d) 38. (c)
39. (d) 40. (a) 41. (b) 42. (a) 43. (d) 44. (a) 45. (b) 46. (c) 47. (c) 48. (c)
49. (c) 50. (c)

Level 3

51. (b) 52. (c) 53. (b) 54. (d) 55. (a) 56. (b) 57. (d) 58. (a) 59. (c) 60. (d)
61. (b) 62. (c) 63. (a) 64. (c) 65. (c) 66. (b) 67. (c) 68. (c) 69. (c) 70. (a)

ASSESSMENT TESTS

Test 1

1. (a)   2. (d)   3. (b)   4. (b)   5. (c)   6. (d)   7. (a)   8. (c)   9. (b)   10. (a)
11. (d) 12. (e) 13. (a) 14. (c) 15. (b)

Test 2

16. (a) 17. (d) 18. (c) 19. (b) 20. (b) 21. (c) 22. (a) 23. (b) 24. (d) 25. (a)
26. (b) 27. (c) 28. (a) 29. (b) 30. (e)

Indices **4.19**

CONCEPT APPLICATION
Level 1

1. $2^{2^2} = 2^9 = 512$
   Hence, the correct option is (d)

2. $0^5 = 0$ and $2^0 = 1$
   $\therefore 32^{2^{0^5}} = 3^1 = 3$
   Hence, the correct option is (c)

3. $200000000 = 2 \times 10^8$
   Hence, the correct option is (a)

4. Base is x.
   Hence, the correct option is (c)

5. $(2^3)^4 = 2^{12} = (2^4)^3$
   Hence, the correct option is (c)

6. $3^4 \left[\left(\dfrac{2}{3}\right)^2 + \left(\dfrac{2}{3}\right) - \left(\dfrac{2}{3}\right)^3\right]$
   $= 3^4 \cdot \dfrac{2^2}{3^2} + 3^4 \cdot \dfrac{2}{3} - 3^4 \cdot \dfrac{2^3}{3^3}$
   $= 36 + 54 - 24 = 66$
   Hence, the correct option is (b)

7. Given,
   $\left(\dfrac{1024}{243}\right)^{\tfrac{3}{5}} = \left(\dfrac{2^{10}}{3^5}\right)^{\tfrac{3}{5}}$
   $= \dfrac{2^6}{3^3} = \dfrac{64}{27}$
   Hence, the correct option is (c)

8. Given,
   $(2^4 + 2^3)^{\tfrac{2}{6}}$
   $= (2^4 + 2^3)^{\tfrac{1}{3}} = [2^3(2+1)]^{1/3}$
   $= 2^{3/3}(3)^{1/3} = 2(3)^{1/3}$
   Hence, the correct option is (b)

9. Given,
   $(25^2 - 15^2)^{\tfrac{3}{2}}$
   $= [(25+15)(25-15)]^{\tfrac{3}{2}}$
   $= [40 \times 10]^{\tfrac{3}{2}} = [2^2 \times 10^2]^{\tfrac{3}{2}} = (20)^3 = 8000$
   Hence, the correct option is (b)

10. $(33^2 - 31^2)^{\tfrac{5}{7}}$
    $= [(33+31)(33-31)]^{\tfrac{5}{7}}$
    $= [64 \times 2]^{\tfrac{5}{7}}$
    $= (2^7)^{\tfrac{5}{7}} = 2^5 = 32$
    Hence, the correct option is (c)

11. Given $abc = 0$
    $\left[(x^a)^b\right]^c$
    $= (x^a)^{bc}$
    $= x^{abc}$
    $= x^0 = 1$
    Hence, the correct option is (a)

12. Given $a + b + c = 0$
    $\sqrt{x^a \cdot x^b \cdot x^c}$
    $= \sqrt{x^{a+b+c}} = \sqrt{x^0} = \sqrt{1} = 1$
    Hence, the correct option is (b)

13. $\dfrac{\sqrt{36} + \sqrt{64}}{2^3 - 1}$
    $= \dfrac{\sqrt{6^2} + \sqrt{8^2}}{2^3 - 1} = \dfrac{6+8}{8-1} = \dfrac{14}{7} = 2$
    Hence, the correct option is (b)

14. Given $a = 25$
    $a^{25^0} + a^{0^{25}} = a^1 + a^0$
    $= a + 1 = 25 + 1 = 26$
    Hence, the correct option is (b)

15. We know $(-1)^x$
    $= -1$, when x is odd.
    $= 1$, when x is even.
    $\therefore (-1)^{301} + (-1)^{302} + \ldots + (-1)^{400}$
    $= (-1) + 1 + (-1) + 1 + \ldots (100 \text{ terms}) = 0$.
    Hence, the correct option is (d)

4.20 Chapter 4

16. $3^x = 6561 = (81)^2 = (3^4)^2 = 3^8$

    $\Rightarrow x = 8$

    $3^{x-3} = 3^{8-3} = 3^5 = 243$.

    Hence, the correct option is (b)

17. (D), (B), (C) and (A) is the required sequential order.

    Hence, the correct option is (b)

18. (C), (B) and (A) is the required sequential order.

    Hence, the correct option is (a)

19. (C), (B) and (A) is the required sequential order.

    Hence, the correct option is (c)

20. (A), (C), (D) and (B) is the required sequential order.

    Hence, the correct option is (c)

21. $\to d: \dfrac{2^{10}}{2^{-10}} = 2^{10} \times 2^{10} = 2^{20}$

22. $\to f: 6^{18} = (2 \times 3)^{18} = 2^{18} \times 3^{18}$

23. $\to e: 2^x = 16 \Rightarrow 2^x = 2^4 \Rightarrow x = 4$

24. $\to b: 5° - 6° = 1 - 1 = 0$

25. $\to e: 5^{-5} = \dfrac{1}{5^5}$

26. $\to b: 6^{-6} \times 6^6 = 6^{-6+6} = 6^0 = 1$

27. $\to d: x^3 = 7^3 \Rightarrow x = 7$

28. $\to f: 2^{123} = 2^1 = 2$

Level 2

29. $4a^{-1}b^{-2}c^3 \times 6a^3b^2c \div 12a^2c^4b$

    $= \dfrac{24a^2c^4}{12a^2c^4b}$

    $= \dfrac{2}{b} = 2b^{-1}$

    Hence, the correct option is (d)

30. Given,

    $3^{-x} = 3000$

    $\Rightarrow 3^x = \dfrac{1}{3000}$ \qquad (1)

    $10^3 \times 3^{(2x+3)} = 1000 \times 3^{2x} \times 3^3$

    $= 1000 \times 3^x \times 3^x \times 3^3$ \qquad (2)

    Substituting the value of 3^x in equation (2), we get

    $= 1000 \times \dfrac{1}{3000} \times \dfrac{1}{3000} \times 3^3$

    $= \dfrac{3}{1000} = 3 \times 10^{-3}$

    Hence, the correct option is (a)

31. $\dfrac{\left(-\dfrac{2}{3}\right)^2}{\left(-\dfrac{2}{3}\right)^3} + \dfrac{\left(-\dfrac{4}{9}\right)}{\left(\dfrac{2}{3}\right)^2}$

    $= \dfrac{\left(-\dfrac{2}{3}\right)^2}{\left(\dfrac{2}{3}\right)^3} - \dfrac{\left(\dfrac{2}{3}\right)^2}{\left(\dfrac{2}{3}\right)^2}$

    $= \dfrac{1}{\left(\dfrac{2}{3}\right)^{3-2}} - 1$

    $= -\dfrac{3}{2} - 1 = -\dfrac{5}{2}$

    Hence, the correct option is (c)

32. Given,

    $(4)^{a+5} \times (16)^{2a} \times (32)^4 = (16)^{3a}$

    $\Rightarrow (2^2)^{a+5} \times (2^4)^{2a} \times (2^5)^4 = (2^4)^{3a}$

    $\Rightarrow (2)^{2a+10} \times 2^{8a} \times 2^{20} = 2^{12a}$

    $\Rightarrow (2)^{2a+10+8a+20} = (2)^{12a}$

    $\Rightarrow (2)^{10a+30} = (2)^{12a}$

    $\Rightarrow 10a + 30 = 12a$

    $\Rightarrow 2a = 30$

    $\Rightarrow a = 15$

    Hence, the correct option is (a)

33. Given.

    $2^n = 4096 = 2^{12}$

    $\therefore 2^{n-5} = \dfrac{2^n}{2^5} = \dfrac{2^{12}}{2^5} = 2^{12-5} = 2^7 = 128$

    Hence, the correct option is (a)

34. Given,

    $2^{3y-x} = 16 = 2^4$

    $\therefore 3y - x = 4$ \qquad (1)

$2^{2y+x} = 2048 = 2^{11}$

$\therefore 2y + x = 11$ (2)

By adding (1) and (2), we get

$5y = 15$

$\Rightarrow y = 3$

Hence, the correct option is (d)

35. Given,

$x^y \times y^x = 72$

Let us factorise 72.

$72 = 2^3 \times 3^2$

$\therefore x = 2, y = 3$ or $x = 3, y = 2$.

$\therefore x + y = 2 + 3 = 5$.

Hence, the correct option is (c)

36. $\left(\dfrac{a^x}{a^y}\right)^z \times \left(\dfrac{a^y}{a^z}\right)^x \times \left(\dfrac{a^z}{a^x}\right)^y$

$= (a^{x-y})^z \times (a^{y-z})^x \times (a^{z-x})^y$

$= (a^{xz-yz}) \times (a^{yz-zx}) \times (a^{zy-xy})$

$= a^{xz-yz+yx-zx+zy-xy} = a^0 = 1$

Hence, the correct option is (a)

37. Given

$(5)^{23}$

$(25)^{12} = (5^2)^{11} = 5^{22}$

$(625)^6 = (5^4)^6 = 5^{24}$

$(3125)^5 = (5^5)^5 = 5^{25}$

$\therefore (3125)^5$ is the greatest.

Hence, the correct option is (d)

38. $(0.01024)^{1/5} = \left[(0.4)^5\right]^{\frac{1}{5}} = 0.4$

39. $\left(\dfrac{a}{b}\right)^{x+y+z} \div \left[\left(\sqrt{\dfrac{a}{b}}\right)^{-x} \times \left(\sqrt{\dfrac{a}{b}}\right)^{-y} \times \left(\sqrt{\dfrac{a}{b}}\right)^{-z}\right]$

$= \left(\dfrac{a}{b}\right)^{x+y+z} \div \left[\left(\dfrac{a}{b}\right)^{-\frac{x}{2}} \times \left(\dfrac{a}{b}\right)^{-\frac{y}{2}} \times \left(\dfrac{a}{b}\right)^{-\frac{z}{2}}\right]$

$= \left(\dfrac{a}{b}\right)^{(x+y+z)} \div \left[\left(\dfrac{a}{b}\right)^{-\left(\frac{x+y+z}{2}\right)}\right]$

$= \left(\dfrac{a}{b}\right)^{\frac{3(x+y+z)}{2}}$

Hence, the correct option is (d)

40. Given, $\dfrac{1}{(512)^{2x \times 2}} = 2^{18}$,

$\Rightarrow \dfrac{1}{(512)^{2x+2}} = 2^{18}$

$\Rightarrow \dfrac{1}{(2^9)^{2x+2}} = 2^{18}$

$\Rightarrow \dfrac{1}{2^{(18x+18)}} = 2^{18}$

$\Rightarrow 2^{-(18x+18)} = 2^{18}$

$\Rightarrow -18x - 18 = 18$

$\Rightarrow -18x = 36$

$\Rightarrow x = -2$

Hence, the correct option is (a)

41. $(2x)^5 = 100000$

$\Rightarrow (2x)^5 = 10^5$

$\Rightarrow 2x = 10$

$\Rightarrow x = 5$

Hence, the correct option is (b)

42. Given

$(x+y)^3 = 1331 = (11)^3$

$\Rightarrow x + y = 11$ (1)

$(x-y)^5 = 243 = 3^5$

$x - y = 3$ (2)

From (1) and (2), $(x+y)(x-y) = (11)(3) = 33$.

Hence, the correct option is (a)

43. Given,

$3^{2x-16} = 4^{3x-24}$

This is possible only, when $2x - 16 = 3x - 24 = 0$.

$\therefore x = 8$

Hence, the correct option is (d)

44. Given $xyz = 0$

$(a^x)^{zy} + (a^y)^{zx} + (a^z)^{xy}$

Chapter 4

$= a^{xyz} + a^{xyz} + a^{xyz} = a^0 + a^0 + a^0$
$= 1 + 1 + 1 = 3$
Hence, the correct option is (a).

45. Marks secured by Varun $= x$
Marks secured by Rahul $= x^2$
Given $x \cdot x^2 = 729$
$x^3 = 729$
$x^3 = 9^3$
$\therefore x = 9$
Marks secured by Varun $= x = 9$.
Marks secured by Rahul $= x^2 = 9^2 = 81$.
Hence, the correct option is (b).

46. Let the required number be x.
$\therefore x \times 2^{-6} = 1$
$\Rightarrow x \left(\dfrac{1}{2^6}\right) = 1$
$\Rightarrow x = 1 \times 2^6 = 2^6 = 64$.
Hence, the correct option is (c).

47. Given $27^{7x-5} = (32)^{x-7}$
$\Rightarrow 2^{7x-5} = 2^{5x-35} \quad (\because 32 = 2^5)$
$\Rightarrow 7x - 5 = 5x - 35$
$\Rightarrow 2x = -30$
$\Rightarrow x = -15$
$\therefore \dfrac{1}{2^x} = \dfrac{1}{2^{-15}} = 2^{15}$
Hence, the correct option is (c).

48. Given $x = \sqrt{100} = \sqrt{10^2}$
$\therefore x = 10$
$\dfrac{x^3 + x^2}{x} = \dfrac{x^3}{x} + \dfrac{x^2}{x}$
$= x^2 + x = (10)^2 + 10 = 110$.
Hence, the correct option is (c).

49. $4^{-2}, 3^{-2}, 2^{-2}, 5^{-2}$
$\Rightarrow \dfrac{1}{4^2}, \dfrac{1}{3^2}, \dfrac{1}{2^2}, \dfrac{1}{5^2}$
$\Rightarrow \dfrac{1}{16}, \dfrac{1}{9}, \dfrac{1}{4}, \dfrac{1}{25}$
Since, the numerators of the given fractions are same, the fraction which has the least denominator has the greatest value.
$\therefore \dfrac{1}{4}$ is the greatest among the given fractions.
Hence, the correct option is (c).

50. The total number of chocolates to be distributed = 2025.
The number of students $= x$.
The number of chocolates each student gets $= x$.
$\therefore 2025 = x \times x$
$2025 = x^2$
$\Rightarrow x = \sqrt{2025} \Rightarrow x = 45$.
The number of students in the class (x) = 45.
Hence, the correct option is (c).

Level 3

51. $\sqrt{32} + \sqrt{252}$
$= 4\sqrt{2} + 6\sqrt{7}$
$= 4 \times 1.414 + 6 \times 2.646$
$= 5.656 + 15.876$
$= 21.532$
Hence, the correct option is (b).

52. $64 > x^3 \Rightarrow 4^3 > x^3$
Clearly 3 is the greatest possible integer value of x.
Hence, the correct option is (c).

53. Let us take G.C.D. of 1152, 768 and 384 which is 384.
$\therefore 2^{1152} = (2^3)^{384} = 8^{384}$
$3^{768} = (3^2)^{384} = 9^{384}$

$5^{384} = (5)^{384}$

∴ The ascending order is as follows.

$5^{384}, 8^{384}, 9^{384}$

i.e., $5^{384}, 2^{1152}, 3^{768}$.

Hence, the correct option is (b)

54. Given $2^a = 4^b = 8^c = 64$

⇒ $2^a = 2^6$ ⇒ $a = 6$

$4^b = 4^3$ ⇒ $b = 3$

$8^c = 8^2$ ⇒ $c = 2$

∴ $a + b + c = 11$.

Hence, the correct option is (d)

55. Given, $\sqrt{2.5} = 1.581$

$\sqrt{0.625} = \sqrt{2.5 \times 0.25}$

$= \sqrt{2.5 \times (0.5)^2}$

$= (0.5)\sqrt{2.5}$

$= \dfrac{1}{2}(1.581) = 0.7905$

Hence, the correct option is (a)

56. $\left(\dfrac{27}{343}\right)^{-\frac{2}{3}} \times \left(\dfrac{343}{729}\right)^{\frac{2}{3}} \div \left(\dfrac{2401}{81}\right)^{\frac{3}{4}}$

$= \left(\dfrac{3^3}{7^3}\right)^{-\frac{2}{3}} \times \left(\dfrac{7^3}{9^3}\right)^{\frac{2}{3}} \div \left(\dfrac{7^4}{3^4}\right)^{\frac{3}{4}}$

$= \left(\dfrac{3}{7}\right)^{-2} \times \left(\dfrac{7^2}{9^2}\right) \div \left(\dfrac{7}{3}\right)^{-3}$

$= \dfrac{7^2}{3^2} \times \dfrac{7^2}{9^2} \times \dfrac{7^3}{3^3}$

$= \dfrac{7^3}{3^2 \times 3^4 \times 3^3} = \dfrac{7^7}{3^9} = \dfrac{1}{9} \times \left(\dfrac{7}{3}\right)^7$

Hence, the correct option is (b)

57. Given,

$(9)^{2x+4} = (243)^{2x-3.2}$

⇒ $(3)^{4x+8} = (3)^{10x-16}$

⇒ $4x + 8 = 10x - 16$

⇒ $6x = 24$ ⇒ $x = 4$.

Hence, the correct option is (d)

58. Given

$2^{2n-3} = 2048$

⇒ $\dfrac{2^{2n}}{8} = 2^{11}$

⇒ $2^{2n} = 2^{11} \times 2^3 = 2^{14}$

⇒ $n = 7$

∴ $(4n + 3n^2)$

$= (4 \times 7 + 3 \times 7^2) = 175$.

Hence, the correct option is (a)

59. Given $a \neq 1$. $7^{a^b} = 2401$ it can be written in the following ways.

$7^{4^1} = 2401$, in the case $a > b$.

$7^{2^2} = 2401$, in this case $a = b$.

Either $a > b$ or $a = b$.

Hence, the correct option is (c)

60. $a^{b^c} = 729$

This can be written in the following ways.

(i) 27^{2^1} (ii) 9^{3^1}

(iii) 729^{1^1} (iv) 3^{6^1}

Clearly, $a + b + c$ will be minimum in case (iv).

∴ Minimum possible value of $a + b + c = 3 + 6 + 1 = 10$.

Hence, the correct option is (d)

61. $(2)^{2x-2} = (8)^{y-1} = (16)^{x-2.5}$

⇒ $(2)^{2x-2} = 2^{3y-3} = 2^{4x-10}$

⇒ $(2)^{2x-2} = 2^{4x-10}$

⇒ $2x - 2 = 4x - 10$

⇒ $x = 4$

⇒ $2^{3y-3} = 2^{2 \times 4 - 2} = 2^{8-2} = 2^6$

⇒ $3y - 3 = 6$

⇒ $y = 3$

∴ The sum of x and y is 7.

Hence, the correct option is (b)

62. $\left(\sqrt{1795} - \sqrt{1170}\right)^{\frac{3}{4}} \times \left(\sqrt{1795} + \sqrt{1170}\right)^{\frac{3}{4}}$

$= \left[\left(\sqrt{1795} - \sqrt{1170}\right)\left(\sqrt{1795} + \sqrt{1170}\right)\right]^{\frac{3}{4}}$

$= \left[\left(\sqrt{1795}\right)^2 - \left(\sqrt{1170}\right)^2\right]^{\frac{3}{4}}$

$= (625)^{\frac{3}{4}} = (5^4)^{\frac{3}{4}} = 5^3 = 125$

Hence, the correct option is (c).

63. Given $a = 3^{-3} - 3^3$ and $b = 3^3 - 3^{-3}$

$\dfrac{a}{b} - \dfrac{b}{a} = \dfrac{a^2 - b^2}{ba} = \dfrac{(a+b)(a-b)}{ba} = 0$

($\because a + b = 0$)

Hence, the correct option is (a).

64. $(4)^{26} = (2^2)^{26} = (2)^{52}$

$(64)^9 = (2^6)^9 = 2^{54}$

$(256)^7 = (2^8)^7 = 2^{56}$

$\therefore (256)^7 > (64)^9 > (4)^{26}$.

Hence, the correct option is (c).

65. $\left(\dfrac{x^{5y-3} \times x^{3-2y}}{x^{4y-6} \times x^{2y-9}}\right)^{-\frac{4}{3}}$

$= \left(\dfrac{x^{5y-3+3-2y}}{x^{4y-6+2y-9}}\right)^{-\frac{4}{3}}$

$= \left(\dfrac{x^{3y}}{x^{6y-15}}\right)^{-\frac{4}{3}}$

$= \left(\dfrac{1}{x^{6y-15-3y}}\right)^{-\frac{4}{3}}$

$= \left(\dfrac{1}{x^{3y-15}}\right)^{-\frac{4}{3}} = \left(\dfrac{1}{x^{-4y+20}}\right) = x^{4y-20}$

Hence, the correct option is (c).

66. If $(xy)^a = z$ (1)

$(yz)^a = x$ (2)

$(xz)^a = y$ (3)

On multiplying equations (1), (2) and (3), we get

$(xyz)^{2a} = xyz$

$\Rightarrow (xyz)^{2a} = (xyz)^1$

$\Rightarrow (xyz)^{2a} = (xyz)^1$

$\Rightarrow 2a = 1 \Rightarrow a = \dfrac{1}{2}$.

Hence, the correct option is (b).

67. $\dfrac{x^{-1} - y^{-1}}{z} + \dfrac{y^{-1} - z^{-1}}{x} + \dfrac{z^{-1} - x^{-1}}{y}$

$= \dfrac{\dfrac{1}{x} - \dfrac{1}{y}}{z} + \dfrac{\dfrac{1}{y} - \dfrac{1}{z}}{x} + \dfrac{\dfrac{1}{z} - \dfrac{1}{x}}{y}$

$= \dfrac{y-x}{xyz} + \dfrac{z-y}{xyz} + \dfrac{x-z}{xyz}$

$= \dfrac{y-x+z-y+x-z}{xyz} = \dfrac{0}{xyz} = 0$

Hence, the correct option is (c).

68. $\dfrac{1}{1+x^{a-b}} + \dfrac{1}{1+x^{b-a}} = \dfrac{1}{1+\dfrac{x^a}{x^b}} + \dfrac{1}{1+\dfrac{x^b}{x^a}}$

$= \dfrac{x^b}{x^b + x^a} + \dfrac{x^a}{x^a + x^b} = \dfrac{x^b + x^a}{x^b + x^a} = 1$

Hence, the correct option is (c).

69. Given $x^5 < 1000$,

The greatest integer value which satisfies the given inequality is 3.

Hence, the correct option is (c).

70. Given x and y are distinct natural numbers and $x^y = y^x$.

$\therefore x$ and y are 2 and 4 since, $2^4 = 4^2$.

$\therefore x + y = 2 + 4 = 6$.

Hence, the correct option is (a).

Indices 4.25

ASSESSMENT TESTS

Test 1

Solutions for questions 1 to 11:

1. (A), (D), (C) and (B) is the required sequential order.
   Hence, the correct option is (a)

2. (C), (A), (B) and (D) is the required sequential order.
   Hence, the correct option is (d)

3. $\dfrac{3^4 x^3 y^6 z^4 \times 27 x^{-2} y^{-3} z^{-6}}{81 xy^2 z^{-3}} = \dfrac{27 xy^3 z^{-2}}{xy^2 z^{-3}} = 27yz$
   Hence, the correct option is (b)

4. $3^x = 243 = 3^5 \Rightarrow x = 5$
   $\therefore 3^{\frac{x+1}{2}} = 3^{\frac{5+1}{2}}$
   $= 3^{\frac{6}{2}} = 3^3 = 27$
   Hence, the correct option is (b)

5. $5^{-5y} = \dfrac{1}{3125}$
   $\Rightarrow 5^{-5y} = 5^{-5} \Rightarrow y = 1$
   $9^x = \dfrac{1}{81}$
   $\Rightarrow 9^x = 9^{-2}$
   $\Rightarrow x = -2$
   $\therefore x - y = -2 - 1 = -3$
   Hence, the correct option is (c)

6. $(\sqrt{3})^{30} = (3^{1/2})^{30} = 3^{15}$
   $(\sqrt{27})^4 = (27^{1/2})^4 = 27^2 = (3^3)^2 = 3^6$
   $(\sqrt{18})^8 = (81^{1/2})^8 = 81^4 = (3^4)^4 = 3^{16}$
   $(\sqrt[3]{9})^{12} = (9^{1/3})^{12} = (3^{2/3})12 = 3^8$
   The greatest among
   $3^{15}, 3^6, 3^{16}, 3^4$ is 3^{16}, i.e., $(\sqrt{81})^8$.
   Hence, the correct option is (d)

7. $\left(\dfrac{a^{x^4}}{a^{y^4}}\right)^{\frac{1}{x^2+y^2}} \times \left(\dfrac{a^{y^4}}{a^{z^4}}\right)^{\frac{1}{y^2+z^2}} \times \left(\dfrac{a^{z^4}}{a^{x^4}}\right)^{\frac{1}{z^2+x^2}}$
   $= \left(a^{x^4-y^4}\right)^{\frac{1}{x^2+y^2}} \times \left(a^{y^4-z^4}\right)^{\frac{1}{y^2+z^2}} \times \left(a^{z^4-x^4}\right)^{\frac{1}{z^2+x^2}}$
   $= a^{\frac{(x^2-y^2)(x^2+y^2)}{x^2+y^2}} \times a^{\frac{(y^2-z^2)(y^2+z^2)}{y^2+z^2}} \times a^{\frac{(z^2-x^2)(z^2+x^2)}{(z^2+x^2)}}$
   $= a^{x^2-y^2+y^2-z^2+z^2-x^2} = a^0 = 1$
   Hence, the correct option is (a)

8. $\left(\sqrt[6]{\dfrac{64}{125}}\right)^{-2} = \left[\left(\dfrac{64}{125}\right)^{\frac{1}{6}}\right]^{-2}$
   $= \left(\dfrac{64}{125}\right)^{-\frac{1}{3}} = \left(\dfrac{5^3}{4^3}\right)^{\frac{1}{3}} = \dfrac{5}{4}$
   Hence, the correct option is (c)

9. $x^y \times y^x = 256 = 16 \times 16 = 2^4 \times 4^2$
   $\Rightarrow x^y \times y^x = 2^4 \times 4^2$
   $\Rightarrow x = 2$ and $y = 4$ ($\because x < y$)
   $\therefore y^2 - x^2 = 16 - 4 = 12.$
   Hence, the correct option is (b)

10. $\left(\sqrt{1024} - \sqrt{24}\right)^{2/3} \left(\sqrt{1024} + \sqrt{24}\right)^{2/3}$
    $= (1024 - 24)^{2/3} = (1000)^{2/3}$
    $= (10^3)^{2/3} = 10^2 = 100.$
    Hence, the correct option is (a)

11. $(8^4 + 8^2)^{1/2}$
    $= [8^2(8^2 + 1)]^{1/2}$
    $= (8 \times 65)^{1/2}$
    $= 8\sqrt{65}$
    Hence, the correct option is (d)

Solutions for questions 12 to 15:

12. → (e): $100^{2^{3^{0^{10}}}} = 100^{2^{3^0}} = 100^{2^1} = 100^2$

13. → (a): $(12^0 - 11^0)(3^1 - 1^3) = (1 - 1)(3 - 1) = 0 \times 2 = 0$

14. → (c): $(128)^{5/7} = (2^7)^{5/7} = 2^5 = 32$

15. → (b): $5^x = 25^y = 625$
    $5^x = 5^{2y} = 5^4$
    $\Rightarrow x = 2y = 4$
    $\Rightarrow x = 4, y = 2$
    $\therefore x^2 - y^2 = 16 - 4 = 12.$

Test 2

Solutions for questions 16 to 26:

16. (D), (A), (B) and (C) is the required sequential order.

    Hence, the correct option is (a)

17. (D), (C), (B) and (A) is the required sequential order.

    Hence, the correct option is (d)

18. $\dfrac{2^3 x^2 y^3 z^5 \times 5^2 x^{-5} y^{-5} z^{-5}}{100 x^{-4} y^{-3} z^{-1}} = \dfrac{2x^{-3} \cdot y^{-2}}{x^{-4} y^{-3} z^{-1}} = 2xyz$

    Hence, the correct option is (c)

19. $2^n = 1024 = 2^{10} \Rightarrow n = 10$

    $\therefore 2^{\frac{n}{2}+2} = 2^{\frac{10}{2}+2}$

    $= 2^{5+2} = 2^7 = 128$

    Hence, the correct option is (b)

20. $2^{2y} = \dfrac{1}{4} \Rightarrow 2^{2y} = 2^{-2}$

    $\Rightarrow 2y = -2 \Rightarrow y = -1$

    $3^{4x} = \dfrac{1}{81} \Rightarrow 3^{4x} = 3^{-4}$

    $\Rightarrow 4x = -4 \Rightarrow x = -1$

    $\therefore x + y = -2$

    Hence, the correct option is (b)

21. 9^6, 27^5, 81^4 and $(243)^2$

    $(3^2)^6$, $(3^3)^5$, $(3^4)^4$ and $(3^5)^2$

    3^{12}, 3^{15}, 3^{16} and 3^{10}

    $\therefore$ The greatest among them is 3^{16}, i.e., $(81)^4$.

    Hence, the correct option is (c)

22. $\left(a^{x^2-y^2}\right)^{\frac{1}{x+y}} \times \left(a^{y^2-z^2}\right)^{\frac{1}{y+z}} \times \left(a^{z^2-x^2}\right)^{\frac{1}{z+x}}$

    $= a^{x-y} \times a^{y-z} \times a^{z-x}$

    $a^{x-y+y-z+z-x} = a^0 = 1$

    Hence, the correct option is (a)

23. $\left(\sqrt[5]{\dfrac{3125}{243}}\right)^{-1} = \sqrt[5]{\dfrac{243}{3125}} = \left(\dfrac{3^5}{5^5}\right)^{\frac{1}{5}} = \dfrac{3}{5}$

    Hence, the correct option is (b)

24. $x^y \times y^x = 800$

    $\Rightarrow x^y \times y^x = 25 \times 32$

    $\Rightarrow 5^2 \times 2^5 = x^y \times y^x \quad (\because x > y)$

    $\Rightarrow x = 5$ and $y = 2$

    $\therefore x - y = 5 - 2 = 3$.

    Hence, the correct option is (d)

25. $\left(\sqrt{2013} - \sqrt{2012}\right)^{\frac{3}{5}} \left(\sqrt{2013} - \sqrt{2012}\right)^{\frac{3}{5}}$

    $= (2013 - 2012)^{\frac{3}{5}} = 1^{\frac{3}{5}} = 1$

    Hence, the correct option is (a)

26. $(3^4 + 3^5)^{3/4}$

    $= [3^4(3 + 1)]^{3/4} = [3^4 \times 4]^{3/4} = 3^3 \times 4^{3/4}$

    $= 27 \times (2^2)^{3/4} = 27(2)^{3/2}$

    Hence, the correct option is (b)

Solutions for questions 27 to 30:

27. → (c): $2^{2^{3^{4^{0^{1^9}}}}} = 2^{2^{3^{4^{0^1}}}} = 2^{2^{3^{4^0}}} = 2^{2^{3^1}} = 2^{2^3} = 2^8 = 256$

28. → (a): $(8^0 + 10^0)(2^1 - 1^2) = (1 + 1)(2 - 1) = 2 \times 1 = 2$

29. → (b): $(512)^{4/9} = (2^9)^{4/9} = 2^4 = 16$

30. → (e): $3^{x-1} = 3^{6-y} = 27$

    $\Rightarrow 3^{x-1} = 3^{6-y} = 3^3$

    $\Rightarrow x - 1 = 6 - y = 3$

    $\Rightarrow x - 1 = 3, 6 - y = 3$

    $\Rightarrow x = 4, y = 3$

    $x^2 + y^2 = 4^2 + 3^2 = 25$

Chapter 5

Geometry

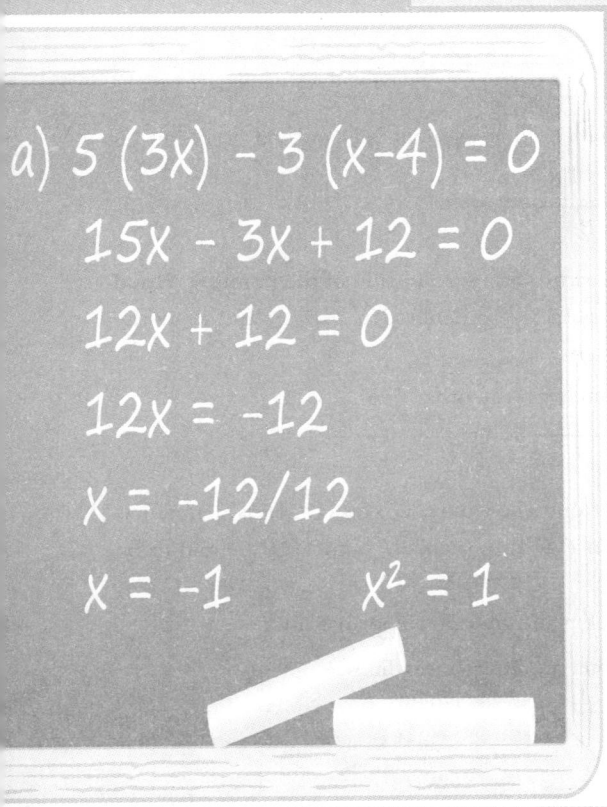

a) $5(3x) - 3(x-4) = 0$
$15x - 3x + 12 = 0$
$12x + 12 = 0$
$12x = -12$
$x = -12/12$
$x = -1 \qquad x^2 = 1$

REMEMBER
Before beginning this chapter, you should be able to:
- Have introductory knowledge of geometrical figures such as point, line, plane, angles
- Name the geometric figures

KEY IDEAS
After completing this chapter, you should be able to:
- Define plane, lines, angles and polygons
- Understand types of triangles, properties of triangles, congruence of triangles, and construction of triangles
- Explain types of quadrilaterals and their properties and constructions
- Learn about chord, secant, tangent, segment and sector of a circle and constructions of circles
- Study line symmetry and point symmetry

INTRODUCTION

Geometry is the study of the properties of lines, angles, polygons and circles. We will begin with some basic concepts.

Plane: A plane is a surface which extends indefinitely in all directions. For example, the surface of a table is part of a plane. A black board is a part of a plane.

Line: A line is a set of infinite points. It has no end points. It is infinite in length. The figure below shows a line l that extends to infinity on either side. If A and B are any two points on line l, we also denote the line l as $\overleftrightarrow{AB}$, read as line AB.

$$\longleftrightarrow$$
$$A \qquad B$$

Line segment: A line segment is a part of a line. The line segment has two end points and it has a finite length.

$$A \qquad B \qquad l$$

In the above figure, AB is a line segment, it is a part of line l, consisting of the points A, B and all the points of l between A and B. Line segment AB is also denoted as $\overline{AB}$. A and B are the end points of $\overline{AB}$.

Ray: A ray has one end point and it extends infinitely on the other side.

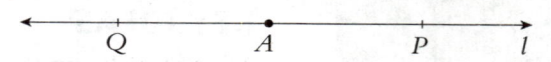

$$Q \qquad A \qquad P \qquad l$$

In the above figure, AP is a ray which has only one end point A. The ray AP is denoted as $\overrightarrow{AP}$. $\overrightarrow{AQ}$ is different from $\overrightarrow{AP}$. If A lies between P and Q, then $\overrightarrow{AP}$ and $\overrightarrow{AQ}$ are said to be opposite rays.

Coplanar lines: Two or more lines lying in a same plane are called coplanar lines.

Intersecting lines: Two lines which meet at a point are intersecting lines.

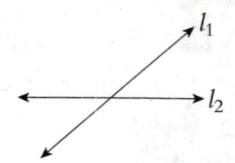

In the figure above, l_1 and l_2 are intersecting lines.

Angle: Two rays which meet at a point are said to form an angle at their meeting point.

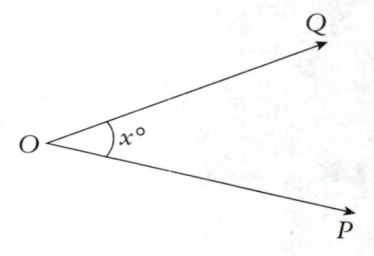

In the figure above, $\overrightarrow{OQ}$ and $\overrightarrow{OP}$ are two rays which have O as their meeting point and an angle with a certain measure is formed. The point O is called the **vertex** of the angle and $\overrightarrow{OQ}$

and $\overrightarrow{OP}$ are called the **sides** or **arms** of the angle. We denote the angle as ∠QOP or ∠POQ (or sometimes simply as ∠O). We see that ∠QOP = ∠POQ. (These are two ways of representing the same angle). A common unit of measurement of angles is a **degree**. This unit is denoted by a small circle placed above and to the right of the number. For example, we write ∠QOP = 40° and read it as 40 degrees.

The angle formed by two opposite rays is called a straight angle. We define the unit of degree such that the measure of a straight angle is 180°.

Let the measure of an angle be $x°$.

1. If $0° < x < 90°$, it is called an **acute angle**.

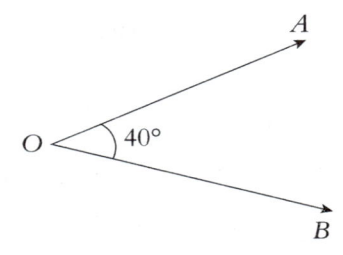

In the figure above, ∠AOB is acute.

2. If $x = 90°$, it is called a **right angle.**

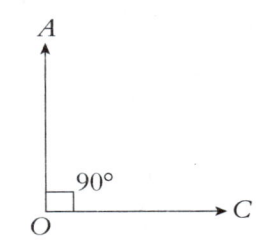

In the figure above, ∠AOC is a right angle.

3. If $90° < x < 180°$, it is called an **obtuse angle**.

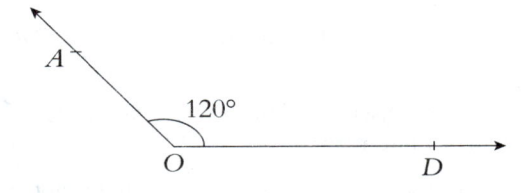

In the figure above, ∠AOD is an obtuse angle.

4. If $x = 180°$, it is the **angle of a straight line** or a **straight angle**.

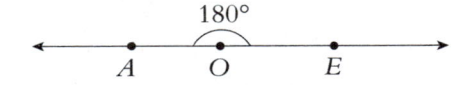

In the figure above, ∠AOE = 180° is a straight angle, $\overleftrightarrow{AE}$ is a straight line and O is a point on the line AE.

Perpendicular lines: Two intersecting lines making an angle of 90° with each other are called perpendicular lines.

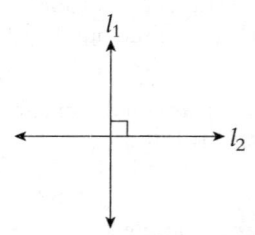

In the figure above, l_1 and l_2 are perpendicular lines. We write $l_1 \perp l_2$ and read this as l_1 is perpendicular to l_2.

Complementary angles: When the sum of two angles is 90°, the two angles are called complementary angles.

For example, if $x + y = 90°$, then $x°$ and $y°$ are called complementary angles.

Supplementary angles: When the sum of two angles is 180°, the two angles are called supplementary angles.

For example, if $a + b = 180°$, then $a°$ and $b°$ are called supplementary angles.

Adjacent angles: If two angles have a common vertex and a common ray and the other two sides lie on opposite sides of the common ray, then they are said to be adjacent angles.

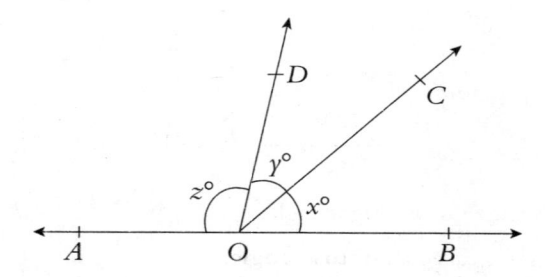

In the diagram above, $\angle AOD$ and $\angle DOC$ are adjacent angles. $\angle DOC$ and $\angle COB$ are also adjacent angles.

Linear pair: If a pair of angles are adjacent and the non-common rays are opposite to each other, then the angles are said to form a linear pair.

For example, if O is a point between A and B and P is a point not on AB, then $\angle AOP$ and $\angle POB$ form a linear pair. The angles of a linear pair are supplementary.

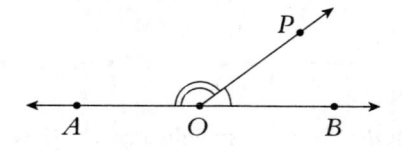

Vertically opposite angles: When two lines intersect each other, four angles are formed.

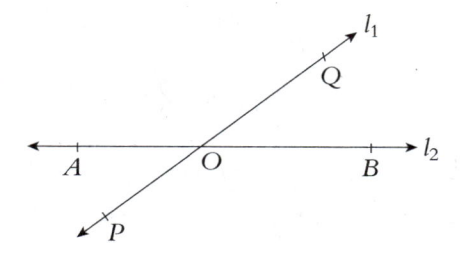

In the figure above, $\overrightarrow{AB}$ and $\overrightarrow{PQ}$ intersect at O. ∠AOP and ∠BOQ are vertically opposite angles. Similarly, ∠POB and ∠QOA are vertically opposite angles. Vertically opposite angles are equal.

Concurrent lines: Two or more lines in a plane passing through a same point are concurrent lines.

Parallel lines: Two coplanar lines that do not meet are called parallel lines.

In the figure above, l_1 and l_2 are parallel lines. Symbolically we write $l_1 \parallel l_2$ and read as l_1 is parallel to l_2.

Properties of Parallel Lines

1. For two given parallel lines, the perpendicular distance between the lines is the same everywhere.
2. If two lines lie in the same plane and are perpendicular to the same line, then they are parallel to each other.
3. If two lines are parallel to the same line, then they are parallel to each other.
4. One and only one parallel line can be drawn to a given line through a given point which is not on the given line.

Transversal: A straight line intersecting a pair of lines in two distinct points is a transversal for the two given lines.

Let l_1 and l_2 be a pair of lines and t be a transversal.

As shown in the figure, totally eight angles are formed.

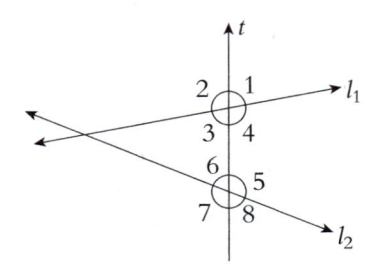

Chapter 5

1. $\angle 1, \angle 2, \angle 7$ and $\angle 8$ are exterior angles and $\angle 3, \angle 4, \angle 5$ and $\angle 6$ are interior angles.
2. $(\angle 1$ and $\angle 5), (\angle 2$ and $\angle 6), (\angle 3$ and $\angle 7)$ and $(\angle 4$ and $\angle 8)$ are pairs of corresponding angles.
3. $(\angle 1$ and $\angle 3), (\angle 2$ and $\angle 4), (\angle 5$ and $\angle 7)$ and $(\angle 6$ and $\angle 8)$ are pairs of vertically opposite angles.
4. $(\angle 4$ and $\angle 6)$, and $(\angle 3$ and $\angle 5)$ are pairs of alternate interior angles.
5. $(\angle 1$ and $\angle 7)$ and $(\angle 2$ and $\angle 8)$ are pairs of alternate exterior angles.

If l_1 and l_2 are parallel, then

1. Corresponding angles are equal, i.e.,
   $\angle 1 = \angle 5, \angle 2 = \angle 6, \angle 3 = \angle 7$ and $\angle 4 = \angle 8$.
2. Alternate interior angles are equal, i.e., $\angle 4 = \angle 6$ and $\angle 3 = \angle 5$.
3. Alternative exterior angles are equal, i.e., $\angle 1 = \angle 7$ and $\angle 2 = \angle 8$
4. Exterior angles on the same side of the transversal are supplementary, i.e.,
   $\angle 1 + \angle 8 = 180°$ and $\angle 2 + \angle 7 = 180°$.
5. Interior angles on the same side of the transversal are supplementary, i.e., $\angle 4 + \angle 5 = 180°$ and $\angle 3 + \angle 6 = 180°$.

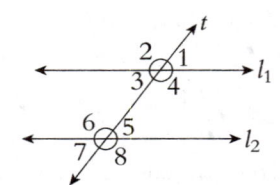

EXAMPLE 5.1

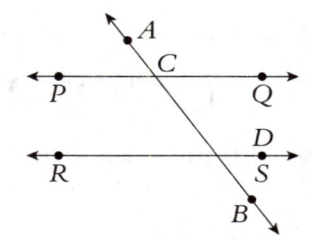

In the figure above, $\overrightarrow{PQ}$ and $\overrightarrow{RS}$ are parallel. $\overrightarrow{AC}$ is a transversal of $\overrightarrow{PQ}$ and $\overrightarrow{RS}$. If $\angle ACP = 5x - 70°$ and $\angle BDR = 4x + 70°$, then find the value of x.

SOLUTION

Given that, $\overrightarrow{PQ}$ // $\overrightarrow{RS}$ and $\overrightarrow{AB}$ is a transversal.

And also given, $\angle ACP = 5x - 70°$ and $\angle BDR = 4x + 70°$.

$\angle ACP + \angle BDR = 180°$ ($\because$ Exterior angles on the same side of the transversal)

$\Rightarrow 5x - 70° + 4x + 70° = 180°$

$\Rightarrow 9x = 180° \Rightarrow x = 20°$.

Triangle

A three sided simple closed figure is a triangle.

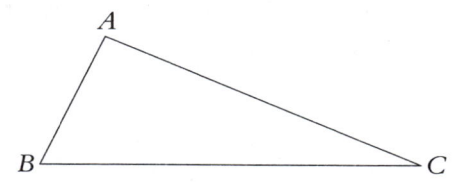

In the above figure, ABC is a triangle. It has three sides AB, BC and CA.

A, B and C are the three vertices of the triangle. $\angle CAB$, $\angle ABC$ and $\angle BCA$ are its three angles. Triangle ABC is denoted as $\triangle ABC$.

Types of Triangles

Based on their sides, triangles can be classified as follows:

1. *Scalene triangle:* A triangle in which no two sides are equal is a **scalene triangle**.

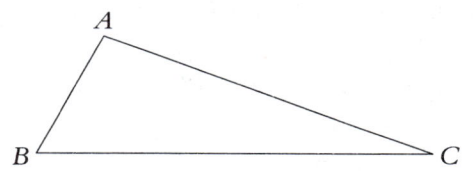

   In the above triangle, $AB \neq BC$, $BC \neq CA$, and $CA \neq AB$. This is a scalene triangle.

2. *Isosceles triangle:* A triangle in which any two sides are equal is an **isosceles triangle**.

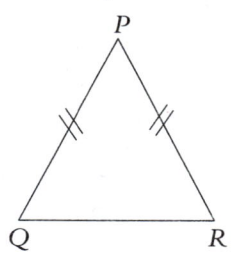

   In the triangle above, $PQ = PR$. Hence, PQR is an isosceles triangle.

3. *Equilateral triangle:* A triangle in which all the three sides are equal is an **equilateral triangle**.

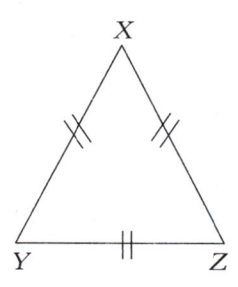

   In the above $\triangle XYZ$, $XY = YZ = ZX$.

   $\therefore$ XYZ is an equilateral triangle.

Based on the angles, triangles can be classified as follows:

1. *Acute-angled triangle:* A triangle in which all the angles are acute is an **acute-angled triangle**. In such a triangle, square of the longest side is less than the sum of the squares of the other two sides.

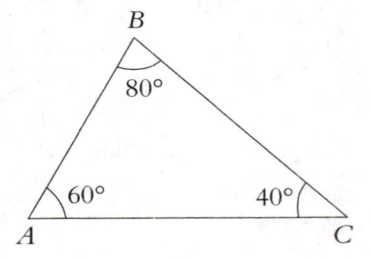

   In the triangle above, each of the angles A, B and C is less than 90°.

   Hence, ABC is an acute-angled triangle.

2. *Right-angled triangle:* A triangle which has a right angle is a right-angled triangle. In such triangle, square of the longest side, the hypotenuse is equal to the sum of the squares of the other two sides.

   In the right triangle, $\angle ABC = 90°$. Hence, ABC is a right-angled triangle.

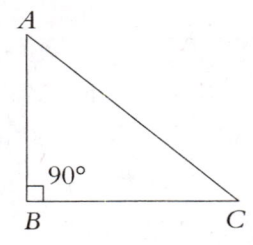

   In the right triangle ABC, if $\overline{AC}$ is the longest side, then $AC^2 = AB^2 + BC^2$.

3. *Obtuse-angled triangle:* A triangle in which one angle is greater than 90° is an **obtuse-angled triangle**. In such a triangle, the square of the longest side is greater than the sum of the squares of the other two sides.

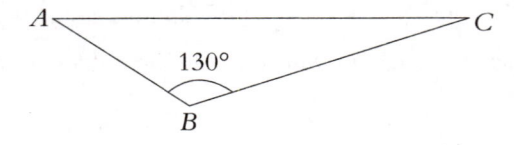

   In the triangle above, $\angle ABC > 90°$.

   Hence, ABC is an obtuse-angled triangle.

   In an obtuse-angled triangle ABC, if $\overline{AC}$ is the longest side, then $AC^2 > AB^2 + BC^2$.

4. *Isosceles right-angled triangle:* A triangle in which two sides are equal and one angle is 90° is an isosceles right-angled triangle.

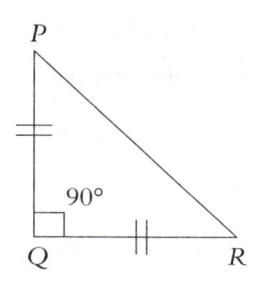

In the triangle PQR, PQ = QR and ∠PQR = 90°.

∴ PQR is an isosceles right triangle.

In such a triangle, the ratio of sides PQ, QR and RP is $1:1:\sqrt{2}$.

Notes (1) A scalene triangle can be acute, right or obtuse angled.

(2) An isosceles triangle can be acute, right or obtuse angled.

(3) An equilateral triangle has to be acute. It cannot contain a right angle or an obtuse angle.

Theorem 1: The sum of the three angles of a triangle is 180°.

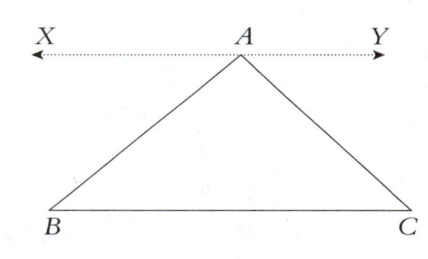

Given: ABC is a triangle.

To prove: ∠A + ∠B + ∠C = 180°

Construction: Draw a line XY through A and parallel to $\overline{BC}$.

Proof: ∠XAB + ∠BAC + ∠CAY = 180° (∵ straight line)   (1)

$\overrightarrow{XY}$ and $\overline{BC}$ are parallel and $\overline{AB}$ is a transversal.

∴ ∠XAB = ∠ABC (∵ alternate angles)   (2)

$\overrightarrow{XY}$ and $\overline{BC}$ are parallel and $\overline{AC}$ is a transversal.

∠YAC = ∠ACB (∵ alternate angles)   (3)

From (1), (2) and (3), we have

∠ABC + ∠BAC + ∠ACB = 180°

Hence proved.

Theorem 2: The exterior angle of a triangle is equal to the sum of the interior angles opposite to it.

Given: ABC is a triangle; BC is produced to the point X.

To Prove: ∠ACX = ∠A + ∠B.

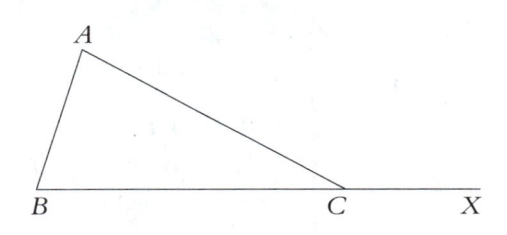

Proof: ∠A + ∠B + ∠BCA = 180° (∵ Angles of △ABC)

∠BCA + ∠ACX = 180° (∵ linear pair)

∴ ∠A + ∠B + ∠BCA = ∠BCA + ∠ACX

⇒ ∠ACX = ∠A + ∠B

Hence proved.

Given below are the statements of some of the properties of triangles:

1. The sum of any two sides of a triangle is greater than the third side.
   In $\triangle PQR$, $PQ + QR > PR$,
   $QR + RP > PQ$ and $RP + PQ > QR$.
2. Difference between any two sides is less than the third side.
   $PQ - QR < PR$, $QR - RP < PQ$ and $RP - PQ < QR$
3. Angles opposite to equal sides are equal and vice versa.
   If $\angle P = \angle Q$, then $QR = PR$.
   If $QR = PR$, then $\angle P = \angle Q$.
4. If the angles are in increasing or decreasing order, then the sides opposite to them also will be in the same order. If $\angle P > \angle Q > \angle R$, then $QR > PR > PQ$.

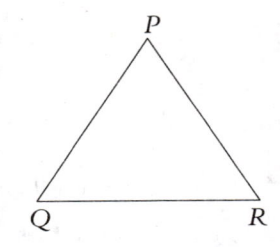

EXAMPLE 5.2

The sides of a $\triangle ABC$ measure 7 cm, 24 cm and 25 cm. What type of a triangle is ABC?

SOLUTION

Since no two sides are equal, ABC is a scalene triangle.
Further, $7^2 + 24^2 = 25^2$.
Since the square of the longest side is equal to the sum of the squares of the other two sides, ABC is a right-angled triangle.

EXAMPLE 5.3

In $\triangle PQR$, $\angle P = 50°$ and $\angle Q = 60°$. Find $\angle R$.

SOLUTION

In a triangle, the sum of the angles is equal to 180°.
$\angle P + \angle Q + \angle R = 180°$
$50° + 60° + \angle R = 180°$
$\Rightarrow \angle R = 70°$.

EXAMPLE 5.4

In $\triangle ABC$, $AB = 5$ cm and $BC = 4$ cm. Find the range of values that CA can take.

SOLUTION

In a triangle, the sum of two sides is greater than the third side and the difference of two sides is less than the third side.
$CA < AB + BC$ and $CA > AB - BC$
$\Rightarrow CA < 9$ cm and $CA > 1$ cm
$\Rightarrow 1$ cm $< CA < 9$ cm.

EXAMPLE 5.5

In $\triangle ABC$, $AC = BC$ and $\angle BAC = 50°$. Find $\angle BCA$.

SOLUTION

Given: $AC = BC$

In a triangle, angles opposite to equal sides are equal.

$\therefore \angle ABC = \angle CAB = 50°$

$\angle ABC + \angle BCA + \angle CAB = 180°$

$50° + \angle BCA + 50° = 180°$

$\Rightarrow \angle BCA = 80°$.

Congruence

Two geometrical figures are congruent if they have the same shape and the same size.

1. Line segments which have the same length are congruent.
2. Circles which have equal radii are congruent.

Congruence of Triangles

The three angles of a triangle determine its shape and its three sides determine its size. If the three angles and the three sides of a triangle are respectively equal to the corresponding angles and sides of another triangle, then the two triangles are congruent. However, to conclude that the triangles are congruent, it is not necessary that all the six corresponding elements of the two triangles are given to be congruent.

Based on the study and experiments, the following results can be used to establish the congruence of two triangles.

1. **S.S.S. congruence property:** By the side-side-side congruence property, two triangles are congruent if the three sides of a triangle are equal to the corresponding sides of the other triangle, i.e., the three angles of the first triangle are bound to be equal to the corresponding angles of the second triangle.

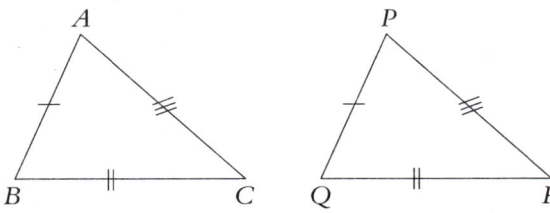

In $\triangle ABC$ and $\triangle PQR$, if $AB = PQ$, $BC = QR$ and $CA = RP$, then $\triangle ABC$ is congruent to $\triangle PQR$.

We write this as $\triangle ABC \cong \triangle PQR$.

$\therefore \angle A = \angle P$, $\angle B = \angle Q$ and $\angle C = \angle R$.

2. **S.A.S. Congruence Property:** By the side-angle-side congruence property, two triangles are congruent if the two sides and the included angle of one triangle are equal to the corresponding sides and the included angle of the other triangle, i.e., the third side and the other two angles of the first triangle are bound to be equal to the corresponding side and angles of the second triangle.

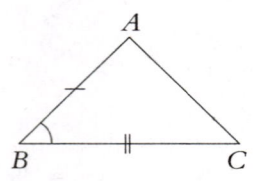

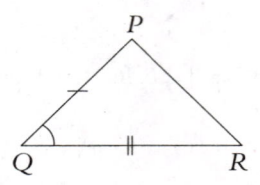

In $\triangle ABC$ and $\triangle PQR$, if $AB = PQ$, $BC = QR$ and $\angle ABC = \angle PQR$, then $\triangle ABC \cong \triangle PQR$.

$\therefore AC = PR$ and $\angle A = \angle P$, $\angle C = \angle R$.

3. **A.S.A. congruence property:** By the angle-side-angle congruence property, two triangles are congruent if any two angles and the side included between them of one triangle are equal to the corresponding angles and the included side of the other triangle, i.e., the third angle and the other two sides of the first triangle are bound to be equal to the corresponding angle and sides of the second triangle.

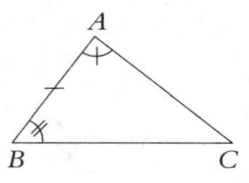

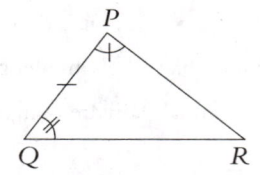

In $\triangle ABC$ and $\triangle PQR$, if $\angle BAC = \angle QPR$ and $\angle ABC = \angle PQR$ and $AB = PQ$, then $\triangle ABC \cong \triangle PQR$.

$\therefore AC = PR$, $BC = QR$ and $\angle C = \angle R$.

4. **R.H.S. congruence property:** The right angle-hypotenuse-side congruence property states that two right triangles are congruent if the hypotenuse and one side of one triangle are equal to the hypotenuse and the corresponding side of the other triangle, i.e., the third side and the other two angles of the first triangle are bound to be equal to the corresponding side and angles of the second triangle.

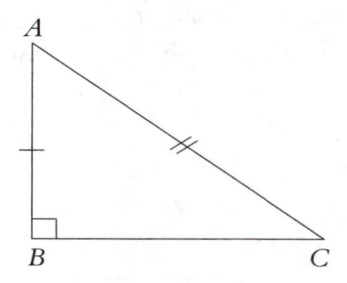

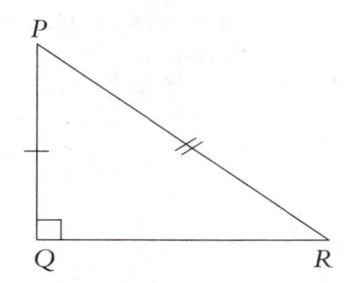

In $\triangle ABC$ and $\triangle PQR$, if $AB = PQ$, $AC = PR$ and $\angle ABC = \angle PQR$ ($= 90°$), then $\triangle ABC \cong \triangle PQR$.

$\therefore BC = QR$, $\angle A = \angle P$ and $\angle C = \angle R$.

Note If two triangles are congruent, then their perimeters are equal and areas are equal.

Similar figures: Geometrical figures which have exactly the same shape but not necessarily the same size are known as similar figures.

Note Two congruent figures are always similar but its converse need not be true.

EXAMPLE 5.6

The ratio of the product of the sides of an equilateral triangle to its perimeter is equal to the ratio of the product of the sides of another equilateral triangle to its perimeter. Then the triangles are

SOLUTION

Let the sides of the triangles be x and y.

Given $\dfrac{x^3}{3x} = \dfrac{y^3}{3y}$

$x^2 = y^2$

$x = y$

∴ Both the triangles are congruent.

∴ They are similar.

Concurrent Lines in Triangles

1. **Median:** A line segment joining the mid-point of a side and its opposite vertex is a median of the triangle.

   In the adjacent triangle, D is the mid-point of side BC and AD is the median.

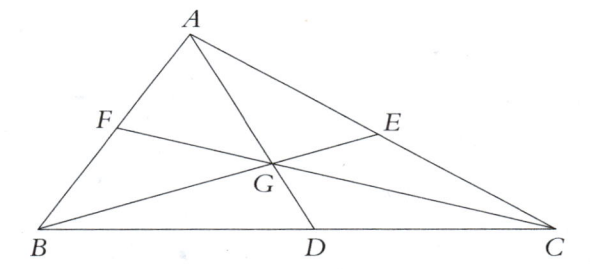

   For each side of the triangle there is a corresponding median.

   $\overline{AD}$ is the median to $\overline{BC}$, $\overline{BE}$ is the median to $\overline{AC}$ and $\overline{CF}$ is the median to $\overline{AB}$.

   The medians of the triangle are concurrent and the point of concurrence of the medians is called the **centroid**. It is denoted by G.

   Notes (1) The centroid divides each median in the ratio 2:1. The smaller part is closer to the side to which the median is drawn.

   ∴ $AG:GD = 2:1$

   (2) A median divides the triangle into two triangles of equal areas.

   (3) The three medians divide the triangle into six triangles of equal area.

2. **Perpendicular bisector:** The perpendicular bisector of a side bisects it and is perpendicular to it. The perpendicular bisector of a side of a triangle need not pass through the opposite vertex.

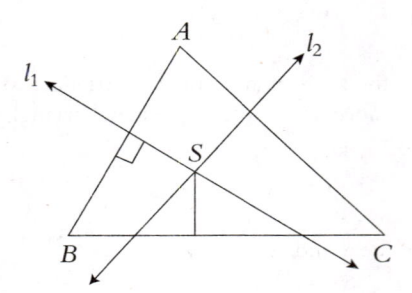

The point of concurrence of perpendicular bisectors is called the circumcentre. It is denoted by S. Since $SA = SB = SC$, taking S as the centre and SA or SB or SC as the radius, a circle can be drawn passing through A, B and C. This circle is called the circumcircle of the triangle. S is the circumcentre and SA (or SB or SC) is the circumradius of the triangle.

Notes (1) In a right-angled triangle, the circumcentre is the mid-point of the hypotenuse.

(2) In an acute angled triangle, the circumcentre lies inside the triangle.

(3) In an obtuse angled triangle, the circumcentre lies outside the triangle.

3. **Angle bisector:** The angle bisector of an angle bisects that angle.

For each angle of a triangle, the bisector can be drawn.

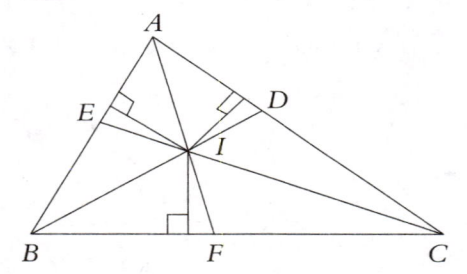

$\overline{BD}$, $\overline{CE}$ and $\overline{AF}$ are the bisectors of $\angle ABC$, $\angle ACB$ and $\angle BAC$ respectively.

The bisectors of the interior angles of a triangle are concurrent. The point of concurrence of the bisectors of the interior angles of triangle is the Incentre of the triangle.

Note For any triangle, the incentre lies inside the triangle.

Altitude: The altitude to a side of a triangle is the perpendicular drawn from its opposite side of the vertex.

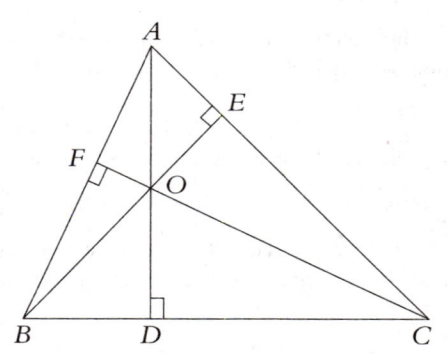

In the above triangle, $\overline{AD}$ is the altitude to side BC.

BE and CF are the altitudes to $\overline{AC}$ and $\overline{AB}$ respectively.

The altitudes of a triangle are concurrent.

The point of concurrence of the altitudes is the orthocentre.

Notes (1) In a right-angled triangle, the orthocentre coincides with the vertex of the right angle.

(2) In an acute-angled triangle, the orthocentre lies inside the triangle.

(3) In an obtuse-angled triangle, the orthocentre lies outside the triangle.

EXAMPLE 5.7

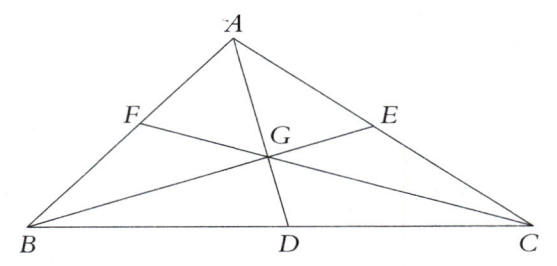

In the above $\triangle ABC$, $\overline{AD}$, $\overline{BE}$ and $\overline{CF}$ are the medians. G is the centroid. What is the ratio of the areas of $\triangle BGD$ and $\triangle GCE$?

SOLUTION

The three medians divide the triangle into six triangles of equal areas. Hence the ratio of the areas of $\triangle BGD$ to that of $\triangle GCE$ is 1:1.

Constructions: The ability to draw accurately is the most important part of the study of geometry.

We shall learn how to construct simple figures using only a straight edge (scale) and a compass.

EXAMPLE 5.8

Draw the perpendicular bisector of the line segment $AB = 6$ cm.

SOLUTION

Steps:

1. Draw a line segment $AB = 6$ cm.
2. With A as the centre and more than half of AB as the radius, draw two arcs on either side of $\overline{AB}$.
3. With B as the centre and with the same radius, draw two arcs which intersect the previous arcs at X and Y.
4. Join XY. XY is the perpendicular bisector of AB.

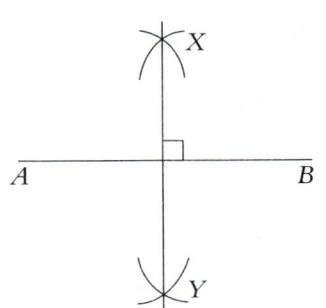

EXAMPLE 5.9

Draw the bisector of ∠AOB = 58°.

SOLUTION

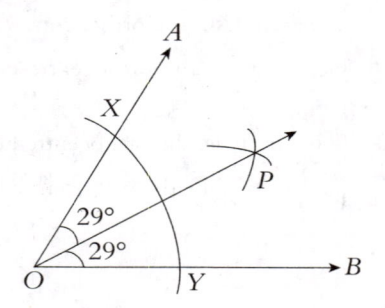

Steps:

1. Draw ∠AOB = 58°.
2. With O as the centre and some convenient radius draw an arc which intersects $\overrightarrow{OA}$ and $\overrightarrow{OB}$ at X and Y respectively.
3. With X as the centre and more than half of XY as radius draw an arc.
4. With Y as the centre and with the same radius draw an arc which cuts the previous arc at P.
5. Join $\overrightarrow{OP}$. $\overrightarrow{OP}$ is the bisector of ∠AOB.
   i.e., ∠POB = ∠POA = 29°.

Construction of triangles: Every triangle has six elements, i.e., three sides and the three angles. We need only three independent elements to construct a triangle. In the following cases we can construct the triangles.

1. All the three sides are given.
2. Two sides and the angle included between them are given.
3. Two angles and a side are given.
4. In the case of a right angled triangle, the hypotenuse and another side are given.

EXAMPLE 5.10

Construct a triangle ABC in which AB = 2.2 cm, BC = 1.9 cm and ∠B = 54°.

SOLUTION

Steps:

1. Draw a line segment BC = 1.9 cm.
2. Draw $\overrightarrow{BX}$ which makes an angle 54° with $\overline{BC}$.
3. Mark point A on $\overrightarrow{BX}$ such that AB = 2.2 cm.
4. Join $\overline{AC}$. ABC is the required triangle.

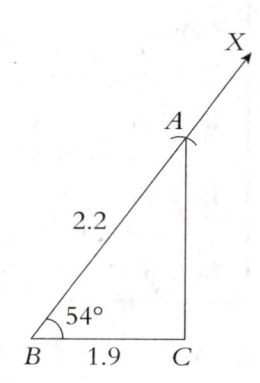

Geometry

EXAMPLE 5.11

Construct a triangle PQR in which $PQ = 2$ cm, $\angle P = 45°$ and $\angle Q = 105°$.

SOLUTION

Steps:

1. Draw a line segment $PQ = 2$ cm.
2. Draw $\overrightarrow{PX}$ which makes an angle 45° with $\overline{PQ}$.
3. Draw $\overrightarrow{QY}$ which makes an angle 105° with $\overline{QP}$.
4. Mark the intersecting point of $\overrightarrow{PX}$ and $\overrightarrow{QY}$ as R.
5. PQR is the required triangle.

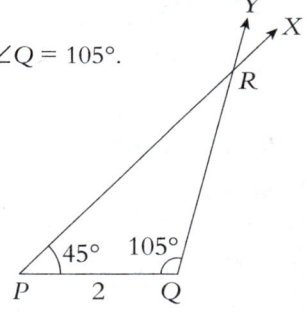

Construction of a circumcircle: A circle which passes through the three vertices of a triangle is its circumcircle. We shall learn the construction of a circumcircle through the following example.

EXAMPLE 5.12

Construct a circumcircle for the triangle ABC in which $AB = 3$ cm, $BC = 3.5$ cm and $AC = 3.5$ cm.

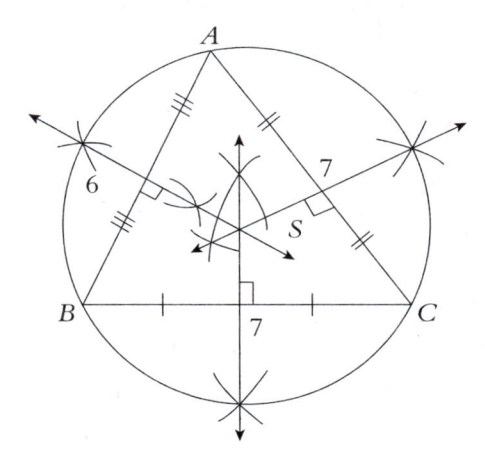

SOLUTION

Steps:

1. Construct the triangle ABC with $AB = 3$ cm, $BC = 3.5$ cm and $AC = 3.5$ cm.
2. Draw the perpendicular bisectors of $\overline{AB}$, $\overline{BC}$ and $\overline{AC}$ and mark their point of concurrence as S.
3. With S as the centre and with SA (or) SB (or) SC as radius, draw a circle. This is the required circumcircle.

Construction of an incircle: The circle which touches the three sides of a triangle internally is called its incircle. We shall learn the construction of an incircle through the following example.

Chapter 5

EXAMPLE 5.13

Construct the incircle for the triangle XYZ in which $\angle Y = 90°$, $XZ = 4.5$ cm and $YZ = 3$ cm.

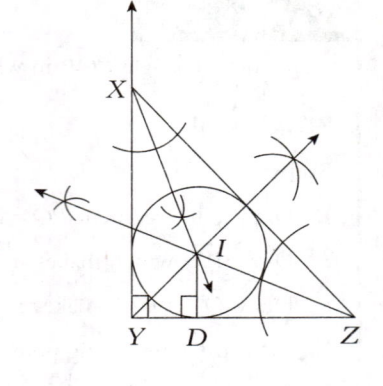

SOLUTION

Steps:

1. Construct the triangle XYZ with $\angle Y = 90°$, $XZ = 4.5$ cm and $YZ = 3$ cm.
2. Draw the bisectors of $\angle X$, $\angle Y$ and $\angle Z$ and mark their point of concurrence as I.
3. Mark the foot of the perpendicular from I on YZ as D.
4. Draw a circle with I as the centre and ID as radius. This is the required incircle.

Construction of an excircle: The circle which touches the three sides of a triangle externally is called its excircle. We shall learn the construction of an excirlce through the following example.

EXAMPLE 5.14

Construct the excircle for the triangle ABC opposite to the vertex A in which $AB = AC = 5$ cm and $BC = 4$ cm.

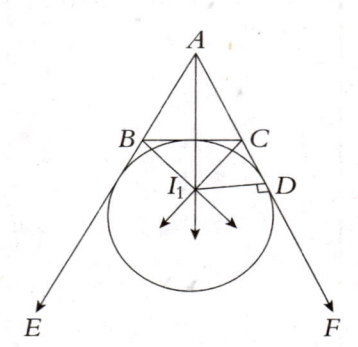

SOLUTION

Steps:

1. Construct the triangle ABC with $AB = AC = 5$ cm and $BC = 4$ cm.
2. Draw the bisector of $\angle A$.
3. Draw the external bisectors of $\angle B$ and $\angle C$.
4. Mark the point of concurrence of the bisectors as I_1.
5. Mark the foot of the perpendicular from I_1 on produced AC as D.
6. Draw a circle with I_1 as the centre and $I_1 D$ as radius. This is the required excircle.

Quadrilaterals

A simple closed figure bounded by four line segments is called a quadrilateral. The different parts of the quadrilateral PQRS are mentioned below.

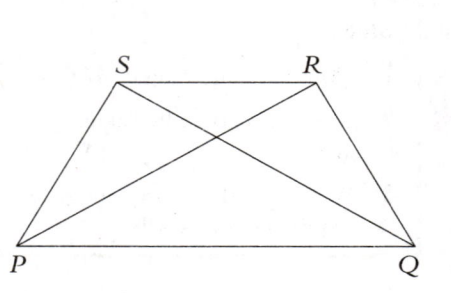

1. Four vertices: P, Q, R and S.
2. Four sides: $\overline{PQ}$, $\overline{QR}$, $\overline{RS}$ and $\overline{PS}$.
3. Four angles: $\angle P$, $\angle Q$, $\angle R$ and $\angle S$.
4. Two diagonals: $\overline{PR}$ and $\overline{QS}$.
5. Four pairs of adjacent sides: ($\overline{PQ}$, $\overline{QR}$), ($\overline{QR}$, $\overline{RS}$), ($\overline{RS}$, $\overline{SP}$) and ($\overline{SP}$, $\overline{PQ}$).

6. Two pairs of opposite sides: ($\overline{PQ}$, $\overline{RS}$) and ($\overline{QR}$, $\overline{PS}$).
7. Four pairs of adjacent angles: (∠P, ∠Q), (∠Q, ∠R), ∠R, ∠S) and (∠S, ∠P).
8. Two pairs of opposite angles: (∠P, ∠R) and (∠Q, ∠S).

   Note The sum of the four angles of a quadrilateral is 360°.

Different Types of Quadrilaterals

1. **Trapezium:** In a quadrilateral, if one pair of opposite sides is parallel to each other, then it is called a trapezium.

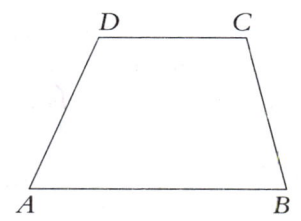

   In the figure above, $\overline{AB}$ || $\overline{CD}$. Hence ABCD is a trapezium.

2. **Isosceles trapezium:** In a trapezium, if the non-parallel sides are equal, then it is called an isosceles trapezium.

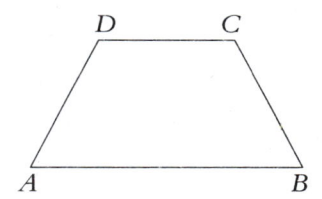

   In the figure given, $\overline{AB}$ || $\overline{CD}$ and BC = AD.

   Hence, ABCD is an isosceles trapezium.

3. **Parallelogram:** In a quadrilateral, if both the pairs of opposite sides are parallel, then it is called a parallelogram.

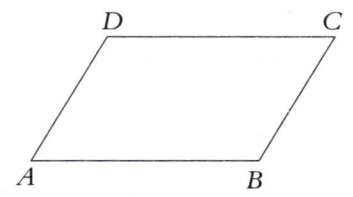

   In the figure given, $\overline{AB}$ || $\overline{CD}$ and $\overline{BC}$ || $\overline{AD}$.

   Hence ABCD is a parallelogram.

   Notes (1) In a parallelogram, opposite sides are equal.

   (2) In a parallelogram, the diagonals need not be equal, but they bisect each other.

4. **Rectangle:** In a parallelogram, if one angle is a right angle, then it is called a rectangle.

   In the given figure, $\angle A = \angle B = \angle C = \angle D = 90°$, $AB = CD$ and $BC = AD$.

   Hence $ABCD$ is a rectangle.

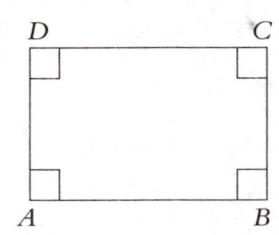

   Note In a rectangle, the diagonals are equal.

5. **Rhombus:** In a parallelogram, if all the sides are equal, then it is called a rhombus.

   In the given figure, $AB = BC = CD = AD$. Hence $ABCD$ is a rhombus.

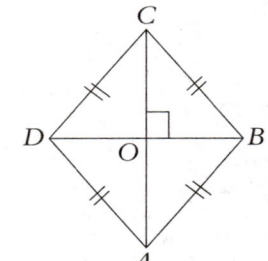

   Notes (1) In a rhombus, the diagonals need not be equal.

   (2) In a rhombus, the diagonals bisect each other at right angles, i.e., $AO = OC$, $BO = OD$ and $AC \perp DB$.

   (3) $AB^2 = OA^2 + OB^2$ (Pythagoras theorem).

6. **Square:** In a rhombus, if one angle is a right angle, then it is called a square.

   (OR)

   In a rectangle, if all the sides are equal, then it is called a square.

   In the given figure, $AB = BC = CD = DA$ and $\angle A = \angle B = \angle C = \angle D = 90°$.

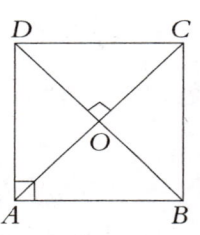

   Hence $ABCD$ is a square.

   Notes (1) In a square, the diagonals bisect each other at right angles.

   (2) In a square, the diagonals are equal.

7. **Kite:** In a quadrilateral, if two pairs of adjacent sides are equal, then it is called a kite.

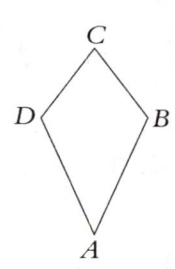

   In the figure above $AB = AD$ and $BC = CD$. Hence, $ABCD$ is a kite.

EXAMPLE 5.15

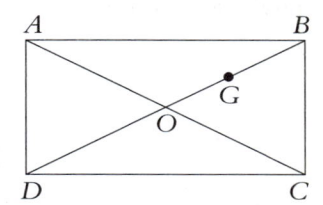

In the figure above, *ABCD* is a rectangle and *G* is the centroid of the triangle *ABC*. If *BG* = 4 cm, then find the length of *AC*.

SOLUTION

We know that, the centroid of a triangle divides each median in the ratio of 2:1 from the vertex.

Given, *BG* = 4 cm.

$OG = \dfrac{1}{2}$ (4 cm) = 2 cm

⇒ *BO* = *BG* + *OG* = 6 cm
⇒ *BO* = *OD* (∵ Diagonals bisect each other)
⇒ *OD* = 6 cm.
BD = *BO* + *OD* = 6 cm + 6 cm = 12 cm
But *AC* = *BD* (∵ Diagonals are equal)
⇒ *AC* = 12 cm.

EXAMPLE 5.16

ABCD is a kite in which *AB* = *AD* and *CB* = *CD*. If ∠*ABD* = 25° and ∠*BDC* = 35°, then find ∠*A* − ∠*C*.

SOLUTION

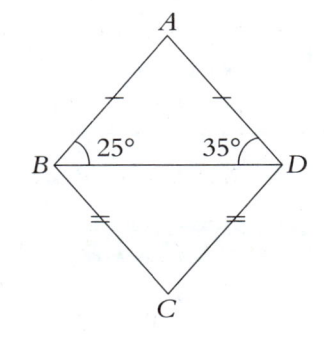

Given ∠*ABD* = 25° and ∠*BDC* = 35°.
In Δ*BCD*, *BC* = *CD*.
⇒ ∠*CDB* = ∠*CBD* ⇒ ∠*CBD* = 35°
⇒ ∠*C* = 180° − (35° + 35°) = 110°
AB = *AD*
And also given,
⇒ ∠*ABD* = ∠*ADB* ⇒ ∠*ADB* = 25°
∠*A* = 180° − (25° + 25°) = 130°
Now
∴ ∠*A* − ∠*C* = 130° − 110° = 20°.

Construction of Quadrilaterals

1. **When four sides and one angle are given.**

EXAMPLE 5.17

Construct a quadrilateral ABCD in which AB = 4.2 cm, $\angle A = 80°$, BC = 2.4 cm, CD = 3.3 cm and AD = 2.4 cm.

SOLUTION

Steps:

1. Draw a line segment AB = 4.2 cm.
2. Draw $\angle BAX = 80°$.
3. Mark D on $\overrightarrow{AX}$, such that AD = 2.4 cm.
4. Taking D as the centre and 3.3 cm as the radius, draw an arc and taking 'B' as the centre and 2.4 cm as radius, draw another arc to intersect the previous arc at C.
5. Join CD and BC. ABCD is the required quadrilateral.

2. **When three consecutive sides and two included angles are given.**

EXAMPLE 5.18

Construct a quadrilateral ABCD with AB = 4 cm, BC = 2.8 cm, CD = 4 cm, $\angle B = 75°$ and $\angle C = 105°$.

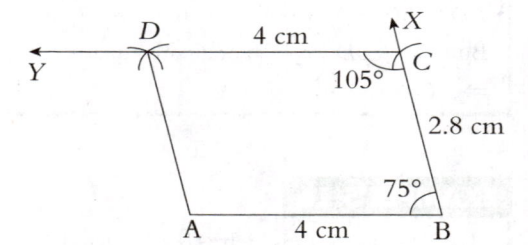

SOLUTION

Steps:

1. Draw a line segment AB = 4 cm.
2. Draw $\overrightarrow{BX}$ such that $\angle ABX = 75°$.
3. With B as centre and a radius of 2.8 cm, draw an arc to cut $\overrightarrow{BX}$ at C.
4. Draw $\overrightarrow{CY}$ which makes an angle 105° with $\overline{BC}$.
5. Mark D on $\overrightarrow{CY}$, such that CD = 4 cm.
6. Join AD. ABCD is the required quadrilateral.

3. **When four sides and one diagonal are given.**

EXAMPLE 5.19

Construct a quadrilateral ABCD in which AB = 4.6 cm, BC = 2.6 cm, CD = 3.5 cm, AD = 2.6 cm and the diagonal AC = 4.9 cm.

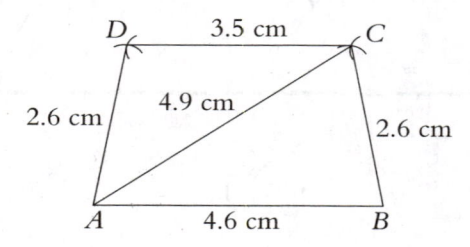

Geometry 5.23

SOLUTION

Steps:

1. Draw a line segment $AB = 4.6$ cm.
2. With A and B as centres, draw two arcs of radii 4.9 cm and 2.6 cm respectively to intersect each other at C.
3. With C and A as centres, draw two arcs of radii 3.5 cm and 2.6 cm respectively to intersect at D.
4. Join BC, CD and AD to form quadrilateral $ABCD$.

   $ABCD$ is the required quadrilateral.

4. To construct a parallelogram, when two consecutive sides and the included angle are given.

EXAMPLE 5.20

Construct a parallelogram $ABCD$, when $AB = 4$ cm, $BC = 2.5$ cm and $\angle B = 100°$.

SOLUTION

Steps:

1. Draw line segment $AB = 4$ cm.
2. Construct line BX such that $\angle ABX = 100°$.
3. Taking B as the centre and the radius = 2.5 cm, cut $\overline{BX}$ at the point C with an arc.
4. Draw two arcs taking C and A as centres and 4 cm and 2.5 cm as radii respectively to intersect at D.
5. Join AD and CD. $ABCD$ is the required parallelogram.

5. Construction of a parallelogram when two adjacent sides and one diagonal are given.

EXAMPLE 5.21

Construct a parallelogram $PQRS$, when $PQ = 3.7$ cm, $QR = 2.3$ cm and $PR = 4.8$ cm.

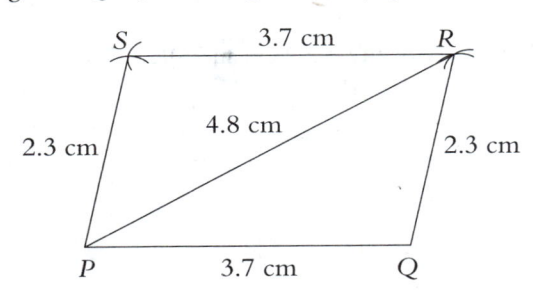

5.24 Chapter 5

SOLUTION

Steps:

1. Draw a line segment $PQ = 3.7$ cm.
2. Draw an arc with P as the centre and a radius of 4.8 cm.
3. With Q as the centre and a radius of 2.3 cm, draw another arc to intersect the previous arc of step 2 at R and join QR.
4. With R as the centre, draw an arc of radius 3.7 cm.
5. With P as the centre, draw another arc of radius 2.3 cm to intersect the arc in step 4 at S. Join RS and PS.

   $PQRS$ is the required parallelogram.

6. **Construction of a parallelogram when both the diagonals and the angle between them are given.**

EXAMPLE 5.22

Construct a parallelogram $PQRS$ with $PR = 3$ cm, $QS = 4.2$ cm and the angle between the diagonals are equal to $75°$.

SOLUTION

Steps:

1. Draw the diagonal $PR = 3$ cm.
2. Bisect PR to mark the midpoint of PR as O.
3. Construct an angle of $75°$ at O, such that $\angle POX = 75°$.
4. Taking O as the centre and radius = $\frac{1}{2}(QS) = \frac{1}{2} \times 4.2$
   = 2.1 cm, draw arcs on the angular line constructed in Step 3 to cut at Q and S.
5. Join PQ, QR, RS and SP to obtain the required parallelogram $PQRS$.

7. **Construction of a rectangle when two adjacent sides are given.**

EXAMPLE 5.23

Construct a rectangle $PQRS$ with $PQ = 5.2$ cm and $QR = 2.6$ cm.

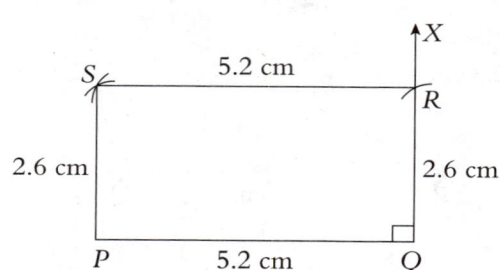

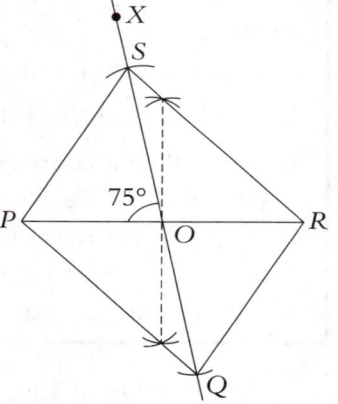

SOLUTION

Steps:

1. Draw $PQ = 5.2$ cm.
2. At Q, construct a right angle, such that $\angle PQX = 90°$.
3. Taking Q as the centre and 2.6 cm as radius, draw an arc to cut $\overrightarrow{QX}$ at R.
4. With R and P as centres draw two arcs with radii 5.2 cm and 2.6 cm respectively to cut each other at S. Join PS and RS. $PQRS$ is the required rectangle.

8. **Construction of a rectangle when a side and a diagonal are given.**

EXAMPLE 5.24

Construct a rectangle $PQRS$ with $PQ = 5.3$ cm and diagonal $PR = 5.8$ cm.

SOLUTION

Steps:

1. Draw a line segment $PQ = 5.3$ cm.
2. At Q, construct $\angle PQX = 90°$.
3. Taking P as the centre and 5.8 cm as radius, draw an arc to cut $\overrightarrow{QX}$ at R.
4. With R and Q as centres, 5.3 cm and 5.8 cm respectively as radii, draw two arcs to intersect each other at S.
5. Join RS and PS to form the required rectangle $PQRS$.

9. **Construction of a rectangle when one diagonal and the angle between two diagonals are given.**

EXAMPLE 5.25

Construct a rectangle $PQRS$ such that $PR = 5.2$ cm and the angle between the diagonals is $50°$.

SOLUTION

Steps:

1. Draw a line segment $PR = 5.2$ cm.
2. Mark the mid-point of PR as O.
3. Draw $\overrightarrow{XY}$ which makes an angle of $50°$ with $\overrightarrow{PR}$ at the point O.
4. With O as the centre and with radius equal to $\frac{1}{2}(PR) = 2.6$ cm cut $\overrightarrow{OX}$ and $\overrightarrow{OY}$ at S and Q respectively.
5. Join PQ, QR, RS and PS to form the required rectangle $PQRS$.

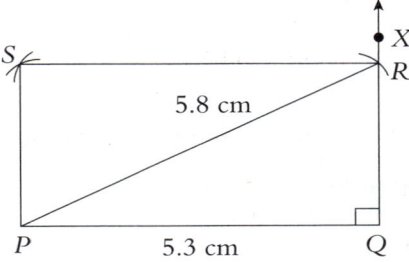

10. Construction of a square when one side is given.

EXAMPLE 5.26

Construct a square of side 3 cm.

SOLUTION

Steps:

1. Draw a line segment $PQ = 3$ cm.
2. Construct $\angle PQX = 90°$.
3. Mark the point R on $\overrightarrow{QX}$ such that $QR = 3$ cm.
4. With R and P as centres and with radii of 3 cm each, draw two arcs to intersect each other at S.
5. Join PS and RS to form the required square $PQRS$.

11. Construction of a square when a diagonal is given.

EXAMPLE 5.27

Construct a square with its diagonal as 4 cm.

SOLUTION

Steps:

1. Draw a line segment $PR = 4$ cm.
2. Draw perpendicular bisector XY of $\overline{PR}$ to bisect $\overline{PR}$ at O.
3. Mark the points Q and S on $\overline{OY}$ and $\overline{OX}$ respectively such that $OQ = OS = 2$ cm.
4. Join PS, RS, PQ and QR to form the required square $PQRS$.

12. Construction of a rhombus when one side and one angle are given.

EXAMPLE 5.28

Construct a rhombus $PQRS$ with $PQ = 3.6$ cm and $\angle P = 50°$.

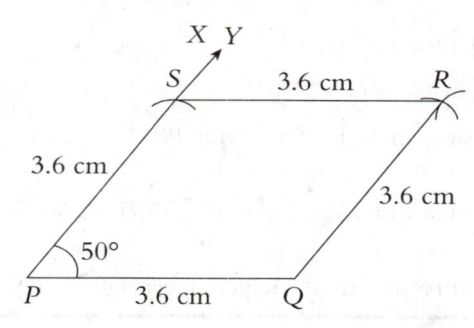

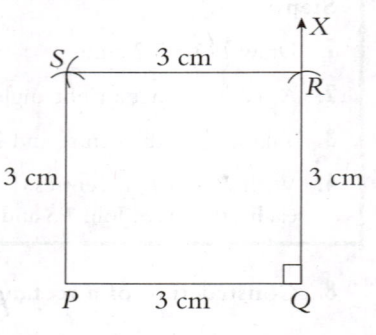

SOLUTION

Steps:

1. Draw a line segment $PQ = 3.6$ cm.
2. Construct $\angle QPX = 50°$.
3. Taking P as the centre and a radius equal to 3.6 cm, draw an arc to cut $\overline{PX}$ at S such that $PS = 3.6$ cm.
4. From Q and S, draw two arcs with radii 3.6 cm each to meet each other at R.
5. Join QR and SR to form the required rhombus $PQRS$.

13. Construction of rhombus when one side and one diagonal are given.

EXAMPLE 5.29

Construct a rhombus $PQRS$ such that $PQ = 3.2$ cm and $PR = 4.2$ cm.

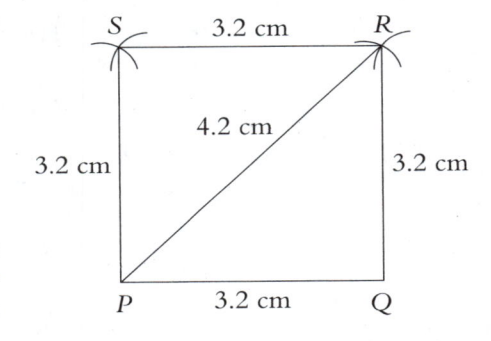

SOLUTION

Steps:

1. Draw a line segment $PQ = 3.2$ cm.
2. Taking P as the centre and radius equal to 4.2 cm, draw an arc and taking Q as the centre, radius as 3.2 cm draw another arc to cut the previous arc at R.
3. With R and P as centres and the radii equal to 3.2 cm each, draw two arcs to meet at S.
4. Join PS, RS and QR to form rhombus $PQRS$.

14. Construction of a rhombus when both the diagonals are given.

EXAMPLE 5.30

Construct a rhombus $PQRS$ with diagonal $PR = 3.4$ cm and $QS = 3.6$ cm.

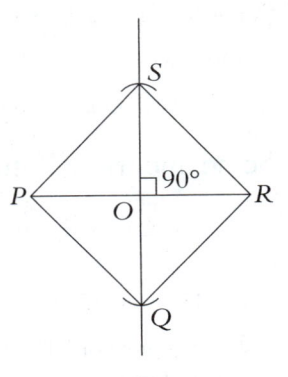

SOLUTION

Steps:

1. Draw a line segment $PR = 3.4$ cm.
2. Bisect PR and mark its midpoint as O.
3. With O as the centre and radii 1.8 cm each, draw arcs on either sides of PR to cut perpendicular bisector of PR at Q and S.
4. Join PS, PQ, QR and RS to form the required rhombus $PQRS$.

Polygons

A simple closed plane figure bounded by three or more line segments is called a polygon.

1. Each line segment is called a side of the polygon.
2. The point at which any two adjacent sides intersect is called a vertex of the polygon.

Different polygons, the number of their sides and their names are given in the table below:

Number of Sides	Name of the Polygon	Corresponding Figure
3	Triangle	
4	Quadrilateral	
5	Pentagon	
6	Hexagon	
7	Septagon	
8	Octagon	
9	Nonagon	
10	Decagon	

Polygons can also be classified as

1. Convex polygons and
2. Concave polygons.

Convex Polygon and Concave Polygon

A polygon in which each interior angle is less than 180° is called a convex polygon.

Otherwise it is called a concave polygon.

Note Unless otherwise mentioned, we refer to convex polygons simply as polygons.

Some Important Results on Polygons

1. The sum of the interior angles of polygon is $(2n - 4)$ 90° or $(2n - 4)$ right angles.
2. The sum of all the exterior angles of a polygon is 360°.
3. In a polygon, if all the sides are equal and all the angles are equal, then it is called a **regular polygon**. In case, all the sides are not equal, then the polygon is called an **irregular polygon**.

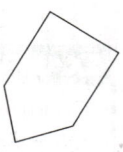

Geometry

4. Each interior angle of a regular polygon of n sides is $\left(\dfrac{(2n-4)90}{n}\right)^\circ$.

5. Each exterior angle of a regular polygon of n sides is $\left(\dfrac{360}{n}\right)^\circ$.

EXAMPLE 5.31

The sum of the interior angles in a polygon is 1980°. Find the number of sides of the polygon.

SOLUTION

Let the number of sides of the polygon be N.

∴ $[180(N-2)]° = 1980°$

⇒ $N - 2 = 11 \Rightarrow N = 13$.

EXAMPLE 5.32

Which of the following angles cannot be an interior angle of any convex polygon?

SOLUTION

Interior angle of any convex polygon cannot be more than 180°.

Circle

A circle is a set of all points in a plane which are at a fixed distance from a fixed point.

The fixed point is the centre of the circle and the fixed distance is the radius of the circle.

In the given figure, O is the centre of the circle and OC is a radius of the circle. AB is a diameter of the circle. OA and OB are also the radii of the circle. The diameter is twice the radius.

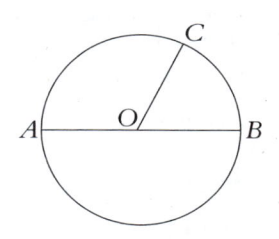

The centre of the circle is generally denoted by O, diameter by d and radius by r.

∴ $d = 2r$.

The perimeter of the circle is called the circumference of the circle. The circumference of the circle is π times the diameter.

In the given circle with centre O, A, B and C are three points in the plane in which the circle lies. We see that $OA < r$, $OB = r$ and $OC > r$. The points O and A are inside the circle. The point B is located on the circle and C is a point outside the circle. The set of all points P in the plane of the circle for which $OP < r$ is called the **interior of the circle** and the set of points Q for which $OQ > r$ is called the **exterior of the circle**.

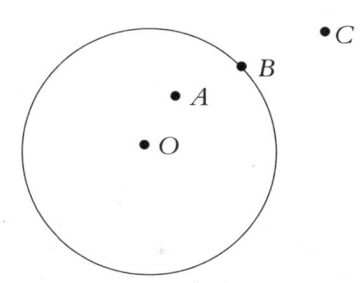

B is a point on the circumference of the circle, $OB = r$.

A is a point in the interior of the circle, $OA < r$.

As C is a point in the exterior of the circle, $OC > r$.

Chord: The line segment joining any two points on a circle is called a chord of the circle.

In the adjacent figure, both $\overline{PQ}$ and $\overline{AB}$ are chords.

AB passes through centre O. Hence, it is a diameter of the circle. A diameter is the longest chord of the circle. It divides the circle into two equal parts.

Secant: A line which intersects a circle in two distinct points is called secant of the circle.

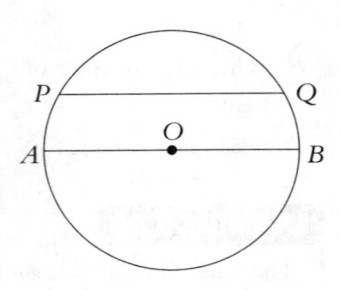

Tangent: A line lying in the plane of the circle that meets the circle in exactly one point is called a tangent to the circle. The point at which it meets the circle is the point of tangency or point of contact.

Arc: A continuous piece of a circle is called an arc of the circle. The end points of a diameter of a circle divide the circle into two equal arcs. Each of these arcs is called semi-circle. The region enclosed by a semicircle and the diameter together including the semicircle and diameter is called a semi-circular region.

If the length of an arc is less than that of a semicircle, it is a **minor arc**. If it is more, it is a **major arc**.

Circumference: The whole arc of the circle is called its circumference.

Angle subtended by a pair of points: The angle subtended by two points A and B at the point P is the angle APB. The angle subtended by an arc or a chord is the angle subtended by the endpoints of the arc or chord.

Segment of a circle: The part of the circular region bounded by an arc AB and the chord AB, including the arc and chord is called a segment of the circle. If the arc is minor, then the segment is also minor. If the arc is major, then the segment is also major.

Sector of a circle: In a circle with centre O, the region enclosed by an arc AB and the bounding radii, OA and OB is called the sector AOB. If the arc is minor, then the sector is minor. If the arc is major, then the sector is major.

Symmetry: Let us examine the following figures drawn on a rectangular piece of paper.

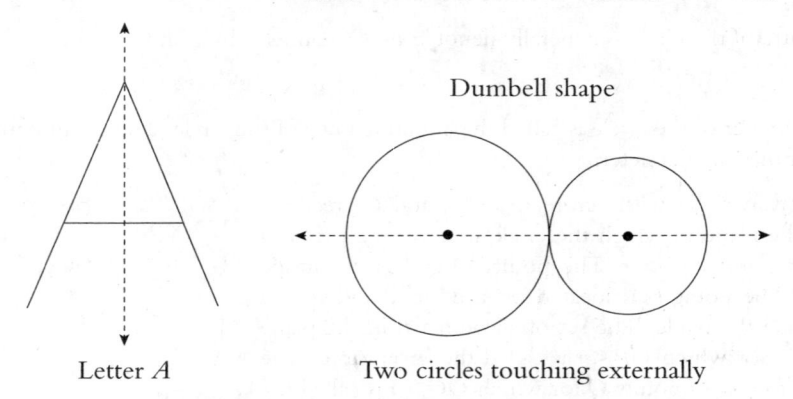

Letter A

Dumbell shape

Two circles touching externally

What do you infer? We can observe that when these figures are folded about the dotted lines, the two parts on either side of the dotted lines coincide. This property of geometrical figures is called **symmetry**.

Geometry

In this chapter, we shall discuss two basic types of symmetry—**line symmetry** and **point symmetry**. Then we shall see how to obtain the image of a point, a line segment and an angle about a line.

Line symmetry: Trace a geometrical figure on a rectangular piece of paper as shown below:

Now fold the paper along the dotted line. You will find that the two parts of the figure on either sides of the line coincide. Thus the line divides the figure into two identical parts .In this case, we say that the figure is symmetrical about the dotted line or 'Line Symmetric'. Also, the dotted line is called 'the Line of Symmetry' or 'the Axis of Symmetry'.

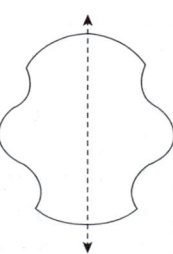

So, a geometrical figure is said to be line symmetric or symmetrical about a line if there exists at least one line in the figure such that the parts of the figure on either sides of the line coincide when it is folded about the line.

EXAMPLE 5.33

Which of the following has only 2 lines of symmetry?

SOLUTION

Rhombus has only 2 lines of symmetry, they are two diagonals.

Point symmetry: Trace the H on a rectangular piece of paper as shown below.

Let P be the mid-point of the inclined line in the figure. Now draw a line segment through the point P touching the two vertical strokes of N. We find that the point P divides the line segment into two equal parts. Thus, every line segment drawn through the point P and touching the vertical strokes of H is bisected at the point P. Also, when we rotate the letter H about the point P through an angle of 180°, we find that it coincides exactly with the initial position. This property of geometrical figures is called 'Point Symmetry'. In this case, we say that the point P is the point of symmetry of the figure.

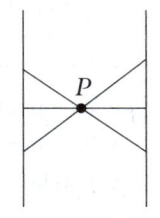

So, a geometrical figure is said to have symmetry about a point P if every line segment through the point P touching the boundary of the figure is bisected at the point P.

OR

A geometrical figure is said to have point symmetry if the figure does not change when rotated through an angle of 180°, about the point P.

Here, point P is called the centre of symmetry.

EXAMPLE 5.34

Which of the following is point symmetric?

SOLUTION

Rectangle is point symmetric.

Points to Remember

- If $x + y = 90°$, then x and y are complementary angles.
- If $a + b = 180°$, then a and b are supplementary angles.
- Linear pair angles are supplementary but supplementary angles need not be linear pair angles.
- The sum of all the three angles in a triangle is 180°.
- An exterior angle of triangle is equal to its interior opposite angles.
- The sum of any two sides of a triangle is greater than the third side.
- The difference of any two sides of triangle is lesser than the third side.
- In a triangle, the angle opposite to longer side is greater.
- In a triangle, the angle opposite to shorter side is smaller.
- In a triangle, if any two sides are equal, then the angles opposite to them are equal.
- In a right triangle, the square of hypotenuse is equal to the sum of the squares of the perpendicular sides.
- The point of concurrence of perpendicular bisectors of the sides of a triangle is the circumcentre of the triangle.
- The point of concurrence of angle bisectors of a triangle is the incentre of the triangle.
- The point of concurrence of altitudes of a triangle is the orthocentre of the triangle.
- The point of concurrence of medians of a triangle is the centroid of the triangle.
- The centroid of a triangle divides its each median in the ratio of 2:1 from its corresponding vertex.
- The point which is equidistant from the vertices of a triangle is its circumcentre.
- The point which is equidistant from the sides of a triangle is its incentre or excentre.
- The number of independent measures required to construct a triangle is three.
- Circles of equal radii are congruent.
- The sum of all the four angles of a quadrilateral is 360°.
- The opposite angles of a parallelogram are equal.
- Adjacent angles of a parallelogram are supplementary.
- The diagonals of a parallelogram bisect to each other.
- The diagonals of a rectangle are equal.
- The diagonals of a rhombus are perpendicular to each other.
- Each diagonal of a parallelogram divides it into two congruent triangles.
- The number of independent measures required to construct a parallelogram is three.
- The number of independent measures required to construct a rectangle or a rhombus is two.
- The number of independent measures required to construct a square is one.
- The number of independent measures required to construct a quadrilateral is five.
- The sum of interior angles of an n-sided polygon is $(2n - 4)\,90°$.
- The sum of exterior angles of a polygon is 360°.
- Each interior angle of an n-sided regular polygon is $\dfrac{(2n-4)90°}{n}$.
- Each exterior angle of an n-sided regular polygon is $\dfrac{360°}{n}$.

TEST YOUR CONCEPTS

Very Short Answer Type Questions

Directions for questions 1 to 6: State whether the following statements are true or false.

1. The point of concurrence of medians of a triangle is called centroid.
2. The point of concurrence of altitudes of a triangle is called orthocentre.
3. Centroid of a triangle divides its median in the ratio of 1:2 from the vertex.
4. The number of independent measurements required to construct a circle is two.
5. The number of independent measurements required to construct an isosceles trapezium is three.
6. Angle made by a longest chord of circle at its centre is 180°.

Directions for questions 7 to 18: Fill in the blanks.

7. The point of concurrence of perpendicular bisectors of the sides of a triangle is called _____.
8. Incentre of a triangle is _____ from all its sides.
9. Each angle in an equilateral triangle is _____.
10. In an isosceles triangle one of its equal angles is 40°. Then the greatest angle is _____.
11. In a $\triangle ABC$, if the exterior angle of C is 135°, $\angle A + \angle B =$ _____.
12. If in a $\triangle ABC$, incentre, circumcentre and orthocentre coincide each other, then $\angle A + \angle B =$ _____.
13. The point which is equidistant from all the points on the circumference of a circle is called _____.
14. Circumference of a circle is _____ times to its radius.
15. Number of independent measurements required to construct a triangle is _____.
16. ABCD is a parallelogram. If $\angle A + \angle C = 120°$, then $\angle B + \angle D =$ ___.
17. If all the sides are equal, then the quadrilateral must be _____.
18. A line which intersects a circle at two distinct points is called a _____ of the circle.

Directions for questions 19 to 25: Select the correct alternative from the given choices.

19. The number of lines of symmetry of a square is _____.
    (a) 2
    (b) 3
    (c) 4
    (d) Infinite

20. The number of lines of symmetry of a rectangle is _____.
    (a) 2
    (b) 3
    (c) 4
    (d) 4

21. The sum of an angle and one third of its supplementary angle is 90°. Find the angle.
    (a) 135°
    (b) 120°
    (c) 60°
    (d) 45°

22. One pair of opposite angles of a parallelogram is $(2x - 50°, x + 20°)$. Then the parallelogram necessarily is
    (a) a rhombus
    (b) a square
    (c) a rectangle
    (d) None of these

23.

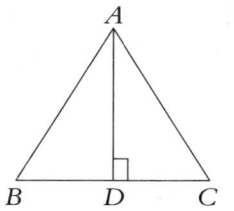

    In the figure above, ABC is a triangle in which BC = 10 cm and AC = 13 cm. If AD is the perpendicular bisector of BC, then find the length of AD.
    (a) 12 cm
    (b) 13 cm
    (c) 10 cm
    (d) 5 cm

5.34 Chapter 5

24. Which of the following is the set of measures of the sides a triangle?
    (a) 8 cm, 4 cm, 20 cm
    (b) 9 cm, 17 cm, 25 cm
    (c) 11 cm, 16 cm, 28 cm
    (d) None of these

25. In which of the following cases, a right triangle cannot be constructed?
    (a) 12 cm, 5 cm, 13 cm
    (b) 8 cm, 6 cm, 10 cm
    (c) 5 cm, 9 cm, 11 cm
    (d) None of these

Short Answer Type Questions

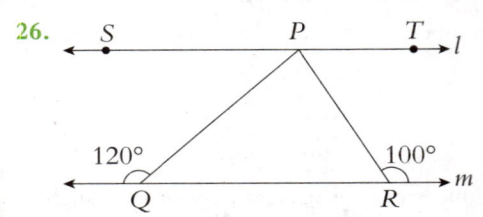

26. In the figure above, if $l // m$, then find $\angle QPS + \angle RPT$.

27. If the supplementary angle of x is 4 times its complementary angle, then find x.

28. Which of the following is/are not Pythagorean triplet(s)?
    (a) 3, 4, 5
    (b) 8, 15, 17
    (c) 7, 24, 25
    (d) 13, 26, 29

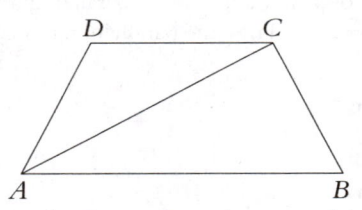

29. In the figure above (not to scale), $ABCD$ is a trapezium in which $AB // DC$. $\angle ACB = 70°$ and $\angle ACD = 30°$. Find $\angle ABC$.

30. Two angles of a triangle are 72° and 38°. Find the third angle.

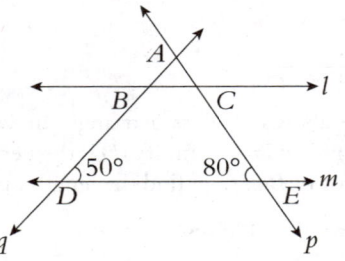

31. In the figure above, if $l // m$, then what type of a triangle is ABC?

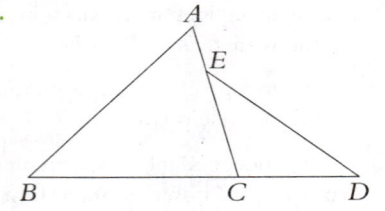

32. In the figure above, $BC = AC$, $CD = CE$. If $\angle ABC = 50°$, then find $\angle CED$.

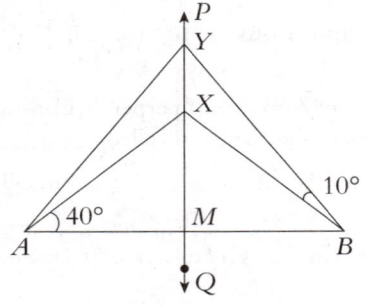

33. In the figure above, $\overrightarrow{PQ}$ is the perpendicular bisector of $\overline{AB}$. If $\angle XAB = 40°$, $\angle XBY = 10°$, then find $\angle AYX$.

34. In a $\triangle ABC$, $\angle B = 90°$ and $AC = 8\sqrt{2}$. If $AB = BC$, then find AB.

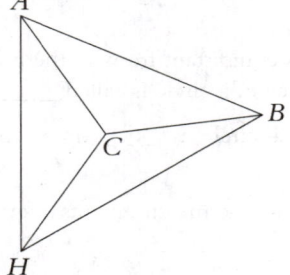

35. In the figure above (not to scale), $\triangle ACB \cong \triangle ACH \cong \triangle BCH$. Find $\angle BCH$.

36.

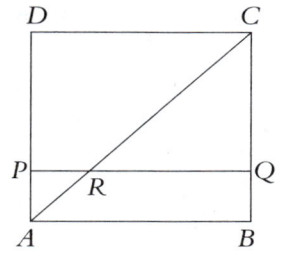

In the figure above, *ABCD* is a square and *PQCD* is a rectangle. Find ∠*PRC*.

37.

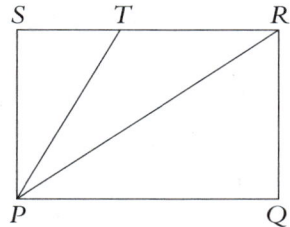

In the figure above, *PQRS* is a square. ∠*PTR* = 110°, then find ∠*TPS*.

38.

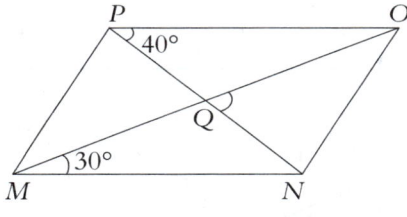

In the figure above, *MNOP* is a parallelogram, diagonals *MO* and *PN* intersect at *Q*, ∠*OPQ* = 40° and ∠*OMN* = 30°. Find ∠*OQN*.

39. In a triangle *ABC*, *AB* = *BC* and ∠*A* = 60°. Find ∠*B*.

40. The angles of quadrilateral are $x - 5°$, x, $x + 5°$ and $x + 10°$. Find the smallest angle of the quadrilateral.

41. Draw all the possible lines of symmetry of the letter *H*.

42. Draw all the possible lines of symmetry of an equilateral triangle.

43.

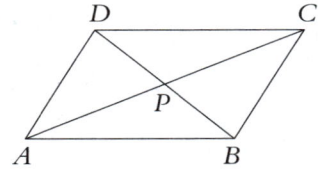

In the figure above, *ABCD* is a parallelogram and if *AC* = 30 cm and *BD* = 20 cm, find *CP* + *DP*.

44. In an *n*-sided regular polygon, each exterior angle is 72°. Find the sum of all the interior angles of the polygon.

45. In the figure given below, *AB* = *AC* and *BC* extended to the point *D*. Find $y - x$.

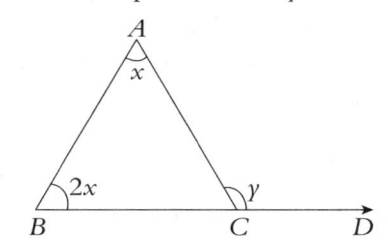

46.

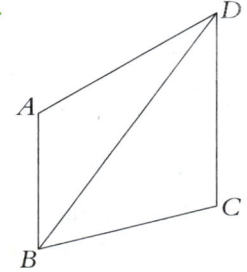

In the above figure, *AB* = *BC* = 8 cm and *AD* = *CD* = 10 cm, which axiom best proves the congruence of △*ABD* and △*CBD*?

47.

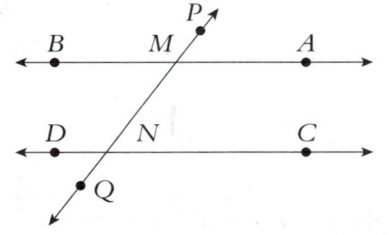

In the figure above, $\overrightarrow{BA}$ is parallel to $\overrightarrow{DC}$ and $\overrightarrow{PQ}$ is a transversal of $\overrightarrow{BA}$ and $\overrightarrow{DC}$. If ∠*PMA* = 70° and ∠*DNM* = $2x + 30°$, then find the value of *x*.

48.

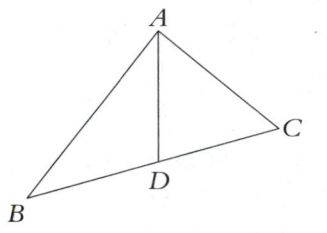

In the figure above, *AD* = *AC* = *BD*. The points *B*, *D* and *C* are collinear. If ∠*CAD* = 80°, then find ∠*DAB*.

5.36 Chapter 5

49.

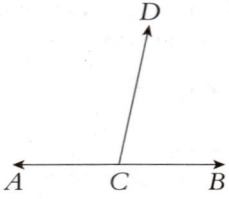

In the above figure, ACB is a straight line and $\angle ACD : \angle DCB = 2:1$. Find $\angle DCB$.

50. In an isosceles right triangle PQR, if $\angle Q = 90°$, then find $\angle PRQ$.

51.

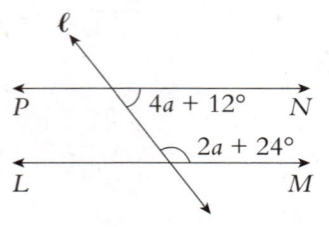

In the given figure, $\overleftrightarrow{LM} \mathbin{//} \overleftrightarrow{PN}$ and the line l is a transversal of $\overleftrightarrow{LM}$ and $\overleftrightarrow{PN}$. Find the value of a.

52. Find the sum of the interior angles of an 8-sided polygon.

53. ABCD is a rhombus, in which the lengths of the diagonals AC and BD are 6 cm and 8 cm. Find the perimeter of the rhombus ABCD.

54. In a triangle ABC, $\angle A = \angle B + \angle C$, then prove that triangle ABC is a right triangle.

55.

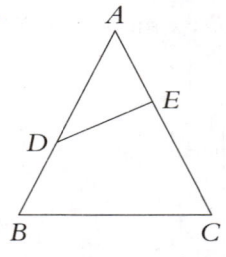

Triangle ABC is an equilateral triangle. If $\angle ADE = 30°$, then find $\angle AED$.

Essay Type Questions

56.

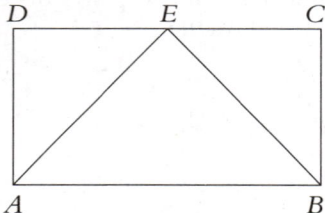

In the figure above (not to scale), ABCD is a rectangle, E is the mid-point of CD. If CD = 24 cm and AD = 5 cm, then find the perimeter of $\triangle ABE$.

57.

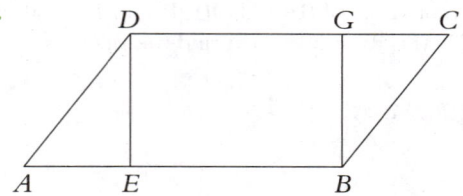

In the figure above, ABCD is a parallelogram, $\overline{DE} \perp \overline{AB}$, $\overline{BG} \perp \overline{CD}$ and EBGD is a square. If BG = 12 cm and BC = 13 cm, then find AB.

58.

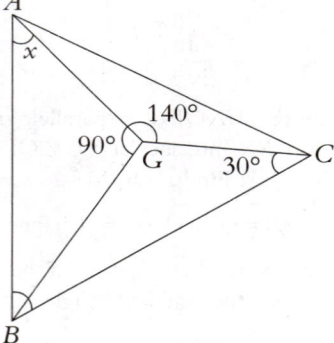

In the figure above, $\angle ABC = 60°$. Find x.

59. In a triangle PQR, if $\angle Q$ is obtuse and S is the orthocentre of $\triangle PQR$, then find the orthocentre of $\triangle PSR$.

60.

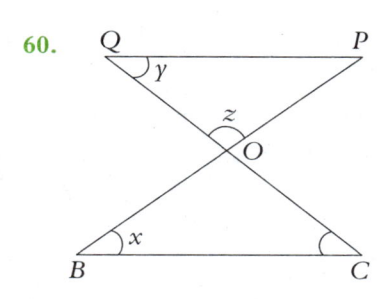

In the figure above, $\overline{BC} \parallel \overline{PQ}$, $\overline{BP}$ and $\overline{CQ}$ intersect at O. If $x + y = 80°$ and $z - y = 55°$, then find x and y.

61.

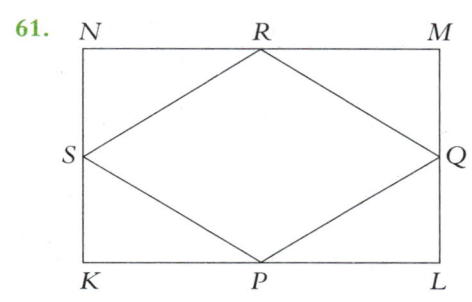

In the figure above, KLMN is a rectangle. P, Q, R and S are the mid-points of $\overline{KL}$, $\overline{LM}$, $\overline{MN}$, and $\overline{NK}$ respectively. If $\angle KPS = 30°$, then find $\angle QRS$.

62.

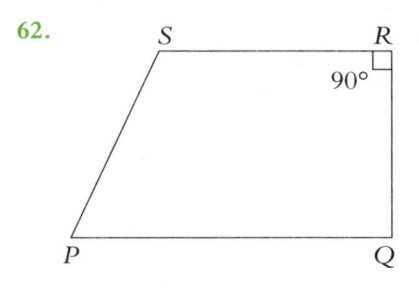

In the figure above, PQRS is a trapezium, PQ // SR, QR = RS and $\angle QRS = 90°$. If QR = 24 cm and PS = 25 cm, then find the length of PQ.

63. The sum of 3 distinct angles is equal to the sum of 2 right angles and the difference between two pairs of the angles is 10°. Find the smallest among the angles.

64. P is an interior point of a square ABCD. Prove that $PA + PB + PC + PD > 2AB$.

65. How can you draw a circle which passes through four vertices of a rectangle? Explain.

66.

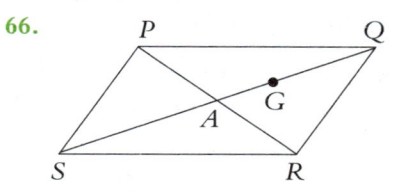

In the figure above, PQRS is a parallelogram and G is the centroid of the triangle PQR. A is the point of intersection of the diagonals PR and SQ. If AG = 3 cm, then find the length of SQ.

67.

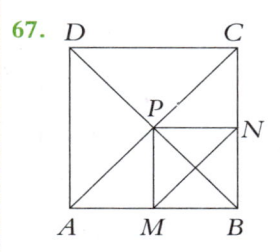

In the figure above, ABCD is a square of side 18 cm and $\overline{PN} \perp \overline{BC}$ and $\overline{PM} \perp \overline{AB}$. Find the length of MN.

68. ABCD is a kite in which AB = AD and CB = CD. If $\angle ABD = 30°$, and $\angle BDC = 40°$, then find $\angle A + \angle C$.

69. Find the complement of an angle whose supplement is 100°.

70.

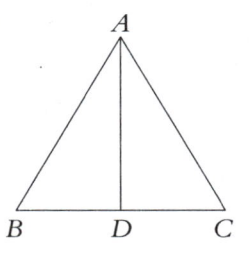

In the given figure, AD is the bisector of $\angle BAC$. Prove that triangles ABD and ADC are congruent, if AB = AC.

CONCEPT APPLICATION
Level 1

Directions for questions 1 to 22: Select the correct alternative from given choices.

1. How many excircles can be drawn for a triangle?
   (a) 3
   (b) 2
   (c) 4
   (d) 1

2. In $\triangle ABC$, if $\angle A = 60°$, $\angle B = 50°$ and $\angle C = 70°$, then find the longest side of the triangle ABC.
   (a) BC
   (b) AB
   (c) AC
   (d) None of these.

3. In a rhombus, if diagonals are equal, then the rhombus necessarily will be
   (a) a rectangle but not square.
   (b) a square.
   (c) a parallelogram but not a square.
   (d) None of these

4. What do you call the triangle whose two of its angles are 40° and 70°?
   (a) Scalene
   (b) Obtuse
   (c) Isosceles
   (d) Equilateral

5. The measures of the sides of a $\triangle PQR$ are integers in cm. If two of its sides are 1 cm each. Find the perimeter of the triangle.
   (a) 3 cm
   (b) 4 cm
   (c) 5 cm
   (d) 6 cm

6. If the angles of a linear pair are equal, then each angle is _____.
   (a) 30°
   (b) 45°
   (c) 60°
   (d) 90°

7. $ABCD$ is rhombus and $\angle BAD = 60°$. The measure of $\angle CAB$ is _____.
   (a) 120°
   (b) 60°
   (c) 30°
   (d) 80°

8. Two complementary angles are in the ratio 2:3. Find the larger angle between them.
   (a) 60°
   (b) 54°
   (c) 66°
   (d) 48°

9. An angle is thrice its supplement. Find it.
   (a) 120°
   (b) 105°
   (c) 135°
   (d) 150°

10.

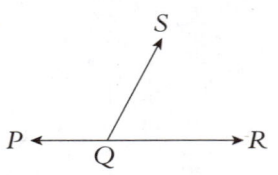

    In the above figure, PQR is a straight line and $\angle PQS : \angle SQR = 7:5$. Find $\angle SQR$.
    (a) 60°
    (b) 62.5°
    (c) 67.5°
    (d) 75°

11.

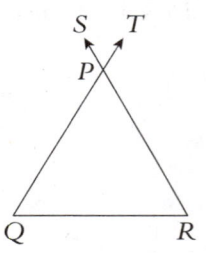

    In the figure above, $\angle SPT = 60°$ and $PQ = PR$. Find $\angle PQR$.
    (a) 50°
    (b) 45°
    (c) 60°
    (d) 55°

12.

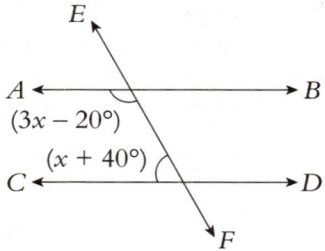

    In the figure above, $\overline{AB} \parallel \overline{CD}$. Find the value of x.

(a) 50°   (b) 45°
(c) 60°   (d) 40°

13.

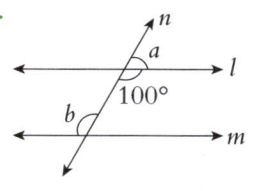

In the figure above, $l \parallel m$. Find the value of $b - a$.

(a) 10°   (b) 20°
(c) 15°   (d) 25°

14. Which of the following is/are point symmetric?
  (a) Rectangle   (b) Square
  (c) Parallelogram   (d) All of these

15. Which of the following has an infinite number of lines of symmetry?
  (a) Equilateral triangle   (b) Isosceles triangle
  (c) Regular hexagon   (d) Circle

16. The sum of an angle and half of its complementary angle is 75°. Find the angle.
  (a) 40°   (b) 50°
  (c) 60°   (d) 80°

17. The adjacent angles of a rhombus are $2x - 35°$ and $x + 5°$. Find x.
  (a) 70°   (b) 40°
  (c) 35°   (d) 45°

18.

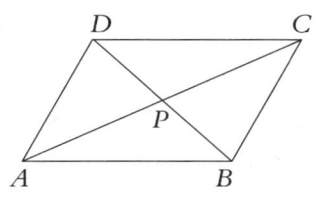

In the figure above, $ABCD$ is a parallelogram, $AC = 14$ cm and $BD = 10$ cm, then $AP + BP =$ _____ cm.
  (a) 5   (b) 7
  (c) 24   (d) 12

19. The following are the steps involved in finding the largest angle of a quadrilateral $PQRS$, if $\angle P : \angle Q : \angle R : \angle S = 1:2:3:4$. Arrange them in sequential order.

(A) $10x = 360° \Rightarrow x = 36°$.
(B) Let the angles be $\angle P = x$, $\angle Q = 2x$, $\angle R = 3x$ and $\angle S = 4x$.
(C) The largest angle = $4(36°) = 144°$.
(D) Given $\angle P : Q : \angle R : S = 1:2:3:4$.
(E) $\angle P + \angle Q + \angle R + \angle S = 360° \Rightarrow x + 2x + 3x + 4x = 360°$.

(a) DBAEC   (b) DBACE
(c) DBECA   (d) DBEAC

20. The following are the steps involved in finding the third side of an isosceles triangle whose two sides are 6 cm and 12 cm. Arrange them in sequential order.

(A) But the difference between two sides is less than the third side.
(B) Since the given triangle is isosceles, the possible measure of the third side is either 6 cm or 12 cm.
(C) $\Rightarrow$ The measure of the third side is 12 cm.
(D) $\therefore$ 6 cm cannot be the measure of the third side.

(a) BDAC   (b) BCAD
(c) BADC   (d) BACD

21. The following are the steps involved in finding the angles of the triangle ABC, when $\angle A : \angle B : \angle C = 1:2:3$. Arrange them in sequential order.

(A) Let the angles be $\angle A = x$, $\angle B = 2x$ and $\angle C = 3x$.
(B) Given $\angle A : \angle B : \angle C = 1:2:3$.
(C) $\angle A = 30°$, $\angle B = 2(30°) = 60°$ and $\angle C = 3(30°) = 90°$.
(D) $\angle A + \angle B + \angle C = 180° \Rightarrow x + 2x + 3x = 180°$.
(E) $6x = 180° \Rightarrow x = 30°$.

(a) BADCE   (b) DBAEC
(c) BADEC   (d) BACDE

22. The following are the steps involved in finding each of interior angle of 10-sided regular polygon. Arrange them in sequential order.

(A) Each exterior angle = 36°.

(B) Each interior angle = 180° − 36° = 144°.

(C) Each exterior angle = $\frac{360°}{n} = \frac{360°}{10}$ (given n = 10)

(a) CAB   (b) BAC

(c) CBA   (d) BCA

Directions for questions 23 to 26: Match the Column A with Column B.

Column A		Column B
23. The supplement of 60° is	( )	(a) Square
24. If the diagonals of a rectangle are perpendicular, then the rectangle is called	( )	(b) 30°
25. The longest side of a right triangle is called	( )	(c) 120°
26. The compliment of 60° is	( )	(d) Rhombus
		(e) Diagonal
		(f) Hypotenuse

Directions for questions 27 to 30: Match the Column A with Column B.

Column A		Column B
27. The compliment of 45° is about a point	( )	(a) Symmetrical
28. The letter B is	( )	(b) 135°
29. The longest chord of a circle is called	( )	(c) 45°
30. A parallelogram in which an angle is 90° is called a	( )	(d) Symmetrical about a line
		(e) Diameter
		(f) Rectangle
		(g) Square

Level 2

Directions for questions 31 to 56: Select the correct alternative from the given choices.

31. In a right triangle, one of the acute angles is four times the other. Find its measure.

(a) 68°   (b) 84°

(c) 80°   (d) 72°

32. In an isosceles triangle ABC, AB = AC and ∠A = 3∠B. Find ∠C.

(a) 36°   (b) 32°

(c) 28°   (d) 40°

33. The lengths of two sides of an isosceles triangle are 5 cm and 12 cm. The length of the third side is _____.

(a) 12 cm   (b) 5 cm

(c) 17 cm   (d) None of these

34. In a triangle, which is not equilateral, the sides (in cm) are integers. The longest side is 3 cm. The perimeter of the triangle is _____.

(a) 5 cm   (b) 6 cm

(c) 8 cm   (d) 7 cm

35. △ABC and △PQR are congruent if

(a) AB = BC = AC and PQ = QR = PR.

(b) ∠A = ∠P, ∠B = ∠Q and ∠C = ∠R.

(c) AB = PQ, BC = QR and ∠B = ∠Q.

(d) None of these

36. In △PQR and △ABC, ∠Q = ∠B = 90°, PQ = AB and QR = BC. Which of the following property can be used to prove the congruence of △PQR and △ABC?

(a) SSS   (b) RHS

(c) ASA   (d) SAS

37. In a triangle TOP, its orthocentre lies at O. Then the circumradius of △TOP is _____.

(a) TO/2   (b) OP/2

(c) TP/2   (d) None of these

38. If in a triangle, the circumcentre does not lie on its longest side, then it must be an/a _____ triangle.
    (a) acute angled   (b) right angled
    (c) obtuse angled   (d) Either (1) or (3)

39. Which of the following must be a square?
    (a) A rhombus whose adjacent angles are equal.
    (b) A rectangle whose adjacent sides are equal.
    (c) Both (a) and (b)
    (d) Neither (a) or (b)

40. A parallelogram in which the diagonals bisect each other at right angles must be
    (a) a rhombus   (b) a rectangle
    (c) a square   (d) Either (b) or (c)

41. A triangle in which the sum of the squares of two sides equals the square of the third side must be a/an _____ triangle.
    (a) right angled   (b) acute angled
    (c) obtuse angled   (d) None of these

42. Which of the following holds true?
    (a) The geometric centre of a triangle equidistant from its sides is called incentre or excentre.
    (b) The centroid divides each median in the ratio 2:1 from the vertex.
    (c) Both (a) and (b).
    (d) Neither (a) nor (b).

43. Which of the following can be one of the angles of a regular polygon?
    (a) 150°   (b) 135°
    (c) 120°   (d) All of these

44. A regular polygon has N sides where $N < 10$. Each of its interior angles is an integer in degrees. How many such polygons are possible?
    (a) 7   (b) 6
    (c) 8   (d) 5

45. The sum of the interior angles of a 10-sided polygon is _____.
    (a) 1260°   (b) 1440°
    (c) 1800°   (d) 1620°

46. In a regular convex polygon, each interior angle is not more than each exterior angle. How many such polygons are possible?
    (a) 2   (b) 3
    (c) 4   (d) 1

47. $ABCD$ is an isosceles trapezium. $\overline{AB} \parallel \overline{CD}$. AE and BF are the perpendiculars drawn to CD. The congruence property used to prove the congruence of triangles AED and BFC is _____.
    (a) RHS   (b) SAS
    (c) SSS   (d) ASA

48.

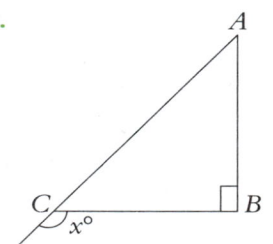

    In the figure above, ABC is a right triangle and $BC = AB$, then find $x°$.
    (a) 45°   (b) 90°
    (c) 120°   (d) 135°

49. Which of the following is not the set of measures of the sides of a triangle?
    (a) 7 cm, 3 cm, 5 cm
    (b) 8 cm, 12 cm, 18 cm
    (c) 5 cm, 6 cm, 14 cm
    (d) 5 cm, 12 cm, 13 cm

50. In which of the following cases can a right triangle ABC be constructed?
    (a) $AB = 5$ cm, $BC = 7$ cm, $AC = 10$ cm
    (b) $AB = 7$ cm, $BC = 8$ cm, $AC = 12$ cm
    (c) $AB = 8$ cm, $BC = 17$ cm $AC = 15$ cm
    (d) None of these

51.

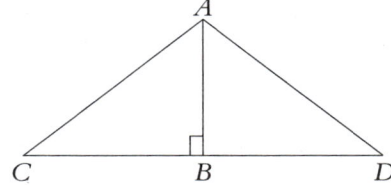

In the figure above, AB is the perpendicular bisector of CD. Which of the following axioms best proves the congruence of △ABC and △ABD?

(a) SSS (b) SAS
(c) RHS (d) ASA

52. In the figure above, AB = AC and BC is extended to D, then find the value of x + y.

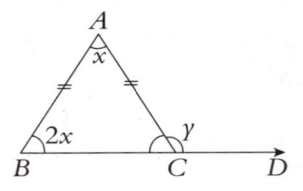

(a) 120° (b) 160°
(c) 40°   (d) 144°

53. In a parallelogram if the diagonals are equal, then the parallelogram necessarily will be

(a) a rhombus   (b) a rectangle
(c) a square   (d) None of these

54.

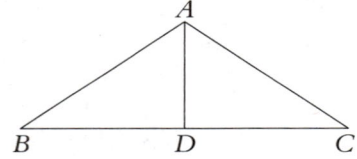

In the figure above, ABC is a triangle in which BC = 24 cm and AC = 13 cm. If AD is the perpendicular bisector of BC, then find the length of AD.

(a) 7 cm   (b) 12 cm
(c) 13 cm (d) 5 cm

55.

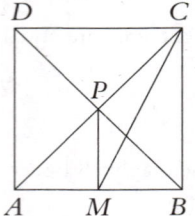

In the figure above, ABCD is a square of side 8 cm and $\overline{PM} \perp \overline{AB}$. Find the length of MC.

(a) $5\sqrt{5}$ cm
(b) $6\sqrt{5}$ cm
(c) $4\sqrt{5}$ cm
(d) $7\sqrt{5}$ cm

56. In an n-sided regular polygon, each interior angle is 144°. Find the number of the sides of the polygon.

(a) 7   (b) 8
(c) 9   (d) 10

Level 3

Directions for questions 57 to 75: Select the correct alternative from given choices.

57. There are three angles. The second angle is one-third of the compliment of the first angle. The third angle is half of the supplement of the first angle. The third angle is 6 times the second angle. Find the first angle.

(a) 45° (b) 60°
(c) 75° (d) 90°

58.

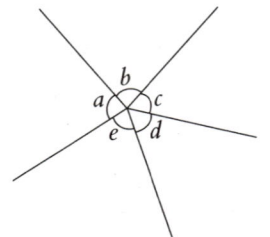

In the figure above, the angles a, b, c, d and e are consecutive integers in degrees. a = _____.

(a) 70°
(b) 74°
(c) Either (a) or (b)
(d) Neither (a) nor (b)

59.

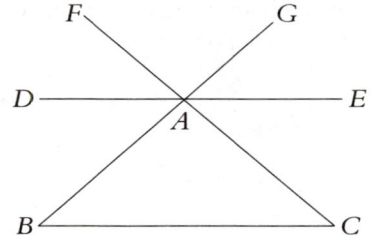

In the figure above (not to scale),
$DAE \parallel BC$, $\angle BAD = (2x - 70)°$,
$\angle ACB = (x - 20)°$ and $\angle FAG = (2x + 20)°$. Find $\angle EAC$.

(a) 40° (b) 25°
(c) 30° (d) 35°

60. In a quadrilateral $ABCD$, $\angle A : \angle B : \angle C : \angle D = 3:4:5:6$. Then $ABCD$ is a _____.

(a) Trapezium (b) Parallelogram
(c) Rhombus   (d) Kite

61. PQR and XYZ are triangles. The perimeter of each triangle is 12 cm. PQR is an equilateral. $XY = 4$ cm and $YZ = ZX$. Both the triangles are

(a) Congruent (b) Similar but not congruent
(c) Similar   (d) Both (a) and (c)

62. In a $\triangle ABC$, $\angle A = \angle B + \angle C$. O and S are the orthocentre and the circumcentre of $\triangle ABC$. If $AB = 12$ cm and $AC = 5$ cm, then find the distance between O and S.

(a) 5.5 cm (b) 5 cm
(c) 6.5 cm (d) 6 cm

63. ABC and DEF are triangles. Consider the following:

I. $\angle A = 40°$; $\angle B = 60°$, $\angle C = 80°$, $AB = 5$ cm and $BC = 6$ cm.

II. $\angle D = \angle F$, $\angle E = 80°$, $DF = 6$ cm and $EF = 8$ cm.

Which of the following can be concluded?

(a) (I) is not possible.
(b) (II) is not possible.
(c) Both (I) and (II) are possible.
(d) Both (I) and (II) are not possible.

64. A is an obtuse angle. The measure of $\angle A$ and twice its supplement differ by 30°. Then $\angle A$ can be

(a) 150° (b) 110°
(c) 140° (d) 120°

65. In the figure below, $2\angle P = \angle QOR$. OQ and OR are bisectors of $\angle Q$ and $\angle R$ respectively. Find $\angle P$.

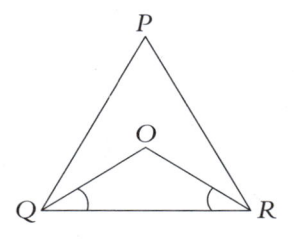

(a) 60° (b) 70°
(c) 40° (d) 80°

66. $ABCD$ is a rhombus in which $\angle B = 120°$ and $BD = 5$ cm. Find the perimeter of the rhombus $ABCD$.

(a) 16 cm (b) 20 cm
(c) 24 cm (d) 30 cm

67.

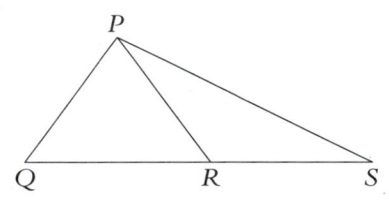

In the figure above, $\angle Q = 2\angle S$ and $\angle QPR = 2\angle RPS$. Find $\angle RPS + \angle S$.

(a) 60° (b) 45°
(c) 72° (d) 54°

68. The angles of a quadrilateral are in the ratio 3:4:5:6. Which of the following can be concluded?

(a) Exactly two angles are acute.
(b) Two pairs of angles are supplementary.
(c) Both (a) and (b)
(d) Neither (a) nor (b)

69. $PQRS$ is a parallelogram in which PR is perpendicular to QS. If $PR = 8$ cm and $QS = 6$ cm. Find PS.

(a) 5 cm (b) 4 cm
(c) 7 cm (d) 6 cm

70. An equilateral triangle has a circumradius of $4\sqrt{3}$ cm. Find its inradius (in cm).

(a) $2\sqrt{3}$ (b) $3\sqrt{3}$
(c) $\sqrt{3}$ (d) $\dfrac{\sqrt{3}}{2}$

5.44 Chapter 5

71.

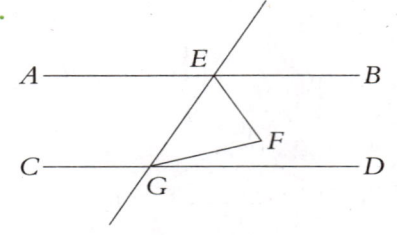

In the figure above, $AB \parallel CD$. EF and FG are the bisectors of $\angle BEG$ and $\angle DGE$ respectively. $\angle FEG = \angle FGE + 10°$. Find $\angle FGE$.

(a) 20° (b) 25°
(c) 40° (d) 35°

72.

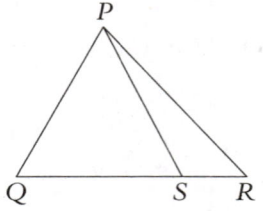

In the figure above, $\angle QPS = 2\angle SPR$, $\angle Q = \angle R + 40°$ and $\angle PSR = 120°$. Find $\angle QPR$.

(a) 50° (b) 55°
(c) 65° (d) 60°

73.

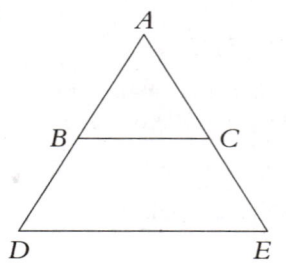

In the figure above, $BC \parallel DE$ and $\angle ABC = \angle CED$. $\angle A = \angle ACB - 30°$. Find $\angle A$.

(a) 40° (b) 50°
(c) 45° (d) 55°

74. In a rhombus $ABCD$, half of angle A exceeds one-sixth of an angle B by 50°. Find the larger of these angles.

(a) 120° (b) 100°
(c) 110° (d) 130°

75.

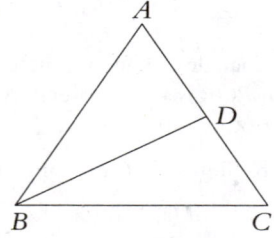

In the figure above, BD is the altitude drawn to AC. Triangles ABD and CBD are congruent if

(a) $AB = BC$
(b) $AD = CD$
(c) Either (a) or (b)
(d) None of these

ASSESSMENT TESTS

Test 1

Directions for questions 1 to 11: Select the correct alternative from the given choices.

1. If the measure of the angles of a triangle is in the ratio of 2:3:4. Find the measure of the angles.

   The following are the steps involved in solving the above problem. Arrange them in sequential order.

   (A) $2x° + 3x° + 4x° = 180°$.
   (B) Let the angles be $2x°$, $3x°$ and $4x°$.
   (C) $x° = 20° \Rightarrow 2x° = 40°$, $3x° = 60°$, $4x° = 80°$.

   (a) BCA (b) BAC
   (c) ABC (d) CBA

2. The measure of one of the angles of a parallelogram is 70°. Find the measures of the angles of the parallelogram.

   The following are the steps involved in solving the above problem. Arrange them in sequential order.

   (A) $70° + x = 180° \Rightarrow x = 110°$.

   (B) Let the angle adjacent to 70° be x.

   (C) The sum of the measures of adjacent angles of a parallelogram is 180°.

   (D) The measures of the angles of the parallelogram are 70°, 110°, 70° and 110°.

   (a) CBDA (b) BCAD
   (c) BCDA (d) CDAB

3. The sum of the measures of the interior angles of a polygon is 540°. Find the number of sides of the polygon.

   (a) 8 (b) 7
   (c) 6 (d) 5

4. In a quadrilateral $ABCD$, $\overline{AC} \perp \overline{BD}$ and $AB = AD$. $ABCD$ is a _____.

   (a) Trapezium
   (b) Rhombus
   (c) Rectangle
   (d) Kite

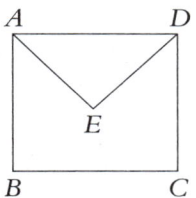

5. In the figure above, $\overline{AB} \parallel \overline{CD}$, $\angle BAE = 30°$ and $\angle CDE = 35°$. If $\overline{AB} \perp \overline{BC}$, then find $\angle AED$.

   (a) 60° (b) 55°
   (c) 65° (d) 85°

6. The measure of one of the exterior angles of a triangle is 100°. Which of the following is definitely the measure of one of the interior angles of that triangle?

   (a) 50° (b) 60°
   (c) 70° (d) 80°

7. In the given figure, if $l \parallel m$, then what type of a triangle is ABC?

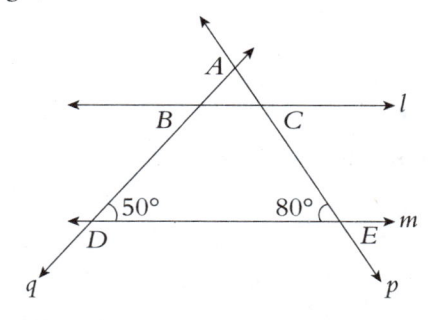

   (a) Equilateral
   (b) Isosceles
   (c) Scalene
   (d) Right angled

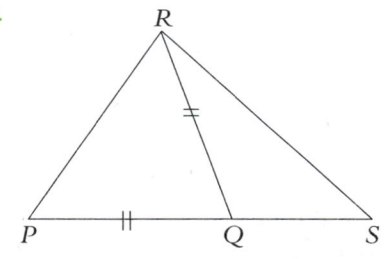

8. In the given figure, $PQ = QR$, $\angle RPQ = 60°$ and $\angle QRS = 20°$. Find the measure of $\angle QSR$.

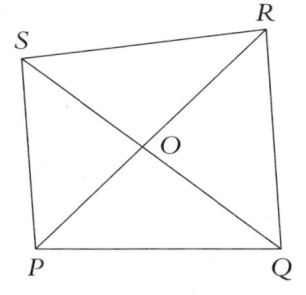

   (a) 20° (b) 30°
   (c) 40° (d) 50°

9. In the given figure, $PQRS$ is a quadrilateral, PR and QS intersect at O. If $\angle PQR = 70°$, $\angle SPQ = 80°$ and $\angle PRQ = 60°$, find $\angle SPR$.

   (a) 30° (b) 50°
   (c) 40° (d) 45°

10. One angle of a parallelogram is 30° more than twice its adjacent angles. Find the measure of adjacent angle.

(a) 50° (b) 60°
(c) 70° (d) 80°

11. Which of the following statements is definitely true?
    (a) In a rhombus, the diagonals are equal.
    (b) In an isosceles trapezium, the diagonals bisect each other.
    (c) In a kite, the diagonals are perpendicular to each other.
    (d) None of these

Directions for questions 12 to 15: Match the Column A with Column B.

Column A		Column B
12. The point of concurrence of altitudes of a triangle is the	( )	(a) Circumcentre
13. The point of concurrence of medians of a triangle is the	( )	(b) Centroid
14. The point of concurrence of the bisectors of the angles of a triangle is the	( )	(c) Incentre
15. The point of concurrence of the perpendicular bisectors of the sides of a triangle is the	( )	(d) Orthocentre

Test 2

Directions for questions 16 to 26: Select the correct alternative from the given choices.

16. The measures of the angles of a triangle are in the ratio of 1:2:3. Find the angles of the triangle.

    The following are the steps involved in solving the above problem. Arrange them in sequential order.

    (A) The angles of the triangle are 30°, 60° and 90°.
    (B) $x = 30°$.
    (C) Let the angles of the triangle be $x°$, $2x°$ and $3x°$.
    (D) $x° + 2x° + 3x° = 180°$.

    (a) CBDA (b) CDBA
    (c) CDAB (d) CBAD

17. The measures of the angles of a quadrilateral are 40°, 80° and 100°. Find the measure of the fourth angle.

    The following are the steps involved in solving the above problem. Arrange them in sequential order.

    (A) $x° = 360° − 220°$.
    (B) Let the measure of the fourth angle be $x°$.
    (C) ∴ The fourth angle, x is 140°.
    (D) $40° + 80° + 100° + x° = 360°$.

    (a) ABDC (b) ABCD
    (c) BADC (d) BDAC

18. Find the number of sides of a regular polygon, if the measure of each of its interior angles is 150°.
    (a) 12 (b) 10
    (c) 8   (d) 6

19. In a quadrilateral PQRS, $\overline{PQ} \parallel \overline{RS}$ and PR = QS. PQRS is a/an _____.
    (a) Square   (b) Rectangle
    (c) Rhombus   (d) Isosceles trapezium

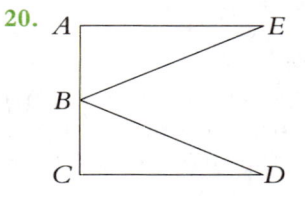

20.

    In the given figure, $\overline{AE} \parallel \overline{CD}$, AB = BC and AE = CD. If $\overline{AC} \perp \overline{CD}$ and $\angle AEB = 35°$, find $\angle DBE$.

(a) 70° (b) 65°
(c) 55° (d) 50°

21. The measure of one of the angles of an isosceles triangle is 94°. Which of the following is definitely the measure of one of the other angles of the given triangle?

    (a) 94° (b) 86°
    (c) 43° (d) 46°

22. In the given figure, if $l \parallel m$, then what type of a triangle is ABC?

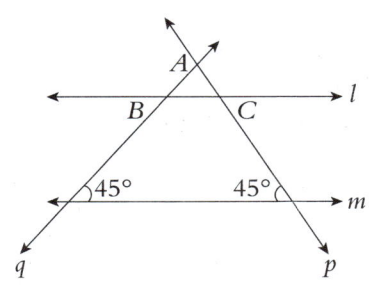

    (a) Scalene   (b) Isosceles
    (c) Right angled (d) Both (b) and (c)

23.

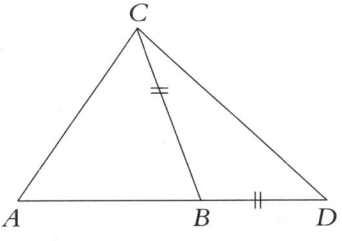

   In the given figure, $\angle ACB = 60°$, $\angle CAB = 50°$ and $BC = BD$. Find $\angle BDC$.

    (a) 35° (b) 60°
    (c) 45° (d) 50°

24.

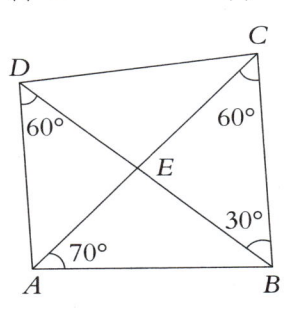

In the given figure, ABCD is a quadrilateral. $\angle ADB = 60°$, $\angle BAC = 70°$, $\angle DBC = 30°$ and $\angle ACB = 60°$. Find $\angle DAC$

(a) 30° (b) 40°
(c) 50° (d) 60°

25. One angle of a parallelogram is thrice its adjacent angle. Which of the following is one of its angles?

    (a) 30°   (b) 45°
    (c) 120° (d) 100°

26. Which of the following statements is true?

    (a) Every trapezium is a parallelogram.
    (b) Every square is a rhombus.
    (c) Every rectangle is a square.
    (d) Every parallelogram is a rectangle.

Directions for questions 27 to 30: Match the Column A with Column B.

Column A		Column B
27. In a right triangle, the orthocentre is	( )	(a) Exterior of any triangle.
28. In a right triangle, the circumcentre is	( )	(b) Interior of any triangle
29. Centroid lies	( )	(c) A point which coincides with one vertex of the triangle, where the right angle in formed.
30. Excentre lies	( )	(d) A midpoint of the longest side of the triangle.

TEST YOUR CONCEPTS

Very Short Answer Type Questions

1. True
2. True
3. False
4. False
5. True
6. True
7. circumcentre
8. equidistant
9. 60°
10. 100°
11. 135°
12. 120°
13. centre of the circle
14. 2π
15. 3
16. 240°
17. rhombus
18. secant
19. 3
20. 1
21. 4
22. 3
23. 1
24. 2
25. 3

Short Answer Type Questions

26. 140°
27. 60°
28. 13, 26 and 29
29. 80°
30. 70°
31. isosceles triangle
32. 40°
33. 40°
34. 8 cm
35. 120°
36. 135°
37. 20°
38. 70°
39. 60°
40. $\left(82\frac{1}{2}\right)^\circ$
43. 25 cm
44. 540°
45. 72°
46. SSS
47. 40°
48. 25°
49. 60°
50. 45°
51. 24°
52. 1080°
53. 20 cm
55. 90°

Essay Type Questions

56. 50 cm
57. 17 cm
58. 50°
59. Q
60. 35°; 45°
61. 120°

62. 31 cm
63. 50°
66. 18 cm
67. $9\sqrt{2}$ cm
68. 220°
69. 10°

CONCEPT APPLICATION

Level 1

1. (a)	2. (b)	3. (b)	4. (c)	5. (a)	6. (d)	7. (c)	8. (b)	9. (c)	10. (d)
11. (c)	12. (d)	13. (b)	14. (d)	15. (d)	16. (c)	17. (a)	18. (d)	19. (d)	20. (c)
21. (c)	22. (a)	23. (c)	24. (a)	25. (f)	26. (b)	27. (c)	28. (d)	29. (e)	30. (f)

Level 2

31. (d)	32. (a)	33. (a)	34. (d)	35. (c)	36. (d)	37. (c)	38. (d)	39. (c)	40. (a)
41. (a)	42. (c)	43. (d)	44. (b)	45. (b)	46. (a)	47. (a)	48. (d)	49. (c)	50. (c)
51. (b)	52. (d)	53. (b)	54. (d)	55. (c)	56. (d)				

Level 3

| 57. (b) | 58. (c) | 59. (c) | 60. (a) | 61. (d) | 62. (c) | 63. (d) | 64. (b) | 65. (a) | 66. (b) |
| 67. (a) | 68. (c) | 69. (a) | 70. (a) | 71. (c) | 72. (d) | 73. (a) | 74. (a) | 75. (c) | |

ASSESSMENT TESTS

Test 1

| 1. (b) | 2. (b) | 3. (d) | 4. (d) | 5. (c) | 6. (d) | 7. (b) | 8. (c) | 9. (a) | 10. (a) |
| 11. (c) | 12. (d) | 13. (b) | 14. (c) | 15. (a) | | | | | |

Test–2

| 16. (b) | 17. (d) | 18. (a) | 19. (d) | 20. (a) | 21. (c) | 22. (d) | 23. (a) | 24. (a) | 25. (b) |
| 26. (b) | 27. (c) | 28. (d) | 29. (b) | 30. (a) | | | | | |

Chapter 5

CONCEPT APPLICATION

Level 1

1. 3

   Hence, the correct option is (a)

2. AB (the side opposite to greatest angle)

   Hence, the correct option is (b)

3. In a rhombus, if the diagonals are equal, then it becomes a square.

   Hence, the correct option is (b)

4. Isosceles triangle

   Hence, the correct option is (c)

5. Perimeter = (1+1+1) cm = 3 cm

   Hence, the correct option is (a)

6. $x + x = 180°$

   $\Rightarrow x = 90°$

   Hence, the correct option is (d)

7. 30°

   Hence, the correct option is (c)

8. Let the angles be $2x°$ and $3x°$.

   $2x° + 3x° = 90°$

   $x° = 18°$

   The larger angle = $(3x)° = 54°$.

   Hence, the correct option is (b)

9. Let the angle be $x°$.

   $\therefore$ Its supplement = $(180 - x)°$

   Given, $x° = 3(180 - x)°$

   $x = 540° - 3x°$

   $x° = 135°$.

   Hence, the correct option is (c)

10. Let $\angle PQS$ and $\angle SQR$ be $7x°$ and $5x°$ respectively.

    $\angle PQS + SQR = 180°$ ($\because$ PQR is a straight line).

    $7x° + 5x° = 180°$

    $x° = 15°$

    $\angle SQR = 5x° = 75°$.

    Hence, the correct option is (d)

11. $PQ = PR \Rightarrow \angle PQR = \angle PRQ$

    $\angle SPT = \angle QPR$

    ($\because$ Vertically opposite angles are equal)

    $\therefore \angle QPR = 60°$

    $\angle PQR = \angle PRQ = \dfrac{180 - \angle QPR}{2} = 60°$

    Hence, the correct option is (c)

12. $AB \parallel CD \Rightarrow (3x - 20)° + (x + 40)° = 180°$

    $\Rightarrow x° = 40°$

    Hence, the correct option is (d)

13. $a° = 180° - 100° = 80°$

    ($\because$ $a°$ and 100° forms linear pair)

    $b° = 100°$ ($\because$ $b°$ and 100° are alternate interior angles)

    $(b - a) = 100° - 80° = 20°$

    Hence, the correct option is (b)

14. All the given figures are point symmetric. The point of symmetry is the point of intersection of their diagonals.

    Hence, the correct option is (d)

15. A circle has an infinite number of lines of symmetry.

    Hence, the correct option is (d)

16. Let the required angle be x.

    $\therefore$ Complement of x is $90° - x$.

    $\Rightarrow x + \dfrac{(90° - x)}{2} = 75°$

    $\Rightarrow 2x + 90° - x = 150°$

    $\Rightarrow x = 60°$.

    Hence, the correct option is (c)

17. The adjacent angles of a rhombus are supplementary.

    $\therefore 2x - 35° + x + 5 = 180°$

    $\Rightarrow 3x = 180° + 30°$

    $\Rightarrow x = \dfrac{210°}{3} = 70°$

    Hence, the correct option is (a)

Geometry 5.51

18. Given

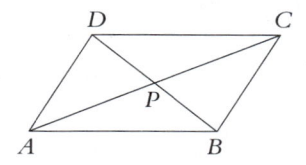

Given that $AC = 14$ cm and $BD = 10$ cm.

We know that, in a parallelogram, the diagonals bisect each other.

∴ P is the midpoint of AC and BD.

$AP = \dfrac{AC}{2}$ and $BP = \dfrac{BD}{2}$

∴ $AP = 7$ cm, $BP = 5$ cm

∴ $AP + BP = 12$ cm

Hence, the correct option is (d).

19. (D), (B), (E), (A) and (C) is the required sequential order.

Hence, the correct option is (d).

20. (B), (A), (D) and (C) is the required sequential order.

Hence, the correct option is (c)

21. (B), (A), (D), (E) and (C) is the required sequential order.

Hence, the correct option is (c)

22. (C), (A) and (B) is the required sequential order.

Hence, the correct option is (a)

23. → c: The supplement of 60° is 180° − 60° = 120°.

24. → a: If the diagonals of a rectangle are perpendicular, then the rectangle is called a square.

25. → f: The longest side of a right triangle is called hypotenuse of the triangle.

26. → b: The compliment of 60° is 90° − 60° = 30°.

27. → c: The compliment of 45° is 90° − 45° = 45°.

28. → d: The letter B is symmetrical about a line.

29. → e: The longest chord of a circle is called diameter of the circle.

30. → f: A parallelogram in which an angle is 90° is called a rectangle.

Level 2

31. Let the require measure be $x°$.

Measure of the other acute angle = $\left(\dfrac{x}{4}\right)°$

$x° + \left(\dfrac{x}{4}\right)° + 90° = 180°$

$x° = 72°$.

Hence, the correct option is (d).

32. Given $AB = AC$

⇒ $\angle B = \angle C$

And also give $\angle A = 3\angle B$

We have, $\angle A + \angle B + \angle C = 180°$

$3\angle B + \angle B + \angle B = 180°$

⇒ $\angle B = 36°$

∴ $\angle c = 36°$.

Hence, the correct option is (a).

33. The lengths of two sides of an isosceles triangle are 5 cm and 12 cm. The third side is either 5 cm or 12 cm long. But only 12 cm satisfies the inequalities of triangle.

Hence, the correct option is (a).

34. Let the unknown sides (in cms) be a and b.

$a, b < 3 \Rightarrow 1 \leq a, b \leq 2$

⇒ $2 \leq a + b \leq 4$.

⇒ $5 \leq$ Perimeter ≤ 7 cm (1)

By triangle inequality $a + b > 3$.

∴ Perimeter > 6 cm.

⇒ Perimeter $= 7$ cm (from (1)).

Hence, the correct option is (d)

35. (a) ⇒ Two equilateral triangles need not be equal.

(b) ⇒ Corresponding sides of the triangles need not be equal.

5.52 Chapter 5

(c) ⇒ By SAS congruence property, $\triangle ABC \cong \triangle PQR$.

Hence, the correct option is (c)

36. Given $PQ = AB$, $QR = BC$, $\angle Q = \angle B = 90°$. Basically by using SAS congruence property, $\triangle PQR \cong ABC$.

Hence, the correct option is (d)

37. Orthocentre $\triangle TOP$ lies at O.

∴ $\angle O = 90°$

∴ Circumradius is half of the hypotenuse. i.e., $TP/2$.

Hence, the correct option is (c)

38. For an acute angled triangle, the circumcentre lies interior of it. For a right angled triangle, it lies on its longest side. For an obtuse angled triangle, it lies exterior of it.

In the given problem, the circumcentre does not lie on the longest side.

∴ It is not right angled. It can be either acute angled or obtuse angled.

Hence, the correct option is (d)

39. A rhombus whose adjacent angles are equal.

∴ Each angle would be 90°.

Such a rhombus would be a square.

A rectangle, whose adjacent sides are equal, would have all sides equal.

∴ Such a rectangle would be a square.

Hence, the correct option is (c)

40. A parallelogram in which the diagonals bisect each other at right angles must be a rhombus.

Hence, the correct option is (a)

41. Choice (a) follows (Pythagoras theorem).

Hence, the correct option is (a)

42. The geometric centre of a triangle equidistant from its sides is called incentre or excentre.

The centroid divides each median in the ratio of 2:1 from the vertex.

Hence, the correct option is (c)

43. Sum of the angles of a regular polygon of N sides $= 180(N - 2)°$.

Each of its interior angle $= \left(\dfrac{180(N-2)}{N}\right)°$

$\left(\dfrac{180(N-2)}{N}\right)° = 150° \Rightarrow N = 12$

$\left(\dfrac{180(N-2)}{N}\right)° = 135° \Rightarrow N = 8$

$\left(\dfrac{180(N-2)}{N}\right)° = 120° \Rightarrow N = 6$

∴ Choice (d) follows.

Hence, the correct option is (d)

44. Sum of the interior angles $= [180(N-2)]°$

Each interior angle $= \left[\dfrac{180(N-2)}{N}\right]°$. For this to be an integer, N must divide $180(N-2)$ exactly.

$N < 10$.

∴ N can be 3 or 4 or 5 or 6 or 8 or 9.

∴ The polygon has 6 possibilities.

Hence, the correct option is (b)

45. The sum of interior angles of an n-sided polygon $= (n-2)180°$.

∴ The sum of interior angles of a 10-sided polygon $= (10-2)180° = 1440°$.

Hence, the correct option is (b)

46. By trial and error method, the equilateral triangle and the square follows the given condition.

∴ There are two such polygons.

Hence, the correct option is (a)

47.

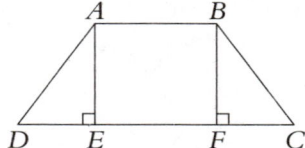

In $\triangle AED$ and $\triangle BFC$, $\angle AED = \angle BFC = 90°$.

$AD = BC$

$AE = BF$ = Height of the trapezium.

By RHS, $\triangle AED$ and $\triangle BFC$ are congruent.

Hence, the correct option is (a) follows.

Hence, the correct option is (a)

48.

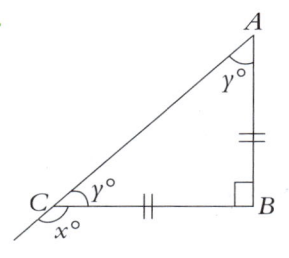

Given $AB = BC$ and $\angle B = 90°$

$\therefore \angle A = \angle C$

Let $\angle A = \angle C = y$

$\therefore \angle A + \angle C + 90° = 180°$

$y + y + 90 = 180°$

$2y + 90° = 180°$

$2y = 180 - 90$

$2y = 90°$

$y = 45°$

$\therefore \angle A = \angle C = 45°$

But $\angle x =$ The sum of opposite interior angles

$\angle x = \angle A + \angle B$

$\quad = 45° + 90° = 135°$

Hence, the correct option is (d)

49. We know that the sum of the lengths of any two sides of a triangle is always greater than the length of the third side and the difference of the lengths of any two sides is less than the length of the third side.

As $5 + 6 < 14$, 5 cm, 6 cm and 14 cm do not follow the above conditions.

Hence, the correct option is (c)

50. In all three given cases, $\triangle ABC$ is constructed, but $AB = 8$ cm, $BC = 17$ cm and $AC = 15$ cm form a right triangle, since $8^2 + 15^2 = 17^2$.

Hence, the correct option is (c)

51.

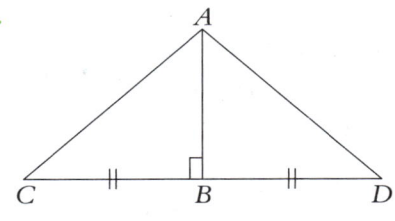

As $\overline{AB}$ is the perpendicular bisector of $\overline{CD}$.

$\angle ABC = \angle ABD = 90°$ and $BC = BD$

In $\triangle ABC$ and $\triangle ABD$, AB is the common side of $BC = BD$ and $\angle ABC = \angle ABD$.

$\therefore$ By SAS congruence property $\triangle ABC \cong \triangle ABD$.

Hence, the correct option is (b)

52.

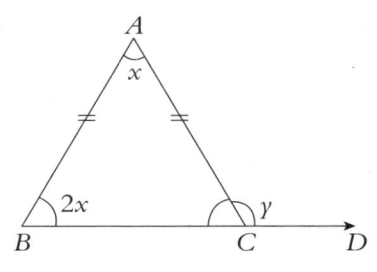

Given $AB = AC$

$\Rightarrow \angle ACB = \angle ABC$

$\Rightarrow \angle ACB = 2x$ (from the figure)

From $\triangle ABC$

$x + 2x + 2x = 180°$

$5x = 180°$

$x = 36°$

But $y = x + 2x = 3x$

$\therefore y = 3 \times 36° = 108°$

$\therefore x + y = 36° + 108° = 144°$.

Hence, the correct option is (d)

53. In a parallelogram, if the diagonals are equal, then it necessarily is a rectangle.

Hence, the correct option is (b)

54.

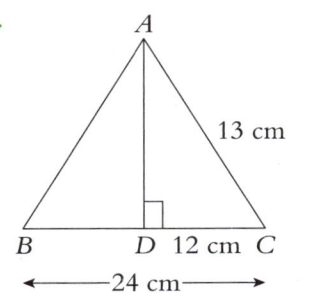

Since AD is the perpendicular bisector of BC, it implies that D is the mid-point of $\overline{BC}$.

5.54 Chapter 5

$DC = \dfrac{BC}{2} = \dfrac{24}{2}$

∴ $DC = 12$ cm.

Now ADC is right triangle and

$AD^2 = AC^2 - DC^2$

$= 13^2 - 12^2 = 169 - 144 = 25$

$AD = \sqrt{25}$

$AD = 5$ cm.

Hence, the correct option is (d)

55.

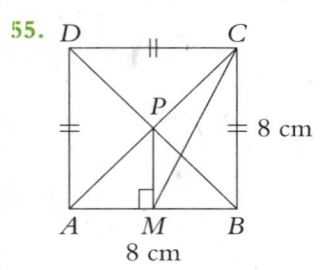

$PM \perp \overline{AB}$

$\Rightarrow AM = BM$

$\Rightarrow BM = \dfrac{AB}{2} = 4$ cm

In $\triangle BMC$, $MC^2 = MD^2 + BC^2 = 4^2 + 8^2$.

$MC^2 = 80$

$MC = \sqrt{80} = 4\sqrt{5}$ cm

Hence, the correct option is (c)

56. Each interior angle = 144°

∴ Each exterior angle = 180° − 144° = 36°.

$\Rightarrow \dfrac{360°}{n} = 36°$ (where n is the number of sides)

$\Rightarrow n = \dfrac{360°}{36°} = 10$.

∴ The number of sides = 10.

Hence, the correct option is (d)

Level 3

57. Let the first angle, the second angle and the third angle be f, s and t respectively.

$s = \left(\dfrac{90° - f}{3}\right) = \left(30° - \dfrac{f}{3}\right)$

$t = \left(\dfrac{180° - f}{2}\right) = \left(90° - \dfrac{f}{2}\right)$

$t = 6s$

∴ $\left(90° - \dfrac{f}{2}\right) = 6\left(30° - \dfrac{f}{3}\right)$

$\Rightarrow \left(90° - \dfrac{f}{2}\right) = (180° - 2f)$

$f = 60°$

Hence, the correct option is (b)

58. $a°$, $b°$, $c°$, $d°$ and $e°$ can be in ascending order or in descending order.

∴ $b = a + 1$, $c = a + 2$, $d = a + 3$ and $e = a + 4$ or $b = a − 1$, $c = a − 2$, $d = a − 3$ and $e = a − 4$.

∴ $a + b + c + d + e = 5a + 10$ or $5a − 10$.

Sum of the angles around a point is 360°.

∴ $a° + b° + c° + d° + e° = 360°$

∴ $(5a + 10)° = 360°$ or $(5a − 10)° = 360°$

∴ $a° = 70°$ or 74°.

Hence, the correct option is (c)

59. DAE || BC ∠DAB = ∠ABC (∴ Alternate interior angles) and ∠EAC = ∠ACB (∴ Alternate interior angles).

FAG = ∠BAC (Q Vertically opposite angles)

$\Rightarrow \angle BAD = \angle ABC = (2x − 70)°$ (given)

Given, ∠BAC = (2x + 20)° and ∠ACB = (x − 20)°

∠ABC + ∠BAC + ∠ACB = 180°

(2x − 70)° + (2x + 20)° + (x − 20)° = 180°

$5x° − 70° = 180°$

$x° = 50° \Rightarrow \angle ACB = (x − 20)° = 30°$

$\Rightarrow \angle EAC = 30°$ (∵ ∠EAC = ∠ACB)

Hence, the correct option is (c)

60. In a quadrilateral ABCD,

∠A + ∠B + ∠C + ∠D = 360°

Given ∠A:∠B:∠C:∠D = 3:4:5:6.

Let, $\angle A = 3x$, $\angle B = 4x$, $\angle C = 5x$ and $\angle D = 6x$.

$\therefore 3x + 4x + 5x + 6x = 360°$.

$\Rightarrow 18x = 360°$

$\Rightarrow x = 20°$

$\therefore \angle A = 60°, \angle B = 80°, \angle C = 100°$ and $\angle D = 120°$

$\therefore \angle A = 60°, \angle B = 80°, \angle C = 100°$ and $\angle D = 120°$

$\angle A + \angle D = \angle B + \angle C = 180°$

$\Rightarrow \overline{AB}$ is parallel to $\overline{DC}$.

$\therefore ABCD$ is a trapezium.

Hence, the correct option is (a)

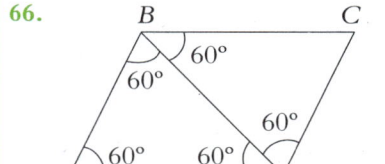

61. Side of $PQR = \dfrac{12}{3} = 4$ cm

Given $XY = 4$ cm and

$XY + YZ + ZX = 12$ cm

$\Rightarrow YZ + ZX = 8$ cm

$YZ = ZX = 4$ cm

$\therefore PQR$ and XYZ are congruent.

Both the triangles are similar. Choice (d) follows.

Hence, the correct option is (d)

62. In $\triangle ABC$, $\angle A + \angle B + \angle C = 180°$.

$\Rightarrow \angle A + \angle A = 180°$ (given $\angle A = \angle B + \angle C$)

$\Rightarrow \angle A = 90°$

In a right triangle, orthocentre (O) lies at the vertex of right angle is at A and circumcentre (S) lies at the midpoint of the hypotenuse.

$\therefore$ Distance between the orthocentre and the circumcentre of a right triangle is circumradius of the triangle, i.e., $\dfrac{BC}{2}$.

Given, $AB = 12$ cm and $AC = 5$ cm.

We have, $(BC)^2 = (AB)^2 + (AC)^2$

$= (12)^2 + (5)^2$

$= 169 = 13^2$

$\Rightarrow BC = 13$ cm

$\Rightarrow \dfrac{BC}{2} = 6.5$ cm

$\therefore$ The required distance $(OS) = 6.5$ cm.

Hence, the correct option is (c)

63. I and II violate the rule that in a triangle the side opposite to greater angle is greater.

Hence, the correct option is (d)

64. Let $A = x°$

Its supplement $= (180 - x)°$

$x°$ and $[2(180 - x)]°$ differ by 30°

$\therefore x° - [2(180° - x)]° = 30°$

or $[2(180 - x)]° - x° = 30°$

$(3x - 360)° = 30°$ or

$(360 - 3x)° = 30°$

$x° = 130°$ or $110°$.

Hence, the correct option is (b)

65. $2\angle P = \angle QOR$

In $\triangle PQR$, $\angle P + \angle Q + \angle R = 180°$ (1)

$\Rightarrow \angle Q + \angle R = 180° - \angle P$

In $\triangle OQR$, $\angle OQR + \angle QOR + \angle ORQ = 180°$.

$\Rightarrow \dfrac{\angle Q}{2} + 2\angle P + \dfrac{\angle R}{2} = 180°$ (2)

$\Rightarrow 2\angle P + \dfrac{180° - \angle P}{2} = 180°$

$\angle P = 60°$

Hence, the correct option is (a)

66.

Given, $\angle B = 120°$

But $\angle A + \angle B = 180°$ (adjacent angles of rhombus)

$\Rightarrow \angle A = 180° - 120° = 60°$

$\angle B = \angle D$ (opposite angles of rhombus)

$\therefore \angle D = 120°$

BD bisects $\angle B$ and $\angle D$

In $\triangle ABD$, $\angle A = \angle ABD = \angle ADB = 60°$

$\therefore \triangle ABD$ is equilateral.

$AB = AD = BD$

∴ AB = 5 cm (given BD = 5 cm)

But AB = BC = CD = AD (side of rhombus)

∴ Perimeter = AB + BC + CD + AD

= (5 + 5 + 5 + 5) cm = 20 cm.

Hence, the correct option is (b)

67. ∠QPR + ∠Q + ∠RPS + ∠S = 180°

2∠RPS + 2∠S + ∠RPS + ∠S = 180°

3(∠RPS + ∠S) = 180°

∠RPS + ∠S = 60°

Hence, the correct option is (a)

68. Let the angles be $(3x)$, $(4x)$, $(5x)$ and $(6x)$.

$3x + 4x + 5x + 6x = 360°$

$x = 20°$

$3x = 60°$, $4x = 80°$, $5x = 100°$ and $6x = 120°$

$3x$ and $4x$ are acute.

And $3x + 6x = 4x + 5x = 180°$.

Choices (a) and (b) follows.

Hence, the correct option is (c)

69. PQRS is a parallelogram and PR ⊥ QS.

∴ PQRS is a rhombus.

Given, PR = 8 cm and

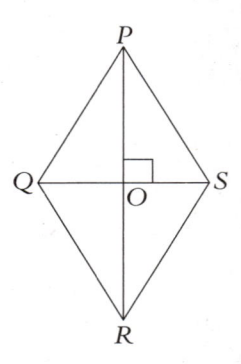

QS = 6 cm. In a rhombus, the diagram bisect to each other.

Let O be the point of intersection of PR and QS.

∴ $PO = \dfrac{PR}{2} = 4$ cm

$OS = \dfrac{QS}{2} = 3$ cm

$(PS)^2 = (OP)^2 + (OS)^2$ (∵ POS form right triangle)

$= 4^2 + 3^2 = 16 + 9 = 25 = 5^2$

⇒ PS = 5 cm.

Hence, the correct option is (a)

70. In any triangle, the centroid divides each median in the ratio 2:1. Such that the part of a median from the centroid to a vertex and the part of it from the centroid to the opposite side are in the ratio of 2 : 1.

In an equilateral triangle, the centroid, the circumcentre, the othocentre and the incentre coincide.

∴ The first part is the circumradius and the second part is the inradius.

∴ Inradius = $\dfrac{\text{Circumradius}}{2}$.

In the given problem, circumradius = $4\sqrt{3}$ cm.

∴ Inradius = $2\sqrt{3}$ cm.

Hence, the correct option is (a)

71. AB || CD

∠BEG + ∠EGD = 180°

EF and FG bisect ∠BEG and ∠DGE respectively.

∴ 2∠FEG + 2∠FGE = 180°

∠FEG + ∠FGE = 90°   (1)

∠FEG = ∠FGE + 10°   (2)

On solving (1) and (2), we get ∠FGE = 40°.

Hence, the correct option is (c)

72.

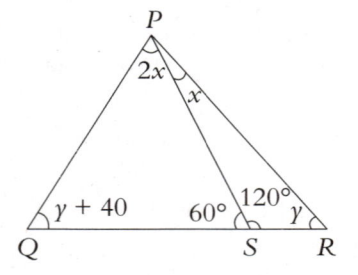

From the figure, in ΔPSR. $x + y = 60°$ (∵ ∠PSR = 120°)

⇒ $y = 60° - x$.   (1)

In ΔPQS, $2x + y + 40° = 120°$.

$2x + 60° - x + 40° = 120$ (from (1))

$x = 20°$

$\therefore \angle QPR = 3x = 3 \times 20° = 60°$.

Hence, the correct option is (d)

73. Given, $BC \parallel DE$, $\angle ABC = \angle CED$ and $\angle A = \angle ACB - 30°$.

$\Rightarrow \angle ACB = \angle CED$

$\Rightarrow \angle ABC = \angle ACB$ ($\because BC \parallel DE$)

In $\triangle ABC$,

$\angle A + \angle ABC + \angle ACB = 180°$

$\angle ACB - 30° + \angle ACB + \angle ACB = 180°$

$\Rightarrow 3\angle ACB = 210° \Rightarrow \angle ACB = 70°$

$\therefore \angle A = 40°$.

Hence, the correct option is (a)

74. $\angle A + \angle B = 180°$ (1)

($\because$ adjacent angles of the rhombus)

Given $\dfrac{\angle A}{2} - \dfrac{\angle B}{6} = 50°$

$\Rightarrow 3\angle A - \angle B = 300°$ (2)

On adding (1) and (2),

$4\angle A = 480°$

$\Rightarrow \angle A = 120°$

$\Rightarrow \angle B = 60°$

$\therefore$ The required angle $= 120°$.

Hence, the correct option is (a)

75.

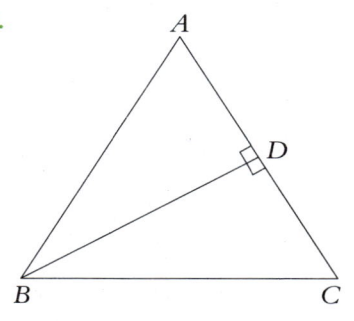

$\overline{BD} \perp \overline{AC} \Rightarrow \angle BDC = \angle BDA = 90°$

BD is the common side of $\triangle ABD$ and $\triangle CBD$.

(i) If $AB = BC$,

$\triangle ABD \cong \triangle CBD$ (by RHS).

(ii) If $AD = CD$,

$\triangle ABD \cong \triangle CBD$ (by SAS).

Hence, the correct option is (c)

ASSESSMENT TESTS

Test 1

Solutions for questions 1 to 11:

1. (B), (A) and (C) is the required sequential order.

   Hence, the correct option is (b)

2. (B), (C), (A) and (D) is the required sequential order.

   Hence, the correct option is (b)

3. $(2n - 4)90° = 540°$

   $\Rightarrow 2n - 4 = 6 \Rightarrow n = 5$

   Hence, the correct option is (d)

4. The given quadrilateral is a kite.

   Hence, the correct option is (d)

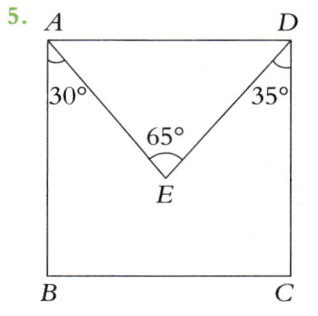

5.

Since $\overline{AB} \parallel \overline{CD}$ and $\overline{AB} \perp \overline{BC}$;

$\angle B = \angle C = 90°$

$\Rightarrow \angle BAD + \angle CDA = 180°$

∴ ∠EAD + ∠EDA = 180° − 65°

Now ∠AED = 180 − (180° − 65°) = 65°.

Hence, the correct option is (c).

6. If the measure of one of the exterior angles is 100°, its supplement is the measure of one of the interior angles of the triangle.

The measure of the required angle = 180° − 100° = 80°.

Hence, the correct option is (d).

7. Given $l \parallel m$,

∠BCA = ∠DEC = 80° (corresponding angles)

∠CBA = ∠EDB = 50° (corresponding angles)

∠BCA + ∠CBA + ∠BAC = 180°

⇒ 80° + 50° + ∠BAC = 180°

⇒ ∠BAC = 50°

∴ In △ABC, ∠A = ∠B = 50

⇒ △ABC is isosceles.

Hence, the correct option is (b).

8.

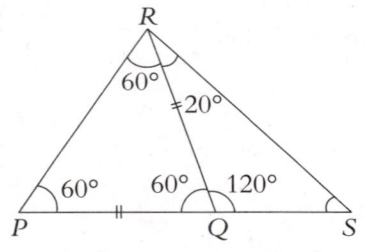

In △PQR, PQ = QR and ∠RPQ = 60°.

∴ ∠PRQ = 60° ⇒ ∠PQR = 60°

⇒ ∠RQS = 180° − 60° = 120°

∠QSR = 180° − (120° + 20°) = 40°.

Hence, the correct option is (c).

9.

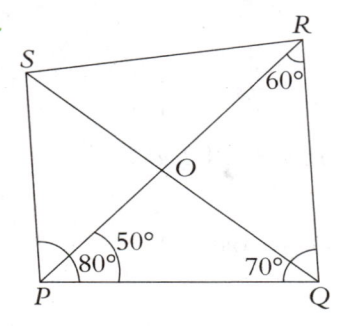

From the figure, in △PQR, ∠PQR = 70°, ∠PRQ = 60°.

⇒ ∠RPQ = 180° − 70° − 60° = 50°

∠SPR = ∠SPQ − ∠RPQ = 80° − 50° = 30°.

Hence, the correct option is (a).

10. Let the adjacent angle be $x°$.

∴ Given angle is $(2x° + 30°)$.

We know that the sum of the adjacent angles of a parallelogram is 180°.

⇒ $(x° + 2x° + 30°) = 180°$

$(3x°) = 180° − 30°$

$x° = \dfrac{150°}{3}$

∴ $x° = 50°$

∴ The adjacent angle = 50°.

Hence, the correct option is (a).

11. In a kite diagonals are perpendicular to each other. It is a true statement.

Hence, the correct option is (c).

Solutions for questions 12 to 15:

12. → (d): The point of concurrence of the attitudes of a triangle is the orthocentre of the triangle.

13. → (b): The point of concurrence of the medians of a triangle is the centroid of the triangle.

14. → (c): The point of concurrence of the bisectors of the interior angles of a triangle is the incentre of the triangle.

15. → (a): The point of concurrence of the perpendicular bisectors of a triangle is the circumcentre of the triangle.

Test 2

Solutions for questions 16 to 26:

16. (C), (D), (B) and (A) is the required sequential order.

    Hence, the correct option is (b)

17. (B), (D), (A) and (C) is the required sequential order.

    Hence, the correct option is (d)

18. Each exterior angle is = 180° − 150° = 30°

    ∴ No. of sides = $\dfrac{360°}{30°}$ = 12.

    Hence, the correct option is (a)

19. Given quadrilateral is an isosceles trapezium.

    Hence, the correct option is (d)

20.

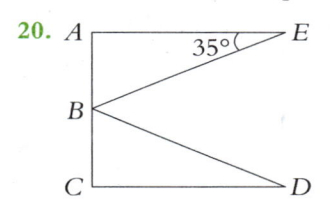

    Since $\overline{AC} \perp \overline{CD}$, ∠BAE = 90°

    ⇒ ∠ABE = 180° − 90° − 35° = 55°

    AB = BC, AE = CD, ∠BAE = ∠BCD = 90°.

    (∵ $\overline{AE} \parallel \overline{CD}$, $\overline{AC} \perp \overline{CD}$)

    ΔABE ≅ ΔCBD

    ∠ABE = ∠CBD = 55°

    ∠DBE = 150° − (55° + 55°) = 70°.

    Hence, the correct option is (a)

21. The sum of the measures of the angles of a triangle = 180°.

    One of the angles is 94° (given)

    ⇒ The sum of the measures of the other two angles = 180° − 94° = 86°.

    ∴ The measure of each of the equal angles is = $\dfrac{86°}{2}$ = 43°.

    Hence, the correct option is (c)

22. Given, $l \parallel m$,

    ∠BCA = ∠DEC = 45° (corresponding angles)

    ∠CBA = ∠EDB = 45° (corresponding angles)

    ∠BCA + ∠CBA + ∠BAC = 180°

    ⇒ 45° + 45° + ∠BAC = 180°

    ⇒ ∠BAC = 90°

    ∴ In ΔABC, ∠A = 90°, ∠B = ∠C = 45°.

    ⇒ ΔABC is a right angled isosceles triangle.

    Hence, the correct option is (d)

23.

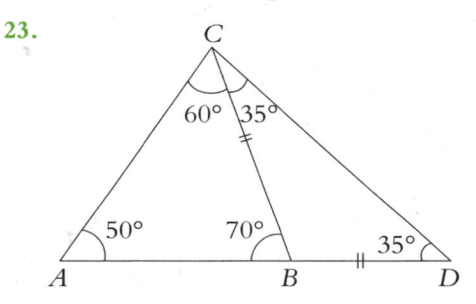

    In ΔABC,

    ∠A + ∠ABC + ∠ACB = 180°

    50° + ∠ABC + 60° = 180°

    ∠ABC = 180° − 110°

    ∠ABC = 70°

    ⇒ ∠BCD + ∠BDC = 70°.

    (∵ an exterior angle of ΔBCD)

    But BC = BD ⇒ ∠BCD = ∠BDC = x° (say)

    ⇒ 2x° = 70°

    ⇒ x° = 35°

    ∠BDC = 35°.

    Hence, the correct option is (a)

24.

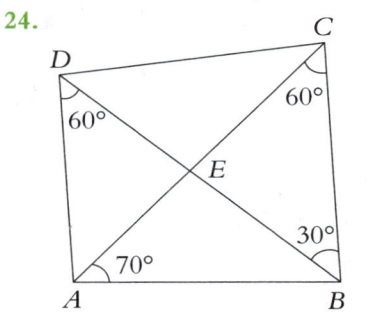

In $\triangle BEC$, $\angle CEB = 180 - 30 - 60 = 90°$.

$\Rightarrow \angle AEB = 90°$

($\because \angle AEB$ and $\angle CEB$ form linear pair)

$\therefore \angle ABE = 180 - 90 - 70 = 20°$.

In $\triangle DEA$,

$\angle DAE = 180° - 60° - 90° = 30°$

$\Rightarrow \angle DAC = \angle DAE = 30°$.

Hence, the correct option is (a)

25. Let (a, b) be a pair of adjacent angles of the parallelogram.

Let $a = 3b$

$\therefore a + b = 180°$.

$3b + b = 180°$ ($\because a = 3b$)

$4b = 180°$

$b = 45°$.

The angles of a parallelogram are 135°, 45°, 135° and 45°.

Hence, the correct option is (b)

26. Every square is a rhombus. It is a true statement.

Hence, the correct option is (b)

Solutions for questions 27 to 30:

27. → (c): In a right triangle, the orthocentre coincides with the vertex of the triangle where the right angle is formed.

28. → (d): In a right triangle, the circumcentre is the midpoint of the hypotenuse of the triangle.

29. → (b): Centroid lies in the interior of any triangle.

30. → (a): Excentre lies in the exterior of any triangle.

Chapter 6

Mensuration

a) $5(3x) - 3(x-4) = 0$
$15x - 3x + 12 = 0$
$12x + 12 = 0$
$12x = -12$
$x = -12/12$
$x = -1 \quad x^2 = 1$

REMEMBER
Before beginning this chapter, you should be able to:
- Describe the plane figures, surface area, volume
- Understand the length, breath and height of an object

KEY IDEAS
After completing this chapter, you should be able to:
- Review the concept of plane figures and units of measurement for geometric figures
- Calculate the areas of rectangles, parallelograms, triangles, trapeziums, rhombuses, squares, circles, circular rings, sectors
- Use standard formulas for parameter and circumference
- Find surface areas and volumes of cubes and cuboids
- Solve numerical using these concepts

INTRODUCTION

Mensuration is a branch of Mathematics that deals with the computation of geometric magnitudes, such as the length of a line, the area of a surface and the volume of a solid. In this chapter, we shall deal with the computation of areas and perimeters of plane figures; surface areas and volume of cube and cuboid.

Plane Figures

A figure lying on a plane is called a plane figure. Triangles, rectangles and circles are some examples of plane figures.

A plane figure is a closed figure if it has no free end. It is called a simple closed figure if it does not cross itself.

A plane figure is made up of lines or curves or both.

A plane figure is called a rectilinear figure if it is made up of only line segments. Triangles, squares, pentagons, etc., are some examples of rectilinear figures. A circle is not a rectilinear figure. The part of the plane which is enclosed by a simple closed figure is called a plane region. The magnitude of a plane region is called its area.

A line segment has one dimension—length. Hence, its size is measured in terms of its length.

A planar region has two dimensions—length and breadth. Hence, its size is measured in terms of its area.

The perimeter of a closed planar figure is the total length of the line segments enclosing the figure.

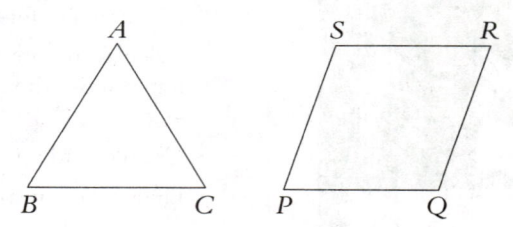

In the figures above, ABC is a triangle and $PQRS$ is a quadrilateral.

The perimeter of the $\triangle ABC = AB + BC + CA$.

The perimeter of the quadrilateral $PQRS = PQ + QR + RS + SP$.

> **Note** The perimeter of a rectangle of length l and breadth $b = 2(l + b)$.
> The perimeter of a square of side $s = 4s$.

Units of Measurement

The length of a line segment is measured in units such as centimetres, inches, metres, feet, etc.

Area is measured in units such as square centimetres, square inches, square metres, square feet, etc.

Estimating Areas

The area of a plane figure can be estimated by finding the number of unit squares that can fit into a whole figure. Observe the following square $ABCD$,

Mensuration

In the square ABCD, there are 16 unit squares. Therefore its area is 16 sq. units.

Observe the adjacent $\triangle PQR$.

In the above figure, PQR is a right angled triangle. All the small squares are not complete squares. Just as the length of a line need not be a whole number, the area of a plane figure also need not be a whole number.

In the above figure, the triangle PQR does not cover 8 unit squares, but its area is equal to that of 8 unit squares (because the partial unit squares when added will be equal to 2 unit squares). Besides that, there are 6 unit squares. Hence its area is equal to 8 unit squares.

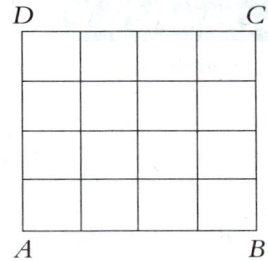

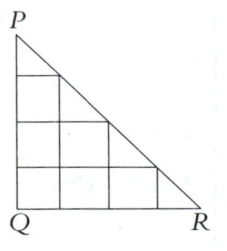

Area of a Rectangle

A rectangle is a closed figure having four sides, in which opposite sides are equal and each angle measures 90°. The following figure ABCD is a rectangle.

In the adjoining rectangle ABCD, AB = 5 cm and BC = 4 cm.

Each of the small squares measures 1 cm by 1 cm and the area occupied by each small square is 1 cm². There are 4 rows, each row consisting of 5 unit squares.

∴ The area occupied by the rectangle = (4) (5) = 20 cm².
The number of unit squares in the rectangle ABCD is equal to the product of the number of unit squares along the length and the number of unit squares along the breadth. Thus, we see that the area of a rectangle = (length) (breadth).

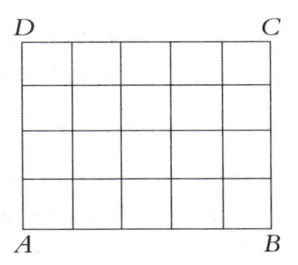

EXAMPLE 6.1

Find the area of the figure given below.

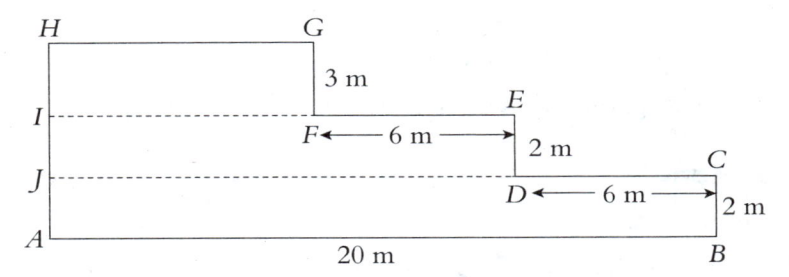

SOLUTION

Join DJ, FI which are represented by the dotted lines above.
The area of the given figure = Sum of the area of the rectangles ABCJ, JDEI and FGHI
= AB × BC + DJ × DE + FI × FG
= 20(2) + 14(2) + 8(3) = 40 + 28 + 24 = 92 m².

6.4 Chapter 6

> **EXAMPLE 6.2**
>
> Find the area of the paths each having a uniform width in the following rectangular field. 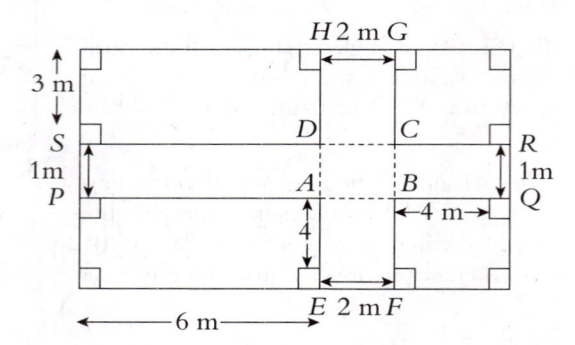
>
> **SOLUTION**
>
> $PQ = RS = 6 + 2 + 4 = 12$ m and $QR = 1$ m
>
> ∴ Area of path $PQRS = PQ \times QR = 12 \times 1 = 12$ m².
>
> $EF = 2$ m and $EH = GF = 3 + 1 + 4 = 8$ m
>
> ∴ Area of path $EFGH = EH \times EF = 8 \times 2 = 16$ m².
>
> ∴ Required area = Area of $PQRS$ + Area of $EFGH$ − Area of $ABCD$
> $= 12 + 16 - 2 \times 1 = 26$ m².

Area of a Parallelogram

Consider the following parallelogram $ABCD$.

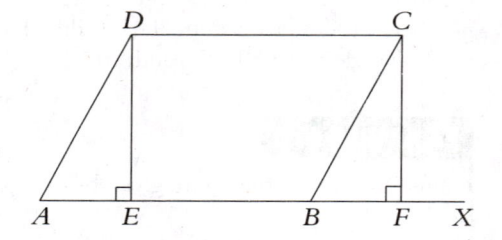

Draw a perpendicular DE from D to AB.

Draw a perpendicular CF to AX (AB extended).

From the ASA congruency property, the area of $\triangle AED$ is equal to the area of $\triangle BFC$.

∴ The area of parallelogram $ABCD$ = The area of rectangle $\triangle EFC = (EF)(FC) = (AB)(DE)$.

∴ **Area of parallelogram = (Base) (Height)**

Area of a Triangle

Consider the following $\triangle ABC$.

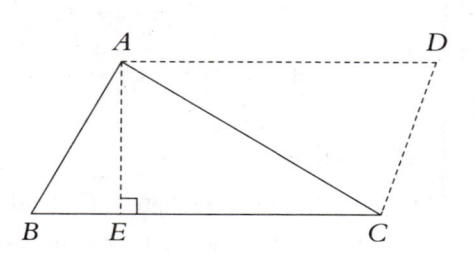

Draw a line AD parallel and equal to BC. Join CD.

$ABCD$ is a parallelogram.

From the ASA axiom, the area of triangle ABC is equal to the area of triangle ADC.

∴ The area of triangle ABC = (1/2) (Area of parallelogram) = (1/2) $(BC)(AE)$.

(Where AE is the height of the triangle or parallelogram)

∴ **The area of a triangle = (1/2) (Base) (Height).**

Area of a Trapezium

Consider the following trapezium $ABCD$ in which BC and AD are the parallel sides and AB and CD are the oblique sides (not parallel).

Join BD

DE is the perpendicular distance between AD and BC.

The area of the trapezium is equal to the sum of the areas of the $\triangle ABD$ and $\triangle BCD$.

$= (1/2)\,(AD)\,(DE) + (1/2)\,(BC)\,(DE)$

$= (1/2)\,(AD + BC)\,(DE)$

∴ **Area of a trapezium = (1/2) (Sum of the parallel sides) (Distance between them).**

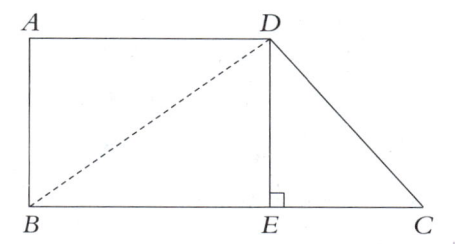

Area of a Rhombus

A rhombus is a parallelogram in which all sides are equal.

Consider the following rhombus $PQRS$.

In a rhombus, the diagonals bisect each other at right angles.

∴ $\angle POS = 90°$ and $\angle POQ = 90°$.

Area of the rhombus $PQRS$ = Area of $\triangle PSR$ + Area of $\triangle PQR$

$= (1/2)\,(PR)\,(OS) + (1/2)\,(PR)\,(OQ)$

$= (1/2)\,(PR)\,(OS + OQ) = (1/2)\,(PR)\,(SQ)$

PR and SQ are the diagonals of the rhombus.

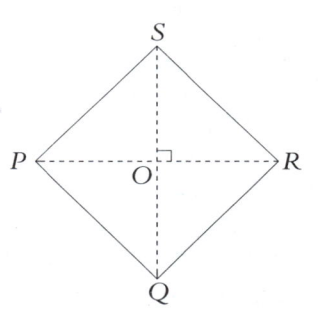

∴ **Area of a rhombus = (1/2) (Product of the diagonals).**

EXAMPLE 6.3

Find the area of rhombus whose one of the diagonals is 32 cm and its perimeter is 80 cm.

SOLUTION

Let AB be the side and E be the point of intersection of diagonals DB and AC.

Given $BD = 32$ cm $\Rightarrow BE = 16$ cm and perimeter of rhombus = 80 cm $\Rightarrow$ side $AB = 80/4 = 20$ cm

In right triangle ABE, $AE^2 = AB^2 - BE^2$

$\qquad\qquad\qquad = 20^2 - 16^2 = 400 - 256$

$AE = \sqrt{144} = 12$ cm.

∴ Other diagonal $AC = 24$ cm.

Area of the rhombus $= \dfrac{1}{2} \times 32 \times 24 = 384$ cm^2.

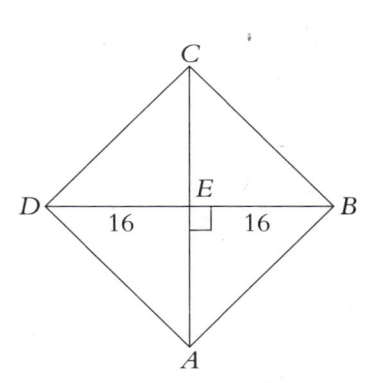

Area of a Square

A square is a rectangle in which all the sides are equal.

Area of a rectangle = (length) (breadth)

In a square, length = breadth

∴ **Area of a square = (Side)^2**

Perimeter of a square = 4 (Side)

Diagonal of a square = $\sqrt{2}$ (Side)

∴ Area of a square = $\dfrac{(\text{Diagonal})^2}{2}$.

EXAMPLE 6.4

A path of 1 m wide runs around and inside a square garden of side 20 m. Find the cost of levelling the path at the rate of ₹2.25 per square metre (in ₹).

SOLUTION

Area of the field
= (20) × (20) − (20 − 2) (20 − 2)
= 400 − 324 = 76 m^2

Cost of levelling the path = 76 × 2.25 = ₹171.

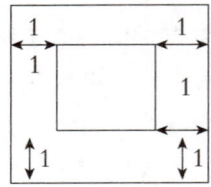

Area of an Equilateral Triangle

In the following figure, ABC is an equilateral triangle.

Let $AB = BC = CA = a$

Draw a perpendicular AD from A to D. (D is a point on $\overline{BC}$)

AD is the height of the triangle.

ADC is a right-angled triangle, where $DC = \dfrac{a}{2}$ and $AC = a$.

$AD = \sqrt{(a)^2 - \left(\dfrac{a}{2}\right)^2} = \dfrac{\sqrt{3}}{2}a$ (Using Pythagoras theorem)

∴ The height of an equilateral triangle = $\dfrac{\sqrt{3}}{2}a$.

The area of an equilateral triangle = $\dfrac{1}{2}(a)\left(\dfrac{\sqrt{3}}{2}a\right) = \dfrac{\sqrt{3}}{4}a^2$

∴ **The area of an equilateral triangle = $\dfrac{\sqrt{3}}{4}$ (side)^2.**

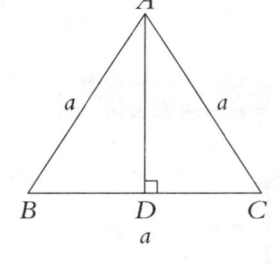

EXAMPLE 6.5

Find the area of an equilateral triangle of side 6 cm.

SOLUTION

Area of an equilateral triangle ABC is $\dfrac{\sqrt{3}}{4}(\text{side})^2 = \dfrac{\sqrt{3}}{4}.6^2 = 9\sqrt{3}$ sq.cm.

Area of a Right-angled Triangle

In the adjacent figure, ABC is a triangle, right angled at B.

Since AB is perpendicular to the base BC of the triangle, AB is the altitude of the triangle.

∴ The area of right-angled $\triangle ABC = (1/2)\,(b)\,(a) = ab/2$.

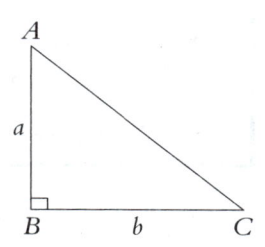

Isosceles Right Triangle

A right triangle in which the two perpendicular sides are equal is an isosceles right triangle.

In the figure, ABC is a right triangle, where $AB = BC$.

The area of the triangle $ABC = (1/2)\,(a)\,(a) = a^2/2$.

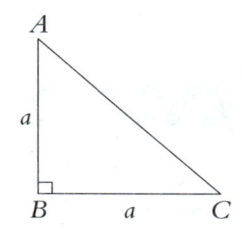

EXAMPLE 6.6

Find the area of a right isosceles triangle whose hypotenuse is $16\sqrt{2}$ cm.

SOLUTION

∴ The area of the right isosceles triangle

$$= \frac{h^2}{4} = \frac{(16\sqrt{2})^2}{4}$$

$$= \frac{256 \times 2}{4}$$

$$= 128 \text{ cm}^2.$$

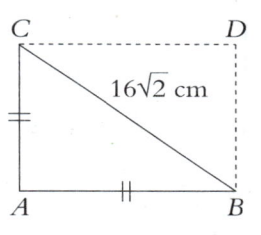

Circumference of a Circle

Since a circle is a closed arc, its perimeter is known as circumference. The circumference of a circle is π (read as pie) times its diameter. π is an irrational number. For numerical work, we use the approximate value of π as 3.14 or 22/7. The circumference of a circle $= \pi d = 2\pi r$. The area of a circle $= \pi r^2$.

Area of a Ring

The region enclosed between two concentric circles is known as a ring.

In the adjoining figure, the shaded region represents a ring. Let the radius of the smaller circle, OA be r and the radius of the bigger circle, OB be R. The area of the ring = (Area of the bigger circle) − (Area of the smaller circle) $= \pi R^2 - \pi r^2 = \pi(R^2 - r^2)$

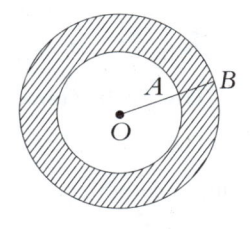

EXAMPLE 6.7

A circular track runs around and inside a circular park. If difference between the circumference of the track and the park is 132 metres, find the width of the track.

SOLUTION

Let R be the radius of the park and r be the radius of the inner circle.
Given $2\pi R - 2\pi r = 132$

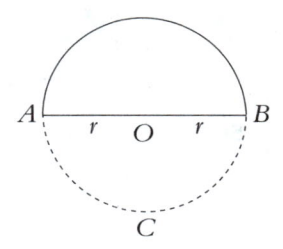

Chapter 6

$2\pi(R - r) = 132$

$R - r = \dfrac{132 \times 7}{2 \times 22} = 21$ cm

∴ Width of the track = 21 cm.

When a circle is cut along one of its diameters into two equal parts then each part is called a semicircle. In the figure, ACB is a semicircle. The perimeter of the semicircle including its diameter is $\pi r + 2r$. Area of the semicircle is $\dfrac{\pi r^2}{2}$.

EXAMPLE 6.8

Find the area and the perimeter of the semicircle whose radius is 14 cm.

SOLUTION

(a) Perimeter of a semicircle is $\pi r + 2r = 22/7 \times 14 + 2 \times 14 = 72$ cm.

(b) Area of semicircle = $\dfrac{\pi r^2}{2} = \dfrac{1}{2} \times \dfrac{22}{7} \times 14 \times 14 = 308$ cm².

Sector of a circle

A sector is a closed figure formed by joining each of the end points of an arc of a circle and the centre of the circle.

In the above circle with centre O, AXB is an arc of the circle. The end points of the arc, A and B are joined with the centre O of the circle. AOB is a sector of the circle. The arc makes an angle θ at the centre of the circle. This is also known as the central angle.

The length of the arc, the circumference of the circle the central angle of the sector and the angle of the circle are in proportion, i.e.,

$$\dfrac{\text{Length of the arc}}{\text{Circumference of the circle}} = \dfrac{\text{Central ange of the sector}}{360}$$

∴ The length of the arc of the sector of the circle $(l) = \dfrac{\theta}{360} \times 2\pi r$.

(Where r is the radius of the circle and θ is the central angle of the sector.)

The perimeter of the sector of the circle = $l + 2r$.

The area of the sector of the circle, the area of the circle and the central angle of the sector, the angle of the circle are in proportion, i.e., $\dfrac{\text{Area of the sector}}{\text{Area of the circle}} = \dfrac{\text{Central angle of the sector}}{360°}$.

∴ The area of a sector of a circle = $\dfrac{\theta}{360}(\pi r^2)$.

EXAMPLE 6.9

Find the area of the sector of a circle of angle 120°, if the radius of the circle is 21 cm.

SOLUTION

The area of the sector = $\dfrac{\theta}{360}(\pi r^2)$

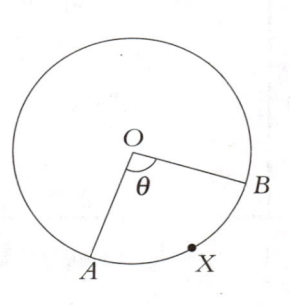

Here $\theta = 120°$ and $r = 21$ cm

$$= \frac{120°}{360°} \times \frac{22}{7} \times 21 \times 21 = 462 \text{ cm}^2.$$

So far we have dealt with the measurement of the perimeters and areas of plane figures such as triangles, quadrilaterals, circles, etc. Now, we shall discuss the methods of measuring the surface areas and volumes of solid figures.

Solids

A plane figure may have one dimension or two dimensions. A circle is said to be uni-dimensional. Triangles and quadrilaterals have two dimensions. For two-dimensional figures, the dimensions are length and breadth or width or height. But many objects such as a brick, a matchbox, a pencil, a marble, a tank, an ice cream cone, etc., have a third dimension. These objects are known as solids.

Thus, a solid is a three dimensional object. In general, any object that occupies space and has a definite shape is called a solid. Solids like prisms, cubes, cuboids, etc., have a plane or flat surfaces.

Cubes and Cuboids

Cuboid

A cuboid is a solid bounded by six rectangular faces and the opposite faces are congruent. In the following figure, faces *ABFE* and *DCGH* are congruent. Similarly, faces *BCGF* and *ADHE* are congruent and faces *ABCD* and *EFGH* are congruent. Each face is a rectangle. Line segments like *AB*, *BC* and *BF* are called edges. A cuboid has 6 faces and 12 edges.

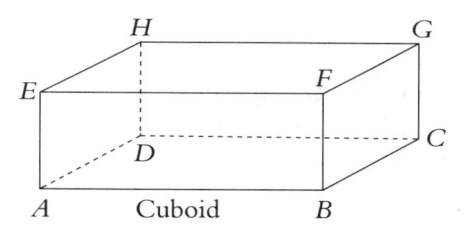

A matchbox, a brick, a room, etc., are in the shape of a cuboid. The three dimensions of the cuboid, length (*l*), breadth (*b*) and height (*h*) are generally denoted by $l \times b \times h$.

1. The lateral surface area of a cuboid = $ph = 2(l + b)h$ sq. units, where p is the perimeter of the base.

2. The total surface area of a cuboid = LSA + 2(base area).

    $= 2(l + b)h + 2\ lb = 2(lb + bh + lh)$ sq. units.

3. The volume of a cuboid = $Ah = (lb)h = lbh$ cubic units, where A is the area of the base.

    Note If a box made of wood of thickness t has inner dimensions l, b and h, then
    the outer length $= l + 2t$,
    the outer breadth $= b + 2t$ and
    the outer height $= h + 2t$.

Cube

In a cuboid, if all the dimensions, i.e., its length, breadth and height are equal, then the solid is called a cube.

All the edges of a cube are equal in length and each edge is called the side of the cube.

Thus, the size of a cube is completely determined by its sides.

If the edge of cube is "*a*" units, then

Chapter 6

1. The lateral surface area of the cube = $4a^2$ sq. units.
2. The total surface area of the cube = LSA + 2(area of base)
   = $4a^2 + 2a^2 = 6a^2$ sq. units.
3. The volume of the cube = a^3 cubic units.

   Note In a cubical box, if the inner edge of a cube is "a" units and the thickness is "t" units, then the outer edge of the cube is given by $(a + 2t)$ units.

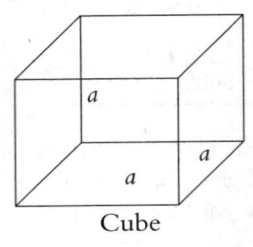

Cube

EXAMPLE 6.10

The dimensions of the cuboid are 24 cm × 12 cm × 5 cm. Find its lateral surface area, total surface area and volume.

SOLUTION

Given, $l = 24$ cm, $b = 12$ cm and $h = 5$ cm.
(a) Lateral surface area of cuboid = $2h(l + b) = 2 \times 5 (24 + 12) = 360$ cm².
(b) Total surface area of cuboid = $2(lb + bh + lh) = 2(24 \times 12 + 12 \times 5 + 24 \times 5) = 936$ cm².
(c) Volume of cuboid = $l \times b \times h = 24 \times 12 \times 5 = 1440$ cm³.

EXAMPLE 6.11

A plot is in the shape of a trapezium. The parallel sides of the plot are 20 m and 16 m long and the distance between them is 8 m. A square plot has the same area as the trapezium. A pit of dimensions 2.4 m long, 1.5 m wide and 1 m deep, is dug inside the trapezium plot and the soil is spread uniformly on the square plot. Find the rise in the level of the plot.

SOLUTION

Area of trapezium = $\frac{1}{2} \times 8 \times (20 + 16)$

= $4 \times 36 = 144$ m²

⇒ Area of the square plot = 144 m²

Volume of the soil dugout = $2.4 \times 1.5 \times 1$

Rise in the level = $\dfrac{\text{Volume of the soil spread}}{\text{Area of the square plot}}$

Rise in level = $\dfrac{3.6}{144}$ m = $\dfrac{360}{144}$ cm = 2.5 cm

EXAMPLE 6.12

The volume of a Harry Potter book is 3600 cubic centimeters. The length, breadth and thickness of each page are 20 cm, 10 cm and 0.006 cm respectively. On an average if each paper costs ₹0.50, then the price of the book is _____.

SOLUTION

Volume of each page = $20 \times 10 \times 0.006 = 1.2$ cm²
Number of pages of book

$$= \frac{\text{Volume of the book}}{\text{Volume of each page}} = \frac{3600}{1.2} = 3000$$

∴ Price of the book = 3000 × 0.50 = ₹1500.

EXAMPLE 6.13

The length and breadth of a rectangular sheet of a paper are 50 cm and 30 cm respectively. A square of side 5 cm is cut and removed from the four corners of the sheet, and the rest of the paper is folded to form a cuboid (without the top face). Find the edge of the cube whose volume is one-fourth the volume of the cuboid so formed.

SOLUTION

Volume of the cuboid formed = (50 − 2 × 5) × (30 − 2 × 5) × 5
$$= 40 \times 20 \times 5$$
$$= 4000 \text{ cm}^3.$$

Let the edge of the cube be x cm.

∴ The volume of the cube = x^3.

Given, $4x^3 = 4000$.

⇒ $x^3 = 1000$.

⇒ $x = 10$ cm.

∴ The edge of the cube whose volume is equal to the cuboid formed = 10 cm.

Points to Remember

- The area of a triangle = (1/2) (Base) (Height).
- Area of a parallelogram = (Base) (Height).
- Area of a trapezium = (1/2) (Sum of the parallel sides) (Distance between them).
- Area of rhombus = (1/2) (Product of the diagonals).
- Area of a square = (Side)^2.
- Perimeter of a square = 4 (Side).
- Diagonal of a square = $\sqrt{2}$ (Side).
- The area of a ring = $\pi(R^2 - r^2)$
- The length of the arc of a sector of a circle (l) = $\dfrac{\theta}{360} \times 2\pi r$.
- The area of a sector of a circle = $\dfrac{\theta}{360}(\pi r^2)$.
- The lateral surface area of a cuboid = ph = $2h(l + b)$ sq units, where p is the perimeter of the base.
- The total surface area of a cuboid = $2(lb + bh + lh)$.
- The volume of a cuboid = lbh.
- The lateral surface area of a cube = $4a^2$.
- The total surface area of a cube = $6a^2$.
- The volume of a cube = a^3.

Chapter 6

TEST YOUR CONCEPTS

Very Short Answer Type Questions

Directions for questions 1 to 5: State whether the following statements are true or false.

1. 1 hectare = 1000 m².
2. If the area of a square is 36 m², then its side is 6 m long.
3. The area of a parallelogram is equal to the product of a side of the parallelogram and the distance between any two sides.
4. The length of a diagonal of a square is $\sqrt{2}$ times its side.
5. If the area of a rhombus is 60 cm², then its diagonals are 10 cm and 6 cm long.

Directions for questions 6 to 11: Fill in the blanks.

6. The area of a sector of a circle is 1/6 of area of the circle, then the angle of the sector is _____.
7. The perimeter of a square is 16 cm, then its area is _____ cm².
8. The length and the breadth of a rectangle are 12 cm and 5 cm respectively, then its diagonal is _____ cm.
9. In a quadrilateral, the length of the altitudes drawn onto a diagonal is 9 cm and 12 cm. If its area is 420 cm², then the length of the diagonal is _____.
10. The area of a circle of radius 7 cm is _____ cm².
11. If the volume of a cube is 27 cm³, then the length of its edge is _____ cm.

Directions for questions 12 to 17: Select the correct alternative from the given choices

12. If the radius of a circle is 7 cm, then the circumference of the circle is _____.
    (a) 22 cm   (b) 44 cm
    (c) 66 cm   (d) 88 cm

13. A cube of edge 9 cm is cut into x cubes each of edge 3 cm. Then x = _____.
    (a) 54   (b) 9
    (c) 27   (d) 81

14. The base of a triangle is 5 cm and its area is 20 cm², then the height of the triangle is _____.
    (a) 8 cm   (b) 9 cm
    (c) 10 cm   (d) 12 cm

15. The angle of a sector is 30° and the radius of the sector is 21 cm, then length of the arc of the sector is _____.
    (a) 9 cm   (b) 11 cm
    (c) 10 cm   (d) 13 cm

16. Find the perimeter of a square whose area is 225 m².
    (a) 15 m   (b) 60 m
    (c) 225 m   (d) None of these

17. The area of a parallelogram is 50 cm². If the base is 10 cm, then find its corresponding height.
    (a) 10 cm   (b) 15 cm
    (c) 5 cm   (d) None of these

Short Answer Type Questions

18. Find the perimeter of a square whose side is 10 cm.
19. Find the perimeter of a rectangle whose length is 9 cm and breadth is 6 cm.
20. Find the length of the rectangle whose area is 120 cm² and breadth is 6 cm.
21. Find the area of a circle whose radius is 2.1 m.
22. The area of a parallelogram is 144 cm² and its height is 18 cm. Find the length of the corresponding base.
23. The area of a rectangular ground is 120 m² and its length is 12 m. Find its breadth.
24. Find the total surface area of a cube whose edge is 8 cm.
25. Find the volume of a cube whose edge is 10 cm.
26. Find the total surface area of a cuboid of length 10 m, breadth 7 m and height 5 m.
27. The total surface area of a cube is 726 m². Find its edge.

28. Find the perimeter of the figure given below.

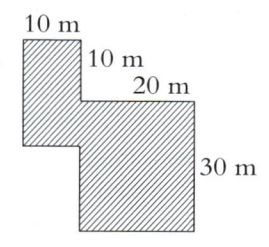

29. From a circular aluminium sheet of radius 14 cm, a sector of angle 45° is removed. Find the percentage of the area of the sector removed.

30. The area of top face of a book, in the shape of a square, is 225 cm². The thickness of the book is 5 cm. Find its volume.

31. The area of a square field is 16 hectares. Find the cost of fencing it with a wire at 2₹ per metre.

32. A rectangular playground is 35 m long and 15 m broad. Find the cost of fencing it at ₹2.50 per metre.

33. The volume of a cube is 343 cm³. Find its total surface area.

34. Find the area of aluminium sheet required to make a closed box of 4 m long, 2 m wide and 1 m high.

35. If the radius of a wheel is 35 cm, then find how many times it should rotate to cover a distance of 33 m.

36. The floor dimensions of a conference hall are 100 ft × 20 ft. Find the number of tiles required for flooring with square tiles of side 2 ft each.

37. Find the length of an arc of a sector whose central angle is 45° and radius is 56 cm.

38. The base and the corresponding height of a parallelogram are 24 cm and x cm respectively. Find the value of x if its area is 216 cm².

39. The length of the parallel sides of a trapezium is 12 cm and 8 cm. Its area is 100 cm². Find the distance between the parallel sides.

40. A square sheet of side 5 cm is cut out from a rectangular piece of an aluminium sheet of length 9 cm and breadth 6 cm. What is the area of the aluminium sheet left over?

41. Find the cost of fencing a semi-circular garden of radius 14 m at ₹10 per metre.

42. The edges of two cubes are 4 cm and 6 cm. Find the ratio of the volumes of the two cubes.

Essay Type Questions

43. Find the area of the figure given below.

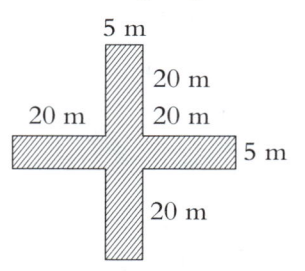

44. Calculate the volume of the given solid.

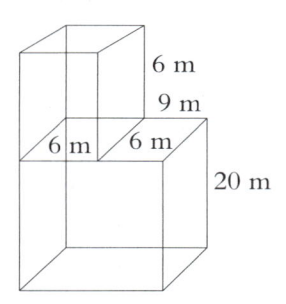

45. The diagonals of a rhombus are 6.5 cm and 14 cm. Find the area of one of the 4 triangles formed by both the diagonals.

46. The circumference of a playground is 132 m. A circular track of width 2 m is laid outside around it. Find the area of the circular track.

47. Find the maximum number of cuboids of dimensions 4 cm, 3 cm and 2 cm that can fit in a box of inner dimensions 36 cm, 16 cm and 8 cm.

48. The length of a cuboid is 2 cm more than its height where as its breadth is 2 cm less than the height. Its height is equal to the side of a square having an area of 100 cm². Find the volume of the cuboid.

49. The inner dimensions of a swimming pool are 50 ft × 20 ft × 10 ft. The inner sides and the floor of the swimming pool have to be fixed with tiles of size 2 ft × 2 ft. Find the cost of fixing the tiles at the rate of ₹15 each.

50. The diagonals of a rhombus are 10 cm and 8 cm. If the area of rhombus is equal to $\frac{1}{10}$th of the area of a square, then find the cost of fencing the square at the rate of ₹10 per metre.

51. In the figure given below, PQRS is a rectangle. Find the area of the shaded portion in the given figure.

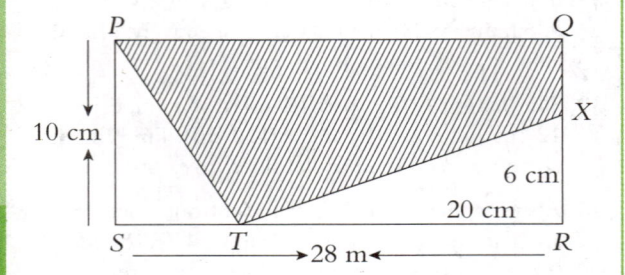

52. Find the total cost of levelling the shaded path of uniform width 2 m, laid in the rectangular field shown below, if the rate per m² is ₹100.

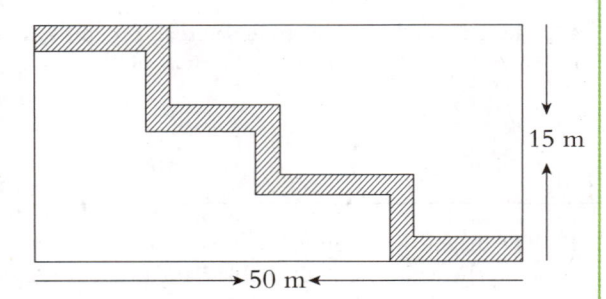

53. The dimensions of a pit are 5 m × 3 m × 2 m. If it is filled with bricks of size 20 cm × 10 cm × 5 cm which costs ₹2.50 each then find the total cost of the bricks.

54. From the four corners of a rectangular plastic sheet of size 34 cm × 24 cm, four squares, each of side 2 cm are cut and removed. The remaining sheet is folded to form a cuboid without the top face. If the cuboid is filled with vanilla ice cream which costs ₹90 per litre, then find the cost of the ice cream in the cuboid.

55. The area of a square field is 36 m². A path of uniform width is laid around and outside of it. If the area of the path is 28 m², then find the width of the path.

56. How many times should a wheel of radius 1.4 m rotate to go around the perimeter of a rectangular field of length 120 m and breadth 100 m?

57. How far does the tip of the minute hand of length 7 cm in a clock move in 30 minutes?

58. A square sheet of maximum area is cut out from a circular piece of cardboard of radius 21 cm. Find the area of the remaining card board.

59. Find the length of the hypotenuse of a right isosceles triangle whose area is 961 cm².

60. The area of a circular play ground is 154 m². A path of width 2 m is laid around and outside of it. Find the cost of levelling the path at the rate of ₹7 per square metre.

61. The parallel sides of a trapezium are 48 cm, 40 cm and the distance between them is 14 cm. Find the radius of a circle whose area is equal to the area of the trapezium.

62. The length and breadth of a rectangular sheet of paper are 60 cm and 30 cm respectively. A square of side 5 cm is cut and removed from the four corners of the sheet. The rest of the paper is folded to form a cuboid (without the top face). Find the volume of the cuboid so formed (in cm³).

63. In the figure given below, PQRS is a square of side 14 cm and two semicircles are drawn inside of it with PQ and SR as diameters. Find the area of the shaded region in the figure.

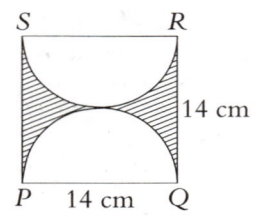

64. The dimensions of a class room are 10 m, 8 m and 5 m. Find the cost of decorating its walls by a wall paper of 1 m width, which costs ₹50 per metre.

65. Find the ratio of the lateral surface area and the total surface area of a cube.

Mensuration 6.15

CONCEPT APPLICATION

Level 1

Directions for questions 1 to 13: Select the correct alternative from the given choices.

1. The volume of a cube is 729 cm³, then the length of its edge is _____.
   (a) 9 m (b) 7 m
   (c) 5 m (d) 3 m

2. Find the maximum length of the side of a square sheet that can cut off from a rectangular sheet of size 8 m × 3 m.
   (a) 3 m (b) 4 m
   (c) 6 m (d) 4 m

3. Inner measurements of a rectangular water tank are 6 m long, 3 m wide and 2 m depth, then the capacity of the tank is _____.
   (a) 36 m³ (b) 18 m³
   (c) 6 m³ (d) None of these

4. Find the total surface area of a cube of edge 5 m.
   (a) 125 m² (b) 250 m²
   (c) 150 m² (d) 300 m²

5. Find the cost of fencing a circular garden of radius 21 m at ₹10 per metre.
   (a) ₹1320 (b) ₹132
   (c) ₹1200 (d) ₹660

6. Find the perimeter of a square whose area is 256 m².
   (a) 24 m (b) 36 m
   (c) 46 m (d) 64 m

7. The area of a parallelogram is 100 cm². If the base is 25 cm, then find its corresponding height.
   (a) 4 cm (b) 6 cm
   (c) 10 cm (d) 5 cm

8. The volume of a cube is 512 cm³, then the length of the edge is _____.
   (a) 7 cm (b) 8 cm
   (c) 4 cm (d) 12 cm

9. The dimensions of a shop are 12 ft × 8 ft × 5 ft. The area of four walls is _____
   (a) 200 ft² (b) 400 ft²
   (c) 500 ft² (d) 800 ft²

10. The side of a square metal sheet is 14 cm. A circular sheet of greatest diameter is cut out of it and removed. Find the area of the remaining metal sheet.

    The following are the steps involved in solving the above problem. Arrange them in sequential order.

    (A) Area of circular sheet = $\pi r^2 = \dfrac{22}{7} \times 7 \times 7 = 154$ cm² and
    Area of square metal sheet = (side)² = (14)² = 196 cm².

    (B) Since the circular sheet has greatest diameter, diameter of the circle = side of the square = 14 cm.

    (C) ∴ Radius of the circular sheet = $\dfrac{14}{2} = 7$ cm.

    (D) ∴ The area of the remaining metal sheet = Area of the square metal sheet − Area of the circular sheet = 196 − 154 = 42 cm².

    (a) BACD (b) ABCD
    (c) BCAD (d) ACBD

11. The inner and the outer radii of a circular track are 7 m and 14 m respectively. Find the cost of levelling the track at the rate of ₹100 per m².

    The following are the steps involved in solving the above problem. Arrange them in sequential order

    (A) Area of circular track = $\pi(R^2 - r^2)$ and given R = 14 m and r = 7 m.

    (B) ∴ Area of circular track = $\dfrac{22}{7}(196 - 49)$.

    (C) Area of the track = $\dfrac{22}{7} \times 147 = 22 \times 21 = 154$ cm²

    (D) Cost of levelling the path = 462 × 100 = ₹46200.

    (a) ABCD (b) ACBD
    (c) CBAD (d) BCAD

Chapter 6

12. The perimeter of an equilateral triangle is 18 cm; find the area of the triangle.

    The following are the steps involved in solving the above problem. Arrange them in sequential order.

    (A) Area of the equilateral triangle (A) $= \frac{\sqrt{3}}{4}a^2 = \frac{\sqrt{3} \times (6)^2}{4}$.

    (B) $A = 9\sqrt{3}$ cm^2

    (C) Let the side of the equilateral triangle be a and its perimeter = $3a$ = 18 cm (given).

    (D) $a = 6$ cm

    (a) CADB
    (b) CDAB
    (c) CBDA
    (d) ABCD

13. The area of the base of a cuboid is 800 cm^2 and its height is 60 cm. Find the cost of the quantity of milk filled in the cuboid at the rate of ₹13 per litre.

    The following are the steps involved in solving the above problem. Arrange them in sequential order.

    (A) Since 1000 cm^3 = 1 l ⇒ Capacity of the cuboid = 48 l.

    (B) ∴ Volume of the cuboid = 800 × 60 = 48000 cm^3.

    (C) The cost of the milk filled in the cuboid = 48 × 13 = ₹624.

    (D) Given area of the base = 800 cm^2 and height h = 60 cm.

    Volume of cuboid = Area of the base × Height.

    (a) BADC
    (b) ADBC
    (c) DACB
    (d) DBAC

Directions for questions 14 to 17: Match the Column A with Column B.

Column A		Column B
14. Area of a trapezium is _____.	( ) (a)	(a) 4 m^2
15. If the volume of a cube is 512 cm^3, then the lateral surface area is _____.	( ) (b)	64 cm
16. If the diagonal of a square is $\sqrt{8}$ m, then its area is _____.	( ) (c)	$\left(\frac{1}{2}\right)$ (Sum of the parallel sides) (Distance between them)
17. If the radius of the circle is 7 cm, then the perimeter of the sector of angle 45° is _____.	( ) (d)	$\left(\frac{1}{2}\right)$ (Product of the diagonals)
	(e)	256 cm^2
	(f)	19.5 m^2

Directions for questions 18 to 21: Match the Column A with Column B.

Column A		Column B
18. If r and R be the radii of inner and outer circles respectively, then area of a ring is	( ) (a)	60
19. The total surface area of a cube of edge $\sqrt{10}$ cm is _____ cm^2.	( ) (b)	$10\sqrt{2}$
20. The diagonal of a square of side 10 cm is _____ cm.	( ) (c)	$\pi(R^2 + r^2)$ sq. units
21. If the radius of the circle is 14 cm, then the area of the sector of angle 45° is _____ cm^2.	( ) (d)	$\pi(R^2 - r^2)$ sq. units
	(e)	$\sqrt{20}$ cm
	(f)	77 cm^2

Level 2

Directions for questions 22 to 65: Select the correct alternative from the given choices.

22. The area of the shaded part in the figure given below is _____.

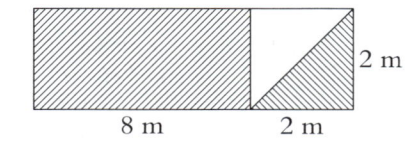

    (a) 16 m² (b) 18 m²
    (c) 14 m² (d) 20 m²

23. The angle of a sector is 120° and its radius is 21 cm. The area of the sector is _____.

    (a) 426 cm² (b) 462 cm²
    (c) 362 cm² (d) 526 cm²

24. The length and breadth of a rectangular field are 4 m and 3 m respectively. The field is divided into two parts by fencing it diagonally. Find the cost of fencing at ₹10 per metre.

    (a) ₹50 (b) ₹30
    (c) ₹40 (d) ₹70

25. A path of 3 metres wide runs around and outside a rectangular playground of length 15 m and width 10 m. Find the area of the path (in m²).

    (a) 150 (b) 200
    (c) 336 (d) 186

26. The length and breadth of a rectangular field are 10 m and 8 m respectively. Two rectangular paths each 3 metres wide run inside the plot, one along the length and the other along the breadth. Find the total area occupied by the paths.

    (a) 30 m² (b) 54 m²
    (c) 90 m² (d) 45 m²

27. The angle of a sector of a circle is 90°. The ratio of the area of the sector and that of the circle is _____.

    (a) 4:1 (b) 3:1
    (c) 1:4 (d) 3:1

28. The radius of a circular field is 105 m. The area obtained by a sector of angle 72° is used for paddy growing. The perimeter of the sector is _____.

    (a) 132 m (b) 342 m
    (c) 210 m (d) None of these

29. A farmer wants to purchase a land which is in the shape of trapezium. The lengths of its parallel sides are 18 ft, 15 ft and perpendicular distance between them is 20 ft. Find the total cost of the field at the rate of ₹2000 per sq. ft.

    (a) ₹666600 (b) ₹666000
    (c) ₹660000 (d) ₹600000

30. The radius of a circular card board is 20 cm. A strip of width 4 cm is removed all along its border. The percentage of the area of the strip removed is _____.

    (a) 50% (b) 36%
    (c) 40% (d) 25%

31. The side of a square metal sheet is 7 m. A circular sheet of maximum radius is cut out of it and removed. Find the area of the circular sheet (in ft²).

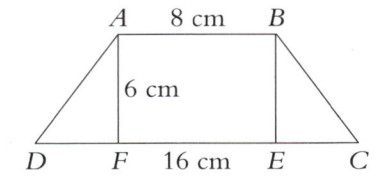

    (a) 49π (b) $\dfrac{49\pi}{2}$
    (c) $\dfrac{49\pi}{4}$ (d) $\dfrac{49\pi}{8}$

32. The side of a square metal sheet is 12 m. Four squares each of side 3 m are cut and removed from the four corners of the sheet and the remaining part is folded to form a cuboid (without top face). Find the volume of the cuboid so formed.

    (a) 108 m³ (b) 36 m³
    (c) 118 m³ (d) None of these

33. The dimensions of a class room are 15 ft × 10 ft × 11 ft. The area of four walls is _____.

(a) 1000 ft² (b) 550 ft²
(c) 500 ft² (d) 600 ft²

34. The parallel sides of a trapezium ABCD are 16 cm and 8 cm and the distance between them is 6 cm. AF and BE are the perpendiculars drawn to $\overline{DC}$. Find the total area of the triangles ADF and BCE (in cm²).

   (a) 36 (b) 24
   (c) 30 (d) 28

35. A circle of radius 7 cm rotates inside and around the circumference of another circle. The smaller circle takes 10 rotations to complete the circumference of the bigger circle. Find the radius of the bigger circle.

   (a) 70 cm (b) 14 cm
   (c) 21 cm (d) None of these

36. A square is inscribed in a circle of radius 7 cm, then find the area of the square (in cm²).

   (a) 49 (b) 98
   (c) 196 (d) 392

37. Find the cost of fencing a semi-circular garden of radius 7 m at ₹10 per metre.

(a) ₹720 (b) ₹360
(c) ₹180 (d) ₹90

38. The edges of two cubes are 2 cm and 4 cm. Find the ratio of the volumes of the two cubes.

   (a) 1:8 (b) 8:1
   (c) 1:9 (d) 9:1

39.

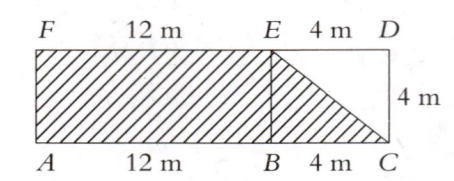

Find the area of the shaded part in the figure given above.

   (a) 48 m² (b) 56 m²
   (c) 64 m² (d) 52 m²

40. The angle of a sector of radius 14 cm is 90°. The area of the sector is _____.

   (a) 144 cm² (b) 154 cm²
   (c) 162 cm² (d) 196 cm²

Level 3

41.

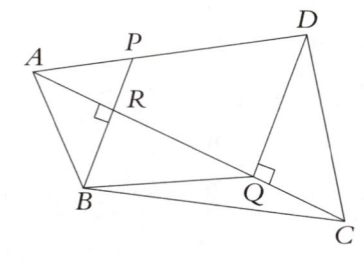

In the figure given above, ABCD is a quadrilateral and BPDQ is a parallelogram. AR = 50 cm, CQ = 70 cm, BR = 60 cm and PR = 40 cm. If the area of the quadrilateral ABCD is 15600 cm², then find the area of the parallelogram BPDQ (in cm²).

   (a) 9460 (b) 8460
   (c) 7500 (d) 9000

42. A path of 1 m wide runs around and inside a rectangular garden of length 40 m and width 20 m. Find the cost of levelling the path at the rate of ₹3.75 per square metre (in ₹).

   (a) 470 (b) 520
   (c) 580 (d) 435

43. The side of a square garden is 150 m. A circular path of width 1.5 m is laid inside the garden touching all the sides. Find the cost of laying the path at ₹28 per m².

   (a) ₹19602 (b) ₹18786
   (c) ₹17754 (d) ₹19888

44. The angle of a sector of a circle is 72° and radius of the circle is 10 m. The ratio of the circumference of the circle and the length of the arc of the sector is _____.

   (a) 5:1 (b) 1:5
   (c) 4:1 (d) 1:4

45. A wire in the form of a circle of radius 3.5 m is bent in the form of a rectangle, whose length and breadth are in the ratio of 6:5. What is the area of the rectangle?

(a) 60 cm² (b) 30 cm²
(c) 45 cm² (d) 15 cm²

46. In the figure given below, ABCD is a rectangle. ED = 16 cm, BF = 20 cm and FG = FC = 8 cm and FG is perpendicular to BC. If CD is half of AD, then find the area of the shaded region (in cm²).

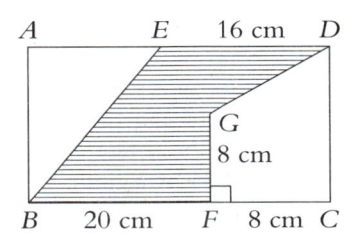

```
A        E    16 cm    D
                       
                  G
                  8 cm
B    20 cm    F  8 cm  C
```

(a) 320 (b) 325
(c) 220 (d) 225

47. Two cubes of edge 10 m each joined together to form a single cuboid. The ratio of the total surface area of one of the cubes and that of the single cuboid formed is ___.

(a) 5:3 (b) 2:1
(c) 3:5 (d) 1:2

48. The inner dimensions of a hall are 40 ft × 20 ft × 30 ft. The hall contains 2 doors of size 5 ft ×10 ft each and 4 windows of size 5 ft × 5 ft each. If the inner walls of the hall are to be fixed by tiles of size 1 ft × 1 ft, then find the number of tiles required.

(a) 8000 (b) 10000
(c) 15400 (d) 3400

49. The ratio of the volumes of two cubes is 729 : 1331. What is the ratio of their total surface areas?

(a) 81 : 121 (b) 9 : 11
(c) 729 : 1331 (d) 27 : 121

50. If the diameter of a circle is equal to the diagonal of a square, then the ratio of their areas is ____.

(a) 7 : 1 (b) 1 : 1
(c) 11 : 7 (d) 22 : 7

51. A rectangular box with lid is made up of certain metal sheet of thickness 1 cm. Its outer dimensions are 50 cm, 40 cm, 20 cm. If the weight of the metal is 4 g per cm³, then the weight of the box is ____.

(a) 26872 g (b) 28762 g
(c) 28672 g (d) 26782 g

52. The sector of a circle has radius of 21 cm and central angle 135°. Find its perimeter.

(a) 91.5 cm (b) 93.5 cm
(c) 94.5 cm (d) 92.5 cm

53. A circle is inscribed in a square of side 14 m. The ratio of the area of the circle and that of the square is _____.

(a) $\pi : 3$ (b) $\pi : 4$
(c) $\pi : 2$ (d) $\pi : 1$

54. The area of a square field is 64 m². A path of uniform width is laid around and outside of it. If the area of the path is 17 m², then find the width of the path.

(a) 1 m (b) 1.5 m
(c) 3 m (d) 0.5 m

55. How many times should a wheel of radius 7 m rotate to go around the perimeter of a rectangular field of length 60 m and breadth 50 m?

(a) 3 (b) 4
(c) 5 (d) 6

56. The minute hand of a clock is 14 cm long. How far does the tip of the minute hand move in 60 minutes?

(a) 22 cm (b) 44 cm
(c) 33 cm (d) 88 cm

57. A square sheet of maximum area is cut out from a circular piece of card board of radius 14 cm. Find the area of the remaining card board (in cm²).

(a) 224 (b) 225
(c) 226 (d) 228

58. The area of a circular play ground is 616 m². A path of width 3 m is laid around and outside of it. Find the cost of levelling the path at the rate of ₹14 per square metre.

(a) ₹4093 (b) ₹4090
(c) ₹4000 (d) ₹4092

6.20 Chapter 6

59. The parallel sides of a trapezium are 24 cm and 20 cm. The distance between them is 7 cm. Find the radius of a circle whose area is equal to the area of the trapezium.
    (a) 9 cm (b) 7 cm
    (c) 14 cm (d) 28 cm

60. In the adjoining figure, ABCD is a square of side 14 cm and two semicircles are drawn inside of it with AB and CD as diameters. Find the area of the shaded region in the figure.
    (a) 144 m² (b) 164 m²
    (c) 184 m² (d) 154 m²

61. The dimensions of a classroom are 12 m, 6 m and 4 m. Find the cost of decorating its walls by a wallpaper of 50 cm width, which costs ₹30 per metre.
    (a) ₹8600 (b) ₹8460
    (c) ₹8640 (d) ₹8600

62. The volume of a cube is 64 cm³. Find the lateral surface area of the cube (in cm²).
    (a) 16 (b) 64
    (c) 144 (d) 256

63. A wire is in the form of a square of side 18 m. It is bent in the form of rectangle, whose length and breadth are in the ratio of 3 : 1. What is the area of the rectangle?
    (a) 81 m² (b) 243 m²
    (c) 144 m² (d) 324 m²

64. The angle of a sector of a circle is 108° and radius of the circle is 15 m. The ratio of the circumference of the circle and the length of the arc of the sector is _____.
    (a) 3 :10 (b) 10 : 3
    (c) 1 : 3 (d) 3 : 1

65. The parallel sides of an isosceles trapezium PQRS are 12 cm and 6 cm and the distance between them is 4 cm. PT and QU are the perpendiculars drawn to SR. Find the area of the triangle PST (in cm²).
    (a) 8 (b) 3
    (c) 6 (d) 12

ASSESSMENT TESTS

Test 1

Directions for questions 1 to 11: Select the correct alternative from the given choices.

1. Find the cost of levelling a ground which is in the shape of a sector with angle 40° and radius 63 m at the rate of ₹15 per square metre.

   The following are the steps involved in solving the above problem. Arrange them in sequential order.

   (A) The area of the sector = $\dfrac{\theta°}{360°}\pi r^2$.

   (B) Cost of levelling the ground per square metre = ₹15.

   (C) ∴ The area of the ground = $\dfrac{40°}{360°} \times \dfrac{22}{7} \times 63 \times 63 = 1386$ m².

   (D) Cost of levelling 1386 m² = 1386 × 15 = ₹20790.

   (a) ADBC (b) ABCD
   (c) CABD (d) ACBD

2. If the radius of a wheel is 42 cm and it makes 20 rotations in one hour, then find the time taken by the wheel to cover a distance of 528 m.

   The following are the steps involved in solving the above problem. Arrange them in sequential order.

   (A) Time taken to complete 200 rotations = $200 \times \dfrac{1}{20}$ = 10 hours

   (B) The number of rotations required to cover a distance of 528m

   = $\dfrac{\text{Distance}}{\text{Circumference}} = \dfrac{528\,\text{m}}{264\,\text{cm}} = \dfrac{52800}{264} = 200$.

(C) The distance covered by the wheel in 1 rotation = Circumference of the wheel = $2\pi r = 2 \times \frac{22}{7} \times 42 = 264$ cm.

(D) Time taken to make 20 rotations = 1 h.

(a) CBDA (b) CDBA
(c) BCDA (d) CABD

3. A rectangular sheet of length 5 cm and breadth 3 cm is cut from each corner of a square piece of an aluminium sheet of side 10 cm. Find the area of the aluminium sheet leftover.

(a) 60 cm² (b) 40 cm²
(c) 85 cm² (d) 75 cm²

4. Find the perimeter (in m) of a square whose area is 225 m².

(a) 15 (b) 60
(c) 225 (d) None of these

5. The area of a square field is 36 m². A path of uniform width is laid around and outside of it. If the area of the path is 28 m², then find the width of the path (in m).

(a) 1 (b) 4
(c) 6 (d) 2

6. The area of a parallelogram is 50 cm². If the base is 10 cm, then find its corresponding height (in cm).

(a) 10 (b) 15
(c) 5 (d) None of these

7. Find the cost of fencing a semi-circular garden of radius 14 m at ₹10 per metre.

(a) ₹1080 (b) ₹1020
(c) ₹700 (d) ₹720

8. In the adjoining figure, PQRS is a square of side 14 cm and two semicircles are drawn inside of it with PQ and SR as diameters. Find the area of the shaded region in the figure (in cm²)

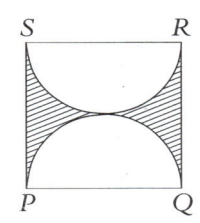

(a) 24
(b) 48
(c) 42
(d) None of these

9. The total surface area of a cube is 726 m². Find its edge.

(a) 8 m (b) 9 m
(c) 11 m (d) 12 m

10. Find the ratio of the lateral surface area and the total surface area of a cube.

(a) 2 : 3 (b) 1 : 2
(c) 4 : 5 (d) 3 : 4

11. The edges of two cubes are 4 cm and 6 cm. Find the ratio of the volumes of the two cubes.

(a) 1 : 3 (b) 2 : 3
(c) 1 : 8 (d) 8 : 27

Directions for questions 12 to 15: Match the Column A with Column B.

Column A		Column B
12. If the area of a square field is 25 hectares, then the sum of the lengths of the diagonals is _____.	( ) (a)	220
13. A playground is in the shape of a rhombus whose diagonals are 10 m and 15 m. The cost of levelling it at the rate of ₹12 per m² is ₹_____.	( ) (b)	1000
14. The volume of a cube of edge 10 cm in cm³ is	( ) (c)	900
15. The length of the arc of a sector with angle 60° and radius 210 cm is _____ cm.	( ) (d)	$1000\sqrt{2}$
	(e)	3000
	(f)	$500\sqrt{2}$

Test 2

Directions for questions 16 to 26: Select the correct alternative from the given choices.

16. Find the cost of fencing a field which is in the shape of a sector with angle 90° and radius 56 m at the rate of ₹5 per metre.

    The following are the steps involved in solving the above problem. Arrange them in sequential order.

    (A) ∴ The cost of fencing = Perimeter × Rate per metre = 200 × ₹5.

    (B) The perimeter of the sector
    $= \frac{\theta°}{360°} \times 2\pi r + 2r = \frac{90°}{360°} \times 2 \times \frac{22}{7} \times 56 + 2 \times 56.$

    (C) The perimeter of the sector = 200 m.

    (D) The cost of fencing = ₹1000.

    (a) CDAB (b) ABCD
    (c) BCAD (d) CBAD

17. If the radius of a wheel is 21 cm, then find how many times it should rotate to cover a distance of 264 m.

    The following are the steps involved in solving the above problem. Arrange them in sequential order.

    (A) The circumference of the wheel = $2\pi r = 2 \times \frac{22}{7} \times 21$ (given $r = 21$ cm).

    (B) ∴ The circumference of the wheel = 132 cm.

    (C) Number of rotations = $\frac{26400 \text{ cm}}{132 \text{ cm}} = 200$.

    (D) Number of rotations
    = $\frac{\text{Distance to be covered}}{\text{Circumference of the wheel}}$.

    (It is given that distance = 264 m and circumference = 132 cm)

    (a) ABDC (b) ACBD
    (c) ABCD (d) BDAC

18. A square sheet of side 2 cm is cut from each corner of a rectangular piece of an aluminium sheet of length 12 cm and breadth 8 cm. Find the area of the leftover aluminium sheet.

    (a) 60 cm² (b) 80 cm²
    (c) 100 cm² (d) 120 cm²

19. Find the perimeter (in m) of a square whose area is 256 m².
    (a) 24 (b) 36
    (c) 46 (d) 64

20. The area of a square field is 64 m². A path of uniform width is laid around and outside of it. If the area of the path is 17 m², then find the width of the path (in m).
    (a) 1 m² (b) 1.5 m²
    (c) 3 m² (d) 0.5 m²

21. The area of a parallelogram is 100 cm². If the base is 25 cm, then find its corresponding height (in cm).
    (a) 4 (b) 6
    (c) 10 (d) 5

22. Find the cost of fencing a semi-circular garden of radius 7 m, at ₹10 per metre.
    (a) ₹720 (b) ₹360
    (c) ₹180 (d) ₹90

23. In the adjoining figure, *ABCD* is a square of side 7 cm and two semicircles are drawn inside of it with *AB* and *CD* as diameters. Find the area of the shaded region (in cm²).
    (a) 77
    (b) 154
    (c) 115.5
    (d) 38.5

24. The volume of a cube is 343 cm³. Find the edge of the cube (in cm).
    (a) 3.5 (b) 7
    (c) 10.5 (d) 14

25. The ratio of the volumes of two cubes is 729 : 1331. What is the ratio of their total surface areas?
    (a) 9 : 1 (b) 81 : 121
    (c) 18 : 11 (d) 9 : 22

26. The edges of two cubes are 2 cm and 4 cm. Find the ratio of the volumes of the two cubes.
    (a) 1 : 8 (b) 1 : 2
    (c) 1 : 9 (d) 1 : 3

Directions for questions 27 to 30: Match the Column A with Column B.

Column A		Column B
27. The TSA of a cube of edge 6 cm is _____ cm^2.	( )	(a) 600 $\sqrt{2}$
28. If the area of a square field is 36 hectares, then its diagonal is _____ m.	( )	(b) 216
29. If the diagonals of a rhombus are 22 cm and 14 cm long, then its area is _____ cm^2.	( )	(c) 1200 $\sqrt{2}$
30. The area of a sector with angle 45° and radius 14 cm is _____ cm^2.	( )	(d) 308
		(e) 154
		(f) 77

Chapter 6

TEST YOUR CONCEPTS

Very Short Answer Type Questions

1. False
2. True
3. False
4. True
5. False
6. 60°
7. 16 cm²
8. 13 cm
9. 40 cm
10. 154 cm²
11. 3 cm
12. (b)
13. (c)
14. (a)
15. (b)
16. (b)
17. (c)

Short Answer Type Questions

18. 40 cm
19. 30 cm
20. 20 cm
21. 13.86 m²
22. 8 cm
23. 10 m
24. 384 m³
25. 1000 cm³
26. 310 m²
27. 11 m
28. 140 m
29. $12\frac{1}{2}$%
30. 1125 m³
31. ₹3200
32. ₹250
33. 294 cm²
34. 28 m²
35. 15
36. 500
37. 44 cm
38. 9 cm
39. 10 cm
40. 29 cm²
41. ₹720
42. 8 : 27

Essay Type Questions

43. 425 m²
44. 2016 m³
45. 11.375 cm²
46. 88 πm²
47. 192
48. 960 cm³
49. ₹9000
50. ₹8
51. 180 cm²
52. ₹12600
53. ₹7500
54. ₹108
55. 1 m
56. 50

Mensuration 6.25

57. 22 cm
58. 504 cm^2
59. 62 cm
60. ₹704
61. 14 cm
62. 5000 cm^3
63. 42 cm^2
64. ₹9000
65. 2 : 3

CONCEPT APPLICATION

Level 1

| 1. (a) | 2. (a) | 3. (a) | 4. (c) | 5. (a) | 6. (d) | 7. (a) | 8. (b) | 9. (a) | 10. (c) |
| 11. (a) | 12. (b) | 13. (d) | 14. (c) | 15. (e) | 16. (a) | 17. (f) | 18. (d) | 19. (a) | 20. (b) |
| 21. (f) |

Level 2

| 22. (b) | 23. (b) | 24. (a) | 25. (d) | 26. (d) | 27. (c) | 28. (b) | 29. (c) | 30. (b) | 31. (c) |
| 32. (a) | 33. (b) | 34. (b) | 35. (a) | 36. (b) | 37. (b) | 38. (a) | 39. (b) | 40. (b) |

Level 3

41. (c)	42. (d)	43. (a)	44. (a)	45. (b)	46. (c)	47. (c)	48. (d)	49. (a)	50. (c)
51. (c)	52. (a)	53. (b)	54. (d)	55. (c)	56. (d)	57. (a)	58. (d)	59. (b)	60. (d)
61. (c)	62. (b)	63. (b)	64. (b)	65. (c)

ASSESSMENT TESTS

Test 1

| 1. (d) | 2. (a) | 3. (b) | 4. (b) | 5. (a) | 6. (c) | 7. (d) | 8. (c) | 9. (c) | 10. (a) |
| 11. (d) | 12. (d) | 13. (c) | 14. (b) | 15. (a) |

Test 2

| 16. (c) | 17. (b) | 18. (b) | 19. (a) | 20. (d) | 21. (c) | 22. (b) | 23. (c) | 24. (b) | 25. (a) |
| 26. (d) | 27. (e) | 28. (a) | 29. (b) | 30. (d) |

CONCEPT APPLICATION

Level 1

1. Given $a^3 = 729 \Rightarrow a = 9$ m.
   Hence, the correct option is (a)

2. The maximum length of side of the square
   = The breadth of the rectangle = 3 m.
   Hence, the correct option is (a)

3. Volume = Length × Breadth × Depth
   = 6 × 3 × 2 = 36 m³.
   Hence, the correct option is (a)

4. Total surface area = $6a^2$
   = 6 × 5² = 6 × 25 = 150 m².
   Hence, the correct option is (c)

5. Circumference of the garden =
   $2\pi r = 2 \times \dfrac{22}{7} \times 21 = 132$ m.
   Total cost = 132 × 10 = `1320.
   Hence, the correct option is (a)

6. Area of the square = 256 m² (given)
   $\Rightarrow$ (side)² = (16)²
   $\Rightarrow$ Side of the square = 16 m.
   ∴ The Perimeter of the square = 4 × side
   = 4 × 16
   = 64 m.
   Hence, the correct option is (d)

7. Area of the parallelogram = 100 cm² (given)
   Area of the parallelogram = Base × Height.
   $\Rightarrow$ Height × 25 = 100
   $\Rightarrow$ Height = 4 cm.
   Hence, the correct option is (a)

8. Let the length of edge of the cube = a cm.
   Then, $a^3 = 512 = 8^3 \Rightarrow a = 8$.
   Hence, the correct option is (b)

9. Area of the four walls = $2h(l + b)$
   = 2 × 5 (12 + 8)

   = 10 × 20 ft² = 200 .ft².
   Hence, the correct option is (a)

10. (B), (C), (A) and (D) is the required sequential order.
    Hence, the correct option is (c)

11. (A), (B), (C) and (D) is the required sequential order.
    Hence, the correct option is (a)

12. (C), (D), (A) and (B) is the required sequential order.
    Hence, the correct option is (b)

13. (D), (B), (A) and (C) is the required sequential order.
    Hence, the correct option is (d)

14. → c: Area of trapezium = $\left(\dfrac{1}{2}\right)$ (Sum of the parallel sides) (Distance between them)

15. → e: Volume of cube $a^3 = 512$.
    $\Rightarrow a^3 = 8^3 \Rightarrow a = 8$ cm.
    Later surface area = $4a^2 = 4 \times 8^2 = 4 \times 64 = 256$ cm².

16. → a: Diagonal of square $(d) = \sqrt{8}$ cm
    ∴ Area = $\dfrac{d^2}{2} = \dfrac{\left(\sqrt{8}\right)^2}{2} = 4$ m².

17. → f: Perimeter of a sector = $l + 2r$
    = $\dfrac{\theta}{360°} 2\pi r + 2r = \dfrac{45°}{360°} \times 2 \times \dfrac{22}{7} \times 7 + 2 \times 7$
    = $\dfrac{1}{8} \times 2 \times 22 + 14$
    = $\dfrac{22}{4} + 14 = 5.5 + 14 = 19.5$ cm.

18. → d: Area of ring is _ $(R^2 - r^2)$.

19. → a: Total surface area of cube = $6(\text{side})^2$ = $6\left(\sqrt{10}\right)^2 = 60$ cm².

20. → b: Diagonal of square = $\sqrt{2} \times$ side = $10\sqrt{2}$ cm.

21. → f: Area of sector = $\dfrac{\theta}{360°} \times \pi r^2$
    = $\dfrac{45°}{360°} \times \dfrac{22}{7} \times 14 \times 14$
    = $\dfrac{1}{8} \times \dfrac{22}{7} \times 14 \times 14$
    = 11 × 7 = 77 cm².

Level 2

22. Area of the shaded part = Area of the rectangle + Area of right angled triangle = $8 \times 2 + \frac{1}{2} \times 2 \times 2$
= 16 + 2 = 18 m².

Hence, the correct option is (b)

23. Area of the sector = $\frac{120°}{360°} \times \pi \times (21)^2$

= $\frac{120°}{360°} \times \frac{22}{7} \times 21 \times 21 = 22 \times 21$

= 462 cm².

Hence, the correct option is (b)

24. Diagonal = $\sqrt{4^2 + 3^2} = \sqrt{16+9}$

= $\sqrt{25}$ = 5 m.

∴ The cost of fencing = 5 × 10 = ₹50.

Hence, the correct option is (a)

25. The required area = 21 × 16 − 15 × 10
= 336 − 150 = 186 m².

26.

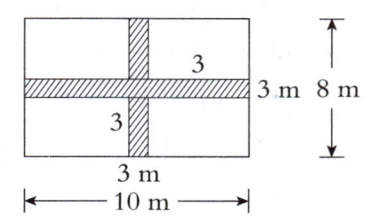

The required area = 10 × 3 + 8 × 3 − 3 × 3
= 30 + 24 − 9 = 45 m².

Hence, the correct option is (d)

27. Area of circle = πr^2

Area of sector = $\frac{90°}{360°} \times \pi r^2$

The required ratio = $\frac{90°}{360°} \times \pi r^2 : \pi r^2 \Rightarrow \frac{1}{4} : 1$

⇒ 1 : 4.

Hence, the correct option is (c)

28. Given, r = 105 m.

Length of the sector (l) = $\frac{72°}{360°} \times 2 \times \frac{22}{7} \times 105$

= 44 × 3 = 132 m.

Perimeter of the sector = l + 2r = 132 + 2 × 105
= 132 + 210 = 342 m.

Hence, the correct option is (b)

29. Area of the field = $\frac{1}{2} h(a+b) = \frac{1}{2} \times 20(18+15)$

= 330 ft².

The total cost of the land = 330 × 2000 = ₹660000.

Hence, the correct option is (c)

30. Area of the strip removed = $\pi(20^2 - 16^2)$
= π (400 − 256) = 144π m².

Area of the circular card board = $\pi \times 20 \times 20 = 400$.

∴ The required percentage = $\frac{144\pi}{400\pi} \times 100\%$

= 36%.

Hence, the correct option is (b)

31. The radius of the circular sheet
= $\frac{1}{2}$ × Side of the square

⇒ $r = \frac{7}{2}$

Area of the circular sheet = $\pi \left(\frac{7}{2}\right)^2 = \frac{49\pi}{4}$.

Hence, the correct option is (c)

32. The required volume of the cuboid
= $(12 - 2 \times 3) \times (12 - 2 \times 3) \times 3 = 6 \times 6 \times 3 = 108$ m³.

Hence, the correct option is (a)

33. The area of four walls = $2(l + b) \times h$
= 2(15 + 10) × 11 = 550 ft².

Hence, the correct option is (b)

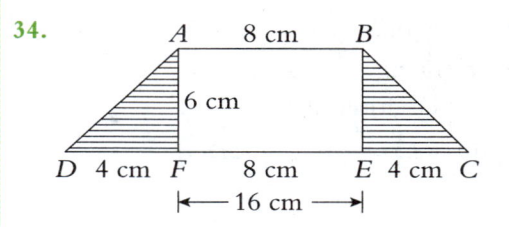

34.

Area of the trapezium
= $\frac{1}{2}(8 + 16) \times 6 = \frac{1}{2} \times 24 \times 6 = 72$ cm².

6.28 Chapter 6

The total area of the triangle = Area of the trapezium − Area of the rectangle.

= 72 − 8 × 6 = 72 − 48 = 24 cm².

Hence, the correct option is (b)

35.

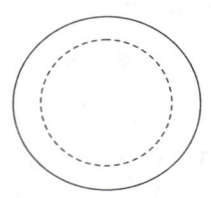

The circumference of the smaller circle = $2\pi r$

= $2 \times \dfrac{22}{7} \times 7 = 44$ cm.

⇒ The distance travelled by the smaller circle in one revolution = 44 cm.

Total distance travelled = 10(44 cm)

Let R be the radius of the bigger circle.

∴ The circumference of the bigger circle = $2\pi R$

= The distance travelled by the smaller circle

⇒ $2\pi R = 10 \times 44$

⇒ $2 \dfrac{22}{7} \times R = 440 \Rightarrow R = \dfrac{440 \times 7}{44} = 70$ cm.

36.

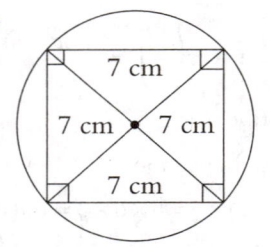

The diagonal of the square = The diameter of the circle = 14 cm.

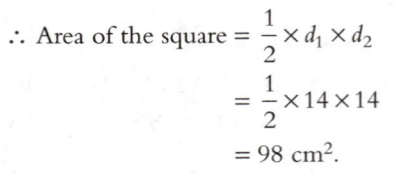

∴ Area of the square = $\dfrac{1}{2} \times d_1 \times d_2$

= $\dfrac{1}{2} \times 14 \times 14$

= 98 cm².

37. Perimeter of a semicircle = $\pi r + 2r$

= $\dfrac{22}{7} \times 7 + 2 \times 7 = 22 + 14 = 36$ m.

∴ The cost of fencing = 36 × 10 = ₹360

Hence, the correct option is (b)

38. The volumes of the given cubes are 2^3 and 4^3, i.e., 8 cm³ and 64 cm³.

∴ Ratio of their volumes = 8 : 64

= 1 : 8

Hence, the correct option is (a)

39.

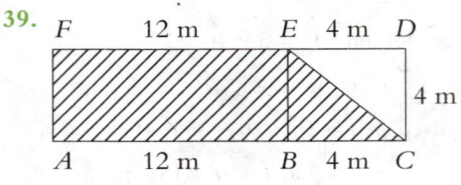

Area of rectangle ABEF = 12 × 4 = 48 m².

Area of triangle BEC = $\dfrac{1}{2} \times 4 \times 4 = 8$ m².

Area of the shaded region = 48 + 8 = 56 m².

Hence, the correct option is (b)

40. Area of the sector = $\dfrac{\theta}{360°} \times \pi r^2$

$\dfrac{90°}{360°} \times \dfrac{22}{7} \times 14 \times 14 = 154$ cm².

Hence, the correct option is (b)

Level 3

41. BPDQ is a parallelogram

BP = DQ = BR + PR = 60 + 40 = 100 cm.

Area of a quadrilateral

= $\dfrac{1}{2} h(d_1 + d_2) = \dfrac{1}{2} AC (BR + DQ)$

$\dfrac{1}{2} \times AC (60 + 100) = 16000$

$AC \times \dfrac{160}{2} = 15600$

AC = 195 cm

AC = AR + RQ + CQ = 50 + RA + 70 = 195

∴ RQ = 195 − 120

RQ = 75 cm

RQ is the perpendicular distance between $\overline{BP}$ and $\overline{DQ}$.

∴ Area of the parallelogram BPDQ = base × height

= BP × RQ = 100 × 75 = 7500 cm².

Hence, the correct option is (c)

42.

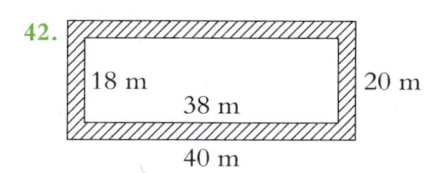

Required path area
= 40 × 20 − [(40 − 2) × (20 − 2)]
= 800 − 38(18) = 116 m².
Cost of levelling the path = 116 × 3.75 = ₹435.
Hence, the correct option is (d)

43.

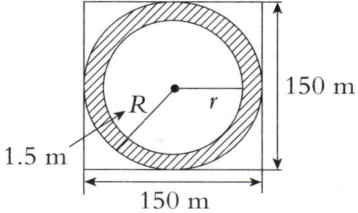

The diameter of the outer circle = 150 m
The diameter of the inner circle = 150 m − 2 × 1.5 m
= 150 m − 3m = 147 m.
∴ Radius of outer circle = $\dfrac{150}{2}$ = 75 and
Radius of inner circle = $\dfrac{147}{2}$ = 73.5
Required area of the path = π(75 + 73.5) (75 − 73.5)
= $\dfrac{22}{7}$ × 148.5 × 1.5
The cost of laying path
= $\dfrac{22}{7}$ × 148.5 × 1.5 × 28 = 19602.

44. Length of the arc = $\dfrac{72°}{360°}$ × 2πr = $\dfrac{2\pi r}{5}$
Circumference of the circle = 2πr
The required ratio = $2\pi r : \dfrac{2\pi r}{5}$ = 5 : 1
Hence, the correct option is (a)

45. The circumference of the circle is equal to the perimeter of the rectangle. Let $l = 6x$ and $b = 5x$
$2(6x + 5x) = 2 \times \dfrac{22}{7} \times 3.5 \Rightarrow x = 1$
∴ $l = 6$ cm and $b = 5$ cm
Area of the rectangle = 6 × 5 = 30 cm².
Hence, the correct option is (b)

46.

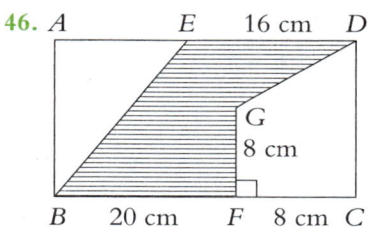

BC = BF + FC = 28 cm
CD = $\dfrac{1}{2}$ AD = $\dfrac{1}{2}$ BC = 14 cm
AE = AD − ED = 28 − 16 = 12 cm
Area of the shaded region = Area of the rectangle
− (Area of △AEB + Area of Trapezium FCDG)
= $28 \times 14 - \left(\dfrac{1}{2} \times 12 \times 14 + \dfrac{1}{2} 8(8+14) \right)$
= 392 − (84 + 88) = 392 − 172 = 220 cm².
Hence, the correct option is (c)

47. Total surface area of original cube = 6 × 10²
= 6 × 100 = 600 m²
The length of the cuboid formed = 10 + 10
= 20 m and breadth = height = 10 m
∴ Total surface area of the cuboid
= 2(20 × 10 + 10 × 10 + 10 × 20) = 2(200 + 100 + 200)
= 2 × 500 = 1000 m²
The required ratio = 600 : 1000 = 6 : 10 = 3 : 5.
Hence, the correct option is (c)

48. Area of the four walls = 2(l + b) × h
= 2(40 + 20) × 30 = 2 × 60 × 30 = 3600 ft²
Total area of 2 doors and four windows
= 2 × 5 × 10 + 4 × 5 × 5 = 100 + 100 = 200 ft²
The required area = 3600 − 200 = 3400 ft².
Number of tiles required = $\dfrac{3400}{1 \times 1}$ = 3400
Hence, the correct option is (d)

49. Ratio of the sides = $\sqrt[3]{729} : \sqrt[3]{1331}$ = 9 : 11
Ratio of surface area = 9² : 11² = 81 : 121
Hence, the correct option is (a)

6.30 Chapter 6

50. Let the diameter of a circle = d cm

    The diagonal of a square = d cm

    Area of a circle = $\pi \dfrac{d^2}{4}$

    Area of a square = $\dfrac{d^2}{2}$

    Ratio of their areas = $\dfrac{\pi d^2}{4} : \dfrac{d^2}{2}$

    $= \pi : 2 = 11 : 7$

    Hence, the correct option is (c)

51. The outer dimensions of the box are 50 cm, 40 cm and 20 cm. The thickness of metal = 1 cm

    ∴ The inner dimensions of the box are (50 − 2) cm, (40 − 2) cm, (20 − 2) cm, i.e., 48 cm, 38 cm, 18 cm.

    Outer volume of the box = 50 × 40 × 20 = 40000 cm².

    Inner volume of the box = 48 × 38 × 18 = 32832 cm³.

    Volume of the metal sheet = 40000 − 32832 = 7168 cm³.

    Weight of the box = 7168 × 4 = 28672 g.

    Hence, the correct option is (c)

52. Perimeter of the sector = length of the arc + 2(radius)

    $= \left(\dfrac{135}{360} \times 2 \times \dfrac{22}{7} \times 21\right) + 2(21)$

    = 49.5 + 42 = 91.5 cm.

    Hence, the correct option is (a)

53. Side of the square = 14 m.

    Side of the square = Diameter of the circle

    ∴ Radius of the circle = 7 m.

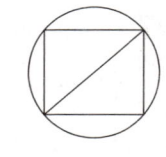

    Area of the circle = $\pi \times 7^2$ = 49 π m².

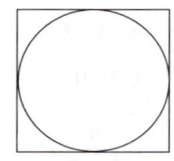

    Area of the square = 14^2 = 196 m².

    Ratio of their areas = 49π : 196 = π : 4.

    Hence, the correct option is (b)

    Hence, the correct option is (d)

54. Area of the square field = 64 m² (given).

    Area of the path = 17 m².

    ∴ The area of the outer square = 64 + 17 = 81 m².

    Side of the outer square = 9 m.

    Let the width of the parallel path be x m,

    ⇒ 8 + 2x = 9 m

    ⇒ x = 0.5 m.

    Hence, the correct option is (d)

55. Perimeter of the rectangular field = 2(60 + 50) = 220 m.

    Circumference of the wheel = $2\pi r$

    $= 2 \times \dfrac{22}{7} \times 7$

    = 44 m.

    Number of rotations

    $= \dfrac{\text{Perimeter of the rectangular field}}{\text{Circumference of the wheel}}$

    $= \dfrac{220}{44} = 5$

    Hence, the correct option is (c)

56. In 60 minutes, the minute hand completes one rotation. The distance travelled by the tip of the minute hand in 60 minutes = The perimeter of the circle

    $= 2\pi r$

    $= 2 \times \dfrac{22}{7} \times 14$

    = 4 × 22 = 88 cm.

    Hence, the correct option is (d)

57. The diagonal of the square of maximum area = the Diameter of the circle

    ∴ Diagonal = 28 cm.

    ∴ Area of the square = $\dfrac{d^2}{2}$

    $= \dfrac{28 \times 28}{2} = 392$ cm².

    Area of the circle

    $= \pi r^2 = \dfrac{22}{7} \times 14 \times 14$

    = 616 cm².

    ∴ Remaining area of the card board

    = 616 − 392 = 224 cm².

    Hence, the correct option is (a)

58. Given that area of the circular field = 616 m².

⇒ $\pi r^2 = 616$

⇒ $\dfrac{22}{7} \times r^2 = 616$

⇒ $r^2 = \dfrac{616 \times 7}{22} = 28 \times 7$

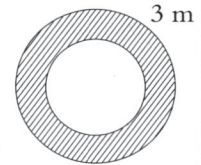

⇒ $r = 2 \times 7 = 14$ m

∴ The outer radius (R) of the circular path
= 14 + 3 = 17 m.

∴ Area of the path = $\pi(R^2 - r^2) = \pi(17^2 - 14^2)$

$= 93 \times \dfrac{22}{7}$

∴ The cost of levelling the path = $93 \times \dfrac{22}{7} \times 14$

= ₹4092

Hence, the correct option is (d)

59. Area of the trapezium = $\dfrac{1}{2}h(a+b)$

$= \dfrac{1}{2} \times 7(24 + 20) = 154\,\text{cm}^2$

Given that area of the trapezium = Area of the circle = $\pi r^2 = 154$

⇒ $\dfrac{22}{7} \times r^2 = 154$

⇒ $r^2 = \dfrac{154 \times 7}{22} = 7 \times 7$

⇒ $r = 7$ cm

Hence, the correct option is (b)

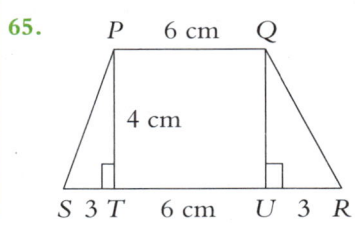

60. The area of the shaded region = Sum of the areas of two semi circles.

= Area of the circle with side of the square as diameter.

Since the side of the square is 14 m.

∴ Radius of the circle = $\dfrac{14}{2} = 7$ m

∴ The required area = πr^2

$= \dfrac{22}{7} \times 7 \times 7$

$= 22 \times 7 = 154\,\text{m}^2$.

Hence, the correct option is (d)

61. Area of 4 walls of the class room = $2h(l + b)$

$= 2 \times 4 \times (12 + 6)$

$= 8 \times 18$

$= 144\,\text{m}^2$.

Area of the wall paper = area of four walls = 144 m².

Length of the wall paper required

$= \dfrac{\text{area of paper in m}^2}{\text{Width of paper in m}}$

$= \dfrac{144}{\dfrac{1}{2}} = 288\,\text{m}$

$\left(\because 50\,\text{cm} = \dfrac{1}{2}\,\text{m}\right)$

∴ The cost of papering = 288 × 30

= ₹8640.

Hence, the correct option is (c)

62. Given that the volume of the cube = 64 cm³ = (4)³

⇒ Side = 4 cm

∴ Lateral surface area = 4 × (side)²

$= 4 \times (4)^2$

$= 64\,\text{cm}^2$.

Hence, the correct option is (b)

63. Total length of wire = Perimeter of square = 4 × 18
= 72 m.

Let the length, breadth of rectangle be $3x$, x.

∴ Perimeter of rectangle = 2(3x + x) = 72 m.

4x = 36.

x = 9

⇒ Length = 3 × 9 = 27 m.

Breadth = 9 m.

Area of rectangle = 27 × 9 = 243 m².

Hence, the correct option is (b)

64. The required ratio = $2\pi r : \dfrac{2\pi r \times 108}{360}$

= 360 : 108 = 10 : 3.

Hence, the correct option is (b)

65.

6.32 Chapter 6

ST = UR

ST + UR + TU = 12 cm (given)

2ST + PQ = 12 cm.

2ST + 6cm = 12 cm

2ST = 6 cm

$ST = \dfrac{6\,cm}{2} = 3\,cm$

PST is a right angled triangle.

Area of triangle $PST = \dfrac{1}{2} \times ST \times PT$

$= \dfrac{1}{2} \times 3 \times 4 = 6\,cm^2.$

Hence, the correct option is (c)

ASSESSMENT TESTS

Test 1

Solutions for questions 1 to 11:

1. (A), (C), (B) and (D) is the required sequential order.

   Hence, the correct option is (d)

2. (C), (B), (D) and (A) is the required sequential order.

   Hence, the correct option is (a)

3. The area of each rectangular sheet cut out = 3 × 5 = 15 cm² and the area of the square piece of aluminium sheet = 10² = 100 cm².

   The total area of the rectangular sheets cut out = 4 × 15 = 60 cm².

   The area of the aluminium sheet leftover = Area of square sheet − Total area of the rectangular sheets cut out.

   ∴ The area of the aluminium sheet leftover = 100 − 60 = 40 cm².

   Hence, the correct option is (b)

4. Given that area of the square = 225 m².

   ⇒ (Side)² = (15)²

   ⇒ Side of the square = 15 m.

   ∴ Perimeter of the square = 4 × side.

   = 4 × 15 = 60 m.

   Hence, the correct option is (b)

5. From the figure, area of the inner square = 36 m².

   The area of the path = 28 m².

   ∴ The area of the outer square = 36 + 28 = 64 m².

   Side of the outer square = 8 m.

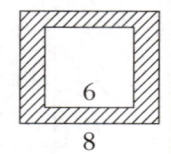

   Let the width of the path be x m.

   ⇒ 6 + 2x = 8

   ⇒ 2x = 2

   ⇒ x = 1

   ∴ The width of the path = 1 m.

   Hence, the correct option is (a)

6. Area of the parallelogram = Base × Height.

   Height × 10 = 50

   Height of the parallelogram = 5 cm.

   Hence, the correct option is (c)

7. Perimeter of a semicircle = $\pi r + 2r$

   $= \dfrac{22}{7} \times 14 + 2 \times 14 = 44 + 28 = 72$ m.

   ∴ The cost of fencing = 72 × 10 = ₹720.

   Hence, the correct option is (d)

8. The area of the shaded region = Area of the square − The sum of the areas of two semi circles.

   = Area of the square − Area of the circle.

   Since the side of the square is 14 cm,

   Radius of the circle = 7 cm.

   ∴ The required area = $14^2 - \pi \times 7^2$

   $= 196 - \dfrac{22}{7} \times 7 \times 7$

   = 196 − 154 = 42 cm².

   Hence, the correct option is (c)

Mensuration 6.33

9. Given $6a^2 = 726$

   $\Rightarrow a^2 = \dfrac{726}{6} \Rightarrow a^2 = 121 \Rightarrow a = 11$

   ∴ Edge of the cube is 11 m.

   Hence, the correct option is (c)

10. The lateral surface area = $4 \times (\text{side})^2$

    Total surface area = $6 \times (\text{side})^2$

    ∴ The required ratio = 4 : 6 = 2 : 3

    Hence, the correct option is (a)

11. Volumes of the given cubes are 4^3 and 6^3,

    i.e., 64 cm^3 and 216 cm^3.

    ∴ The required ratio = 64 : 216 = 8 : 27

    Hence, the correct option is (d)

12. → d: Area of the square = 25 hectares = 250000 m^2.

    Let a be the side of the square

    $\Rightarrow a^2 = 250000 \Rightarrow a = \sqrt{250000} = 500$

    $a = 500$ m.

    Length of the diagonal = $\sqrt{2} \times a$

    $= 500\sqrt{2}$

    Sum of the lengths of the diagonals

    $= 500\sqrt{2} + 500\sqrt{2} = 1000\sqrt{2}$ m

13. → c: Area of the playground = $\dfrac{1}{3} \times 10 \times 15$

    $= 75$ m^2

    ∴ The cost of levelling = 75×12 = ₹900.

14. → b: Volume of a cube of edge, 10 cm = 1000 cm^3.

15. → a: Length of the arc of a sector

    $= \dfrac{\theta°}{360°} \times 2\pi r$

    $= \dfrac{60°}{360°} \times 2\dfrac{22}{7} \times 210$

    $= \dfrac{1}{6} \times 2 \times \dfrac{22}{7} \times 210$

    $= 220$ cm

Test 2

Solutions for questions 16 to 26:

16. (B), (C), (A) and (D) is the required sequential order.

    Hence, the correct option is (c)

17. (A), (B), (D) and (C) is the required sequential order.

    Hence, the correct option is (a)

18. Area of each square cut out from the rectangular sheet = $2 \times 2 = 4$ cm^2.

    Area of the rectangular sheet = $12 \times 8 = 96$ cm^2.

    The required area = Area of the rectangular sheet − Total area of the squares cut out.

    The required area = $96 − 16 = 80$ cm^2.

    Hence, the correct option is (b)

19. Area of the square = 256 m^2 (given)

    $\Rightarrow (\text{side})^2 = (16)^2$

    $\Rightarrow$ Side of the square = 16 m

    ∴ The perimeter of the square = $4 \times$ side.

    $= 4 \times 16 = 64$ m.

    Hence, the correct option is (d)

20. Area of the square field = 64 m^2 (given)

    Area of the path = 17 m^2

    ∴ The area of the outer square = $64 + 17 = 81$ m^2.

    Side of the outer square = 9 m.

    Let the width of the path be x m,

    $\Rightarrow 8 + 2x = 9$ m

    $\Rightarrow x = 0.5$ m.

    Hence, the correct option is (d)

21. Area of the parallelogram = 100 cm^2 (given)

    Area of the parallelogram = base × height.

    $\Rightarrow$ Height $\times 25 = 100$

    $\Rightarrow$ Height = 4 cm.

    Hence, the correct option is (a)

22. Perimeter of a semicircle = $\pi r + 2r$

    $= \dfrac{22}{7} \times 7 + 2 \times 7 = 22 + 14 = 36$ m

    ∴ The cost of fencing = 36×10 = ₹360

    Hence, the correct option is (b)

23. The area of the shaded region = Sum of the areas of two semi circles.

    Since the side of the square is 7 cm.

    ∴ Radius of the circle = $\frac{7}{2}$ cm

    ∴ The required area = $\pi r^2 = \frac{22}{7} \times \frac{7}{2} \times \frac{7}{2}$

    = 38.5 cm^2

    Hence, the correct option is (d)

24. Given volume = 343 cm^3 ⇒ a^3 = 343

    ⇒ a = 7 cm

    Hence, the correct option is (b).

25. Ratio of the sides = $a : a_1 = \sqrt[3]{729} : \sqrt[3]{1331}$

    = 9 : 11

    Ratio of surface area = $6a^2 : 6a_1^2$

    = $9^2 : 11^2$ = 81 : 121

    Hence, the correct option is (b)

26. The volumes of the given cubes are 2^3 and 4^3, i.e., 8 cm^3 and 64 cm^3.

    ∴ The ratio of their volumes = 8 : 64 = 1 : 8

    Hence, the correct option is (a)

Solutions for questions 27 to 30:

27. → b: The TSA of a cube of edge, 6 = 6 × 6^2 = 216.

28. → a: Let a be the side of the square.

    It is given that the area of the square = 36 hectares
    = 36 × 10000 m^2 = 360000 m^2

    i.e., a^2 = 360000 ⇒ $a = \sqrt{360000}$

    a = 600 m.

    Diagonal = $\sqrt{2}$ × side = 600$\sqrt{2}$ m

29. → e: Area of a rhombus

    $\frac{1}{2}$ (The product of diagonals)

    = $\frac{1}{2} \times 22 \times 14 = 22 \times 7$

    = 154 cm^2.

30. → f: Area of a sector = $\frac{\theta°}{360°} \times \pi r^2$

    = $\frac{45°}{360°} \times \frac{22}{7} \times 14 \times 14$

    = $\frac{1}{8} \times 22 \times 28$

    = $\frac{616}{8}$ = 77 cm^2

Chapter 7

Equations and their Applications

a) $5(3x) - 3(x-4) = 0$
$15x - 3x + 12 = 0$
$12x + 12 = 0$
$12x = -12$
$x = -12/12$
$x = -1 \qquad x^2 = 1$

REMEMBER

Before beginning this chapter, you should be able to:
- Know the numbers and symbols, and to write simple equations
- Solve simple equations using trial and error method

KEY IDEAS

After completing this chapter, you should be able to:
- Explain numerical expression and algebraic expression
- Identify the mathematical sentence and mathematical statement
- Study linear equation, simple equation, properties of equations
- Solve equations with one variable
- Solve word problems on simple equations
- Understand inequalities, their notation, inequation, continued inequation, linear inequation, solving of inequations

INTRODUCTION

In the previous class, you have learnt to solve simple equation using trial and error method or by guessing. In this class we shall discuss solving simple equations. Using properties of equality, we shall discuss the methods of solving word problems. Besides, we shall discuss the method of solving inequations. First, we have to understand the meaning of certain terms which are associated with equations like numbers, symbols, knowns, unknowns, constants, variables, expressions, sentences, statements, etc.

Numbers and Symbols

In lower classes, we worked with numbers like 1, 2, 3, 1.2, −2.3 as well as letters like a, b, c or x, y, z which can be used instead of numbers. These letters can be used for some known or unknown numbers. Accordingly, they are called knowns or unknowns. We'll also come across situations in which the letters represent some particular numbers or a whole set of numbers. Accordingly, we call them constants or variables.

Numerical Expressions

Expressions of the form 3×5, $(2 + 6)$, $5 \div (-4)$, $3^2 + 4^{1/2}$, $\sqrt{2} + 5 \div 3$ are numerical expressions. Numerical expressions are made up of numbers, the basic arithmetical operations $(+, -, \times, \div)$, involution (raising to a power) and evolution (root extraction).

Algebraic Expression

Expressions of the form $2x$, $(3x + 5)$, $(4x - 2y)$, $2x^2 + 3\sqrt{y}$, $3x^3/2\sqrt{y}$ are algebraic expressions. $3x$ and 5 are the terms of $(3x + 5)$, $4x$ and $2y$ are the terms of $4x - 2y$. Algebraic expressions are made up of numbers, symbols and the basic arithmetical operations.

Mathematical Sentence

Two expressions joined by the equality sign (=) or an inequality sign ($<, >, \leq, \geq$) are mathematical sentences.

For example, $2 + 3 = 4 + 1$, $5 \times 3 > 2 \times 4$, $5 \times 6 = 10 \times 3$, $15 - 8 < 3 \times 3$, $3x + 5 \leq 10$ and $5x - 4 \geq 15$ are some mathematical sentences. Those which have the equality sign are equations or those which have an inequality sign are inequations.

Mathematical Statement

A mathematical sentence that can be verified as either true or false is a mathematical statement.

For example, $5 + 6 = 3 + 8$. This is a true statement.
$5 - 3 > 2 \times 2$. This is a false statement.

All sentences involving only numerical expressions can be verified as either true or false. Hence, they are statements. But sentences involving a variable can be of three different types. They may be true for all values of the variable (Example $x + 1 = 1 + x$, $x < x + 1$), or only for some values (Example $x + 1 = 4$, $x < 4$) or for no values (Example $x = x + 1$, $x > x + 1$).

Open Sentences

Sentences which are true for some values of the variable and false for the other values of the variable are called open sentences. When a certain value is substituted for the variable, the sentence becomes a statement, either true or false.

Equations and their Applications 7.3

For example,

1. $x + 1 = 4$. This is an open sentence. When we substitute 3 for x, we get a true statement. When we substitute any other value, we get a false statement.

2. $x + 1 < 3$. This is another open sentence. If we substitute any value less than 2 for x we get a true statement. If we substitute 2 or any number greater than 2, we get a false statement.

Equation

An open sentence containing the equality sign is an equation. In other words, an equation is a sentence in which there is an equality sign between two algebraic expressions.

For example, $2x + 5 = x + 3$, $3y - 4 = 20$, $5x + 6 = x + 1$ are equations. Here x and y are unknown quantities and 5, 3, 20, etc., are known quantities.

Linear Equation

An equation, in which the highest index of the unknowns present is one, is a linear equation.

$x + 5 = 10$, $2x - 3 = 23$, $x + y = 10$ and $3x - 4y = 15$ are some linear equations.

Simple Equation

A linear equation which has only one unknown is a simple equation.

$3x + 4 = 16$ and $2x - 5 = x + 3$ are examples of simple equations. The part of an equation which is to the left side of the equality sign is known as the left hand side, abbreviated as LHS. The part of an equation which is to the right side of the equality sign is known as the right hand side, abbreviated as RHS. The process of finding the value of an unknown in an equation is called solving the equation. The value/values of the unknown found after solving an equation is/are called the solution(s) or the root(s) of the equation. Before we learn how to solve an equation, let us review the basic properties of equality.

Properties of Equality

1. **Reflexive property:** Every number is equal to itself.

    For example, $5 = 5$, $2 = 2$ and so on.

2. **Symmetric property:** For any two numbers, if the first number is equal to the second, then the second number is equal to the first. If x and y are two numbers and $x = y$, then $y = x$.

    For example, $3 + 4 = 5 + 2 \Rightarrow 5 + 2 = 3 + 4$.

3. **Transitive property:** If x, y and z are three numbers such that $x = y$ and $y = z$, then $x = z$.

    For example, $9 + 3 = 12$, $12 = 3 \times 4$

    $\therefore 9 + 3 = 3 \times 4$.

4. If x, y and z are three numbers such that $x = y$ and $x = z$, then $y = z$.

    For example, $24 = 8 \times 3$ and $24 = 14 + 10 \Rightarrow 8 \times 3 = 14 + 10$.

5. (a) **Addition property:** If equal numbers are added to both sides of an equality, the equality remains the same. If $x = y$, then $x + z = y + z$.

    (b) **Subtraction property:** If equal numbers are subtracted from both sides of an equality, the equality remains the same. If $x = y$, then $x - z = y - z$.

(c) **Multiplication property:** If both sides of an equality are multiplied by the same number, the equality remains the same. If $x = y$, then $(x)(z) = (y)(z)$.

(d) **Division Property:** If both sides of an equality are divided by a non-zero number, the equality remains the same. If $x = y$, then, $x/z = y/z$ where $z \neq 0$.

Solving an Equation in One Variable

The following steps are involved in solving an equation.

Step 1: Always ensure that the unknown quantities are on the LHS and the known quantities or constants on the RHS.

Step 2: Add all the terms containing the unknowns on the LHS and all the knowns on the RHS so that each side of the equation contains only one term. On the LHS, the number with which the unknown is multiplied is called the coefficient.

Step 3: Divide both sides of the equation by the coefficient of the unknown.

EXAMPLE 7.1

If $3x + 20 = 65$, then find the value of x.

SOLUTION

Step 1: Group the known quantities as the RHS of the equation, i.e., $3x = 65 - 20$.
Step 2: Simplify the numbers on the RHS. $\Rightarrow 3x = 45$.
Step 3: Since 3 is the coefficient of x, divide both sides of the equation by 3.

$$\frac{3x}{3} = \frac{45}{3} \Rightarrow x = 15.$$

EXAMPLE 7.2

If $5x - 8 = 3x + 22$, then find the value of x.

SOLUTION

Step 1: $5x - 3x = 22 + 8$
Step 2: $2x = 30$
Step 3: $\dfrac{2x}{2} = \dfrac{30}{2} \Rightarrow x = 15$.

EXAMPLE 7.3

Solve for m: $\dfrac{m-3}{2} + \dfrac{m-2}{3} = -13$

SOLUTION

Given, $\dfrac{m-3}{2} + \dfrac{m-2}{3} = -13$

$$\Rightarrow \frac{3(m-3) + 2(m-2)}{6} = -13$$
$\Rightarrow 3m - 9 + 2m - 4 = -78$
$\Rightarrow 5m = -65$
$\Rightarrow m = -65/5$
$\therefore m = -13.$

Word Problems and Application of Simple Equations

Let us understand with the help of the following examples as to how word problems can be solved using simple equations.

EXAMPLE 7.4

The sum of the digits of a two-digit number is 9. If 27 is subtracted from the number, the digits interchange their places. Find the number.

SOLUTION

Let the digit in the units place be x. Then the digit in the tens place is $(9 - x)$.
The number is $10(9 - x) + x = 90 - 9x$.
The number formed by interchanging the digits is $10x + 9 - x = 9x + 9$.
Given that $90 - 9x - 27 = 9x + 9$
$\Rightarrow x = 3 \therefore$ Units digit is 3, tens digit is $9 - 3 = 6$ and the number is 63.

EXAMPLE 7.5

Ram and Rahim have ₹60,000 together. If Ram has ₹8000 more than Rahim, then find how much money Ram has.

SOLUTION

Let the amount with Rahim be x.
So the amount with Ram is $x + 8000$.
$\therefore x + x + 8000 = 60000$
$\Rightarrow x = 26,000$.
$\therefore$ The amount with Ram $= 8000 + 26,000 = ₹34,000$.

EXAMPLE 7.6

Sixteen years hence, a man's age will be 9 times his age 16 years ago. Find his age 5 years hence.

SOLUTION

Let the present age be x.
$x + 16 = 9(x - 16)$
$\Rightarrow x = 20$
$\therefore$ Man's age 5 years hence will be 25 years.

EXAMPLE 7.7

In a bag, there are 50 paise coins, ₹1 coins and ₹2 coins. The total value of these coins is ₹30. The number of ₹2 coins is half the number of ₹1 coins which is half the number of 50 paise coins. Find the number of ₹1 coins.

SOLUTION

Let the number of 50 p coins, ₹1 coins and ₹2 coins in the bag be f, θ and t respectively.

Total value (in ₹) = $\dfrac{50}{100} f + 1\theta + 2t = 30$.

$\dfrac{f}{2} + \theta + 2t = 30.$ (1)

$t = \dfrac{\theta}{2}$ and $\theta = \dfrac{f}{2}$

$\therefore 2t = \theta = \dfrac{f}{2}.$

(1) $\Rightarrow \theta + \theta + \theta = 30.$

$3\theta = 30 \Rightarrow \theta = 10.$

EXAMPLE 7.8

The cost of three chairs and four tables is ₹2800. If the cost of each chair is ₹600, then find the cost of each table (in ₹).

SOLUTION

The cost of each chair is ₹600.

The cost of three chairs = 3 × 600 = ₹1800.

$\Rightarrow$ Let the cost of each table = T.

₹1800 + 4T = ₹2800 (given)

$\Rightarrow 4T = ₹1000$

$T = ₹250.$

Inequalities

If a is any real number, then a is either positive or negative or zero. When a is positive, we write $a > 0$, which is read as 'a is greater than zero'. When a is negative, we write $a < 0$, which is read as 'a is less than zero'. If a is zero, we write $a = 0$ and in this case, a is neither positive nor negative. The two signs > and < are called the 'signs of inequalities'.

Notation

1. '>' denotes 'greater than'.
2. '<' denotes 'less than'.
3. '≥' denotes 'greater than or equal to'.
4. '≤' denotes 'less than or equal to'.

Equations and their Applications 7.7

Definition

For any two non-zero real numbers a and b,

1. a is said to be greater than b when $a - b$ is positive, i.e., $a > b$ when $a - b > 0$ and
2. a is said to be less than b when $a - b$ is negative, i.e., $a < b$ when $a - b < 0$.

Listed below are some properties/results which are needed to solve problems on inequalities. The letters a, b, c, d, etc. represent real numbers.

1. For any two real numbers a and b, either $a > b$ or $a < b$ or $a = b$.
2. If $a > b$, then $b < a$.
3. Transitive property: If $a > b$ and $b > c$, then $a > c$. If $a < b$ and $b < c$, then $a < c$.
4. If $a > b$, then $a + c > b + c$ and $a - c > b - c$.
5. If $a > b$ and $c > 0$, then $ac > bc$ and $a/c > b/c$.
6. If $a > b$ and $c < 0$, then $ac < bc$ and $a/c < b/c$.
7. If $a > b$ and $c > d$, then $a + c > b + d$.
8. The square of any real number is always greater than or equal to 0.
9. The square of any non-zero real number is always greater than 0.
10. If $a > 0$, then $-a < 0$ and if $a > b$, then $-a < -b$.
11. If a and b are positive numbers and $a > b$, then $1/a < 1/b$.
12. If a and b are negative numbers and $a > b$, then $1/a < 1/b$.
13. If $a > 0$ and $b < 0$, $1/a > 1/b$.

Inequation

An open sentence which consists of one of the symbols $>, <, \geq, \leq$ is called an inequation.

For example, (i) $5x - 3 > 17$
(ii) $6y + 4 < 28$

Continued Inequation

Two inequations of the same type (i.e., both consisting of $>$ or $\geq$ or both consisting of $<$ or $\leq$) can be combined into a continued inequation as explained below.

For example, (i) If $a < b$ and $b < c$, we can write $a < b < c$.
(ii) If $a \geq b$ and $b > c$, we can write $a \geq b > c$.

Linear Inequation

An inequation in which the highest degree of the variables present is one is called a linear inequation.

For example, (i) $5x - 6 \leq 12 - 4x$
(ii) $3x + 14 \geq x + 20$

Note   A linear inequation having one variable is called a linear inequation in one variable.

Solving Inequations

We are familiar with solving linear equations. Now let us look at solving some linear inequations in one variable.

EXAMPLE 7.9

Solve the following inequations.

(a) $x + 5 < 7$, $x \in R$.
(b) $4x - 3 \geq 17$, $x \in Z$.
(c) $3x - 2 < 1$, $x \in N$.

SOLUTION

(a) Subtracting 5 from both sides,
$x + 5 - 5 < 7 - 5 \Rightarrow x < 2$.
The set of all the numbers which are less than 2 is the solution set of the given inequation.

(b) Adding 3 on both sides,
$4x - 3 + 3 \geq 17 + 3$
$4x \geq 20$.
Dividing both sides by 4, $\dfrac{4x}{4} \geq \dfrac{20}{4} \Rightarrow x \geq 5$.
The set of all the integers which are greater than or equal to 5 is the solution set of the given inequation.

(c) Adding 2 on both sides $3x - 2 + 2 < 1 + 2$.
$3x < 3$.
Dividing both sides by 3, $\dfrac{3x}{3} < \dfrac{3}{3} \Rightarrow x < \dfrac{3}{3} \Rightarrow x < 1$.
There is no natural number which is less than 1.
∴ There is no solution for the given inequation in the set of natural numbers.

EXAMPLE 7.10

If $\dfrac{18 - 2m}{5} + \dfrac{4m + 3}{7} \geq \dfrac{m}{5} + \dfrac{8}{7}$, then

SOLUTION

$\dfrac{18 - 2m}{5} + \dfrac{4m + 3}{7} \geq \dfrac{m}{5} + \dfrac{8}{7}$

$\dfrac{18}{5} - \dfrac{2m}{5} + \dfrac{4m}{7} + \dfrac{3}{7} \geq \dfrac{m}{5} + \dfrac{8}{7}$

$\dfrac{18}{5} + \dfrac{3}{7} - \dfrac{8}{7} \geq \dfrac{m}{5} + \dfrac{2m}{5} - \dfrac{4m}{7}$

$\dfrac{18}{5} - \dfrac{5}{7} \geq \dfrac{7m + 14m - 20m}{35}$

$\dfrac{126 - 25}{35} \geq \dfrac{m}{35}$

$\dfrac{101}{35} \geq \dfrac{m}{35} \Rightarrow m \leq 101$

EXAMPLE 7.11

Represent the following inequations on number line.

(a) $x \leq 3$
(b) $y \geq -1$
(c) $z < -4$

SOLUTION

(a) In the figure above, the ray drawn above the number line represents the solution set of the inequation. The end point of the ray is part of the solution set. This is indicated by placing a solid dot at the end point.

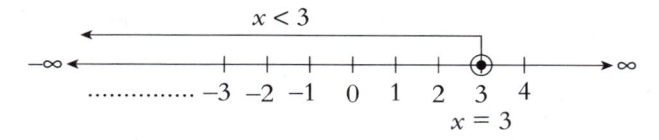

(b) In the figure above, the ray drawn above the number line represents the solution set of the inequation. The end point of the ray is part of the solution set. This is indicated by placing a solid dot at the end point.

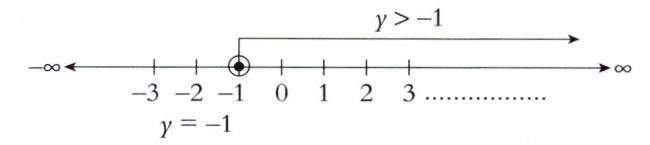

(c) The ray drawn above the number line represents the solution set of the given inequation. The end point of the ray is not a part of the solution set. This is indicated by placing a hollow dot at the end point. We see that the graph of an inequality in one variable is a ray.

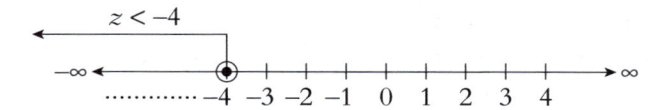

Points to Remember

- A mathematical sentence that can be verified as either true or false is a mathematical statement.
- An open sentence containing the equality sign is an equation.
- An equation, in which the highest index of the unknowns present is one, is a linear equation.
- An inequation, in which the highest degree of the variables present is one, is called a linear inequation.

TEST YOUR CONCEPTS

Very Short Answer Type Questions

Directions for questions 1 to 5: State whether the following statements are true or false.

1. An open sentence containing the sign 'is equal to' is called an equation.
2. The root of the equation $\dfrac{2x+3}{4} = x + 8$, is 7.
3. For the inequation $\dfrac{4}{3}x - 2 < 0$, $x = 2$ is a solution.
4. $3p - 15 > p + 25$, $p = 20$ is a solution of the given inequation.
5. If 68 is divided into two parts such that one part is one third of the other, then the smallest part is 17.

Directions for questions 6 to 19: Fill in the blanks.

6. If the number of variables present in the equation is ____, then it is called a simple equation.
7. If $5m + 18 = 8$, then $m =$ ____.
8. If $x = 5$, then $\dfrac{x}{2} + \dfrac{1}{2} =$ ____.
9. The root of the equation $0.8x + 9 = 17$, is ____.
10. In a two-digit number, if p is the ten's digit and q is the units digit, then the two-digit number is ____.
11. If one-third of a number x is subtracted from two-third of the number x, then the result is 6. Then $x =$ ____.
12. If $0.2y + 10.2 > 11$, then $y >$ ____.
13. If $\dfrac{x}{2} - 3 = 6$, then $x =$ ____.
14. The solution of $\dfrac{3x}{4} + \dfrac{x}{4} \leq 4$ is ____.
15. If $\dfrac{x}{5} \geq 25$, then $x = 125$ is a ____ of $\dfrac{x}{5} \geq 25$.
16. If $2x - 3 = 13$, then $x =$ ____.
17. If $x + 3^{1/2} = 5$, then $x =$ ____.
18. If $\dfrac{x}{5} - 9 = 1$, then $x =$ ____.
19. If $\dfrac{x}{2} + 1 = 3$, then $x =$ ____.

Directions for questions 20 to 28: Select the correct alternative from the given choices.

20. Solve for z: $\dfrac{2z}{5} + 6 = z - 3$.
    (a) 12 (b) 15
    (c) 11 (d) 10

21. Solve for m: $3(4m + 5) - 4(3 - 2m) = 13$.
    (a) 1/2 (b) 2
    (c) 1/4 (d) 1

22. Solve for x: $3x + 5 > 7$
    (a) $x > 3/2$ (b) $x > 2/3$
    (c) $x < 3/2$ (d) $x < 2/3$

23. Solve for y: $\dfrac{4y}{3} - 5 < 10$
    (a) $y < 27/4$ (b) $y < 33/4$
    (c) $y < 45/4$ (d) $y < 59/4$

24. Solve for z: $\dfrac{z}{3} - 7 \geq z - 19$
    (a) $z \leq 12$ (b) $z \leq 15$
    (c) $z \leq 18$ (d) $z \leq 21$

25. Solve for x: $2x + 3 = 9$
    (a) 4 (b) 3
    (c) 2 (d) 1

26. Solve for y: $\dfrac{y}{3} - 7 = -4$
    (a) 15 (b) 12
    (c) 6 (d) 9

27. Solve for x: $12x - 7 = 7x - 13$
    (a) -1.2 (b) -1.6
    (c) -1.8 (d) -1.4

28. Solve for m: $1.2m + 2.6 = 5$
    (a) 4 (b) 3
    (c) 2 (d) 1

Equations and their Applications 7.11

Short Answer Type Questions

29. Divide ₹98 into two parts such that one part is 6 times to the other part.

30. One third of a number is equal to 24, then find the value of $1\frac{1}{2}$ times of the number.

31. The sum of three consecutive integers is 24. Find the smallest number.

32. Solve: $\frac{x}{3} + \frac{x}{6} + \frac{x}{9} = 11$

33. Solve: $0.5y + 0.75y = 125$

34. Solve: $2.5t + 7.3t = 21.6 - t$

35. Equal parts of a flag pole are painted with saffron, white and green colours. If one-third of the portion painted green is 2 m long, then find the length of the pole.

36. A person covered 5/8 of his journey and he has to walk 240 m more to complete his journey. Find the total distance of the journey.

37. A boy has given 7/12 of his marbles to his friend and is left with only 20 marbles. Find how many marbles he had with him initially.

38. A number is multiplied by 5 and 25 is subtracted from the product. The result is equal to four times the number itself. Find the number.

39. One-fourth of a certain number exceeds it's one seventh by 3. Find the number.

40. Solve: $\frac{5x-2}{3} + \frac{4x+3}{2} = \frac{3x+19}{2}$

41. Solve: $0.3(3y - 4.5) + 2.9(5.5 - 5y) = 1$

42. Solve: $\frac{9m+4}{5} - \frac{27m+1}{8} - \frac{1}{2} = 0$

43. Solve: $2m + 5 > 9 - 4m$, $m \in Q$.

44. Solve: $\frac{z}{2} - \frac{z}{3} - \frac{z}{4} = -1$

45. Solve: $\frac{y+2}{3} + \frac{y+3}{2} = y + 1$

46. Solve: $\frac{2n+3}{6n-5} = 1$

47. Solve: $2(k + 3) + 3(k - 4) = 24$

48. One-fifth of a number is 5 more than one-tenth of the number. Find the number.

49. Twice a number is added to half the number, the result is 250. Find one-tenth of the number.

50. Solve for x, $8x + 4 \leq 20$ in the set of natural numbers.

Essay Type Questions

51. Five years ago, the age of a person was half of his present age. How old is he now?

52. The present age of A is twice that of B. The sum of their present ages is 33 years. Find the present age of A (in years).

53. In a two-digit number, the units digit is twice the ten's digit and the difference between the number and the number formed by reversing the digits is 18. Find the original number.

54. Thirty years ago, the age of a man was three-fifth of his present age. Find his present age (in years).

55. The present age of Shobha is equal to one-fifth of her mother Sudha's age. Twenty five years later, the age of Shobha will be 4 years less than half the age of her mother Sudha. Find their present ages.

56. In a two-digit number, the tens digit is one more than twice the units digit. The sum of the digits is 36 less than the number formed by reversing the digits. Find the product of the digits.

57. A man's age 15 years hence would be two times his age six years ago. Find his present age.

58. In a two-digit number, the sum of the digits is 5 more than the units digit. The difference between the original number and the sum of digits is 10 more than the number formed by reversing the digits. Then find the difference between the digits.

59. Solve: $\frac{4t}{5} - \frac{5}{3} < \frac{t}{4} + \frac{3}{2}$, $t \in Q$.

60. Solve: $8a - 7 < \frac{6a}{5} + 27$; $a \in Q$.

Chapter 7

61. Solve: $\dfrac{14y}{3} + \dfrac{3}{2} \leq \dfrac{20y}{3} - \dfrac{83}{4}$, where $y \in Z$.

62. Solve: $2x - 5 > 4x - 3$

63. In a two-digit number, the sum of the digits is 9. If 9 is subtracted from the number, then the digits get reversed. Find the product of the digits.

64. A purse contains a certain number of coins of denominations ₹1 and 25 paise. The total value of the coins (in ₹) is 6 less than the total number of coins. Find the number of 25 paise coins.

65. ₹x is divided among Mr Bilhari, Mr Narahari and Mr Murahari. The share of Bilhari is one-fourth of the total money, the share of Murahari is one-third of the remaining money and the share of Narahari is ₹1200. Find x.

CONCEPT APPLICATION

Level 1

Directions for questions 1 to 22: Select the correct alternative from the given choices.

1. The root of the equation $\dfrac{3}{4}x + 5 = 8$ is _____.
   (a) 5   (b) 4
   (c) 2   (d) 1

2. Which of the following is the solution of the equation $5p - 10 = 5$?
   (a) $p = 5$ (b) $p = 4$
   (c) $p = 2$ (d) $p = 3$

3. If $\dfrac{x}{2} + \dfrac{x}{3} = 5$, then $x = $ ____.
   (a) 2   (b) 3
   (c) 4   (d) 6

4. If $x = 10$, then $0.2x + 0.2 = $ ____.
   (a) 12.2   (b) 10.2
   (c) 2.2   (d) 22

5. If $\dfrac{0.2}{x} + 0.1 \geq 2.1$, then $x \leq $ ____.
   (a) 1   (b) 0.1
   (c) 2   (d) 0.2

6. If $\dfrac{x}{12} + \dfrac{1}{2} = x - 5$, then $x = $ ____.
   (a) 2   (b) 4
   (c) 8   (d) 6

7. If $1.7y + 2.3y = 2$, then $y = $ ____.
   (a) 1/4   (b) 1/2
   (c) 8   (d) 6

8. If $5\left(\dfrac{x}{6} + \dfrac{1}{2}\right) = 5x - 10$, then $x = $ ____.
   (a) 3   (b) 1/3
   (c) 6   (d) 1/6

9. If $\dfrac{1}{x} + \dfrac{3}{x} = \dfrac{11}{3}$, then $x = $ ____.
   (a) 11/3   (b) 2/3
   (c) 12/11   (d) 3/11

10. If $9.1x + 3x + 1.9x \leq 42$, then $x \leq$ ____.
    (a) 1   (b) 3
    (c) 5   (d) 4

11. If $5z + 1.5(0.5z + 10) \geq 590$, then $z \geq$ ____.
    (a) 15   (b) 100
    (c) 150   (d) 200

12. If $8.3x - 9.8 = x + 26.7$, then $x = $ ____.
    (a) 2   (b) 3
    (c) 4   (d) 5

13. If $3(a - 2) - 2(a + 9) = 1$, then $a = $ ____.
    (a) 20   (b) 22
    (c) 25   (d) 27

14. If $2(p - 5) = \dfrac{p}{2} + 5$, then $p = $ ____.
    (a) 8   (b) 9
    (c) 10   (d) 11

15. Solve for a: $3a - 4 = -16$
    (a) -3 (b) -4
    (c) -5 (d) 4

16. Solve for b: $\dfrac{b}{5} - 5 = 7$

    (a) 30 (b) 60
    (c) 90 (d) 120

17. Solve for x: $5x - 6 = 8x - 4$

    (a) $-1/3$ (b) $-2/3$
    (c) $1/3$ (d) $2/3$

18. Solve for k: $1.5k - 3.7 = 0.8$

    (a) 7 (b) 6
    (c) 5 (d) 3

19. Solve: $\dfrac{3x+4}{7} - \dfrac{x+5}{14} = \dfrac{x}{28} + \dfrac{x+1}{14}$

    The following are the steps involved in solving the above problem. Arrange them in sequential order.

    (A) $5x + 3 = \dfrac{3x+2}{2} \Rightarrow 10x + 6 = 3x + 2$

    (B) $\Rightarrow 7x = -4$

    (C) Given $\dfrac{3x+4}{7} - \dfrac{x+5}{14} = \dfrac{x}{28} + \dfrac{x+1}{14}$
    $\Rightarrow \dfrac{6x+8-x-5}{14} = \dfrac{x+2x+2}{28}$

    (D) $\Rightarrow x = -\dfrac{4}{7}$

    (a) ABCD (b) CADB
    (c) CABD (d) BCAD

20. Two-third of a certain number exceeds one-third of the number by 10. Find the number. The following are the steps involved in solving the above problem. Arrange them in sequential order.

    (A) $\dfrac{2x}{3} - \dfrac{x}{3} = 10$ (given)

    (B) $\dfrac{2x-x}{3} = 10 \Rightarrow \dfrac{x}{3} = 10$

    (C) $\Rightarrow x = 30$

    (D) Let the number be x.

    (a) DACB (b) BDAC
    (c) ADBC (d) DABC

21. Solve: $\dfrac{7x+3}{4} + \dfrac{9x-5}{8} = \dfrac{16x-3}{16}$

    The following are the steps involved in solving the above problem. Arrange them in sequential order.

    (A) $x = \dfrac{-5}{30} = \dfrac{-1}{6}$

    (B) $\dfrac{7x+3}{4} + \dfrac{9x-5}{8} = \dfrac{16x-3}{16}$
    $\Rightarrow \dfrac{14x+6+9x-5}{8} = \dfrac{16x-3}{16}$

    (C) $\dfrac{23x+1}{8} = \dfrac{16x-3}{16} \Rightarrow 23x + 1 = \dfrac{16x-3}{2}$

    (D) $46x + 2 = 16x - 3 \Rightarrow 30x = -5$

    (a) BCDA (b) CBDA
    (c) BCAD (d) BDCA

22. One-third of a certain number exceeds $\dfrac{1}{9}$ th of the number by 20. Find the number.

    The following are the steps involved in solving the above problem. Arrange them in sequential order.

    (A) Let the number be x.

    (B) $\dfrac{2x}{9} = 20 \Rightarrow \dfrac{x}{9} = 10 \Rightarrow x = 90$

    (C) Given $\dfrac{x}{3} - \dfrac{x}{9} = 20$

    (D) $\dfrac{3x-x}{9} = 20$

    (a) ADCB (b) ACDB
    (c) DACB (d) CADB

Directions for questions 23 to 26: Match the column A with column B.

Column A	Column B
23. If $\dfrac{17}{3}x + 20 = 71$, then $x =$ ____ . ( )	(a) $y < \dfrac{36}{5}$
24. The root of the equation $\dfrac{4}{5}x + 9 = 2x - 3$ is ____ . ( )	(b) 9
25. The solution set of $\dfrac{y}{6} + \dfrac{4}{5} < 2$ is ____ . ( )	(c) $y > \dfrac{36}{5}$
26. If 69 is divided into two parts such that one part is twice the other, then the greater part is ____ . ( )	(d) 10
	(e) 46
	(f) 23

Chapter 7

Directions for questions 27 to 30: Match the Column A with Column B.

Column A		Column B
27. If $\frac{15}{2}y + 10 = -5$, then $y =$ ___.	( )	(a) $x > 5$
28. The root of the equation $\frac{7}{2}x + 2 = 14 + 3x$ is ___.	( )	(b) 64
29. The solution set $\frac{x}{5} + 4 > 5$ of is ___.	( )	(c) –2
30. If 96 is divided into two parts such that one part is twice the other, then the greater part is ___.	( )	(d) 2
		(e) 24
		(f) $x < 5$

Level 2

Directions for questions 31 to 70: Select the correct alternative from the given choices.

31. If $\frac{9x-5}{7} + \frac{6-3x}{2} = 2$ then $x =$ ___.
    (a) 2/3 (b) 3/4
    (c) 5/4 (d) 4/3

32. If $3y + 1\frac{1}{2} + 6(4 - 5y) = 12$ then $y =$ ___.
    (a) 1/6 (b) 11
    (c) 1/2 (d) 7

33. If $\frac{z+5}{7} + \frac{4(z-11)}{9} + 3 = 0$, then $z =$ ___.
    (a) 1 (b) 2
    (c) –4 (d) 5

34. If $\frac{0.3(3x-4)}{5} + \frac{0.4x+3.6}{5} = 3.5x$, then $x =$ ___.
    (a) 0.1 (b) 0.3
    (c) 0.5 (d) 0.2

35. If $\frac{7t+13}{15} + 7\left(\frac{2t-1}{5}\right) = 6$, then the value of t is ___.
    (a) 2 (b) 3
    (c) 1 (d) –2

36. Two-third of a number exceeds one third of the number by 10. Find the number.
    (a) 10 (b) 20
    (c) 30 (d) 40

37. A number is doubled and half of the number is added to it. If 10 is subtracted from the result, then we get a number which is one less than the original number. Find the original number.
    (a) 5 (b) 6
    (c) 7 (d) 8

38. If a number is multiplied by 5 and 5 is added to it, then the result is equal to 50. Find the number.
    (a) 9 (b) 8
    (c) 7 (d) 6

39. If seven times a number is added to one-fifth of itself, then five-sixth of the sum is equal to 30. Find the number.
    (a) 5 (b) 6
    (c) 15 (d) 10

40. If one-fourth, half and one-third of a number are added to the number itself, then the result is equal to 25. Find the number.
    (a) 10 (b) 11
    (c) 12 (d) 14

41. Solve for x: $\dfrac{x}{5} + \dfrac{x}{7} = 12$
    (a) 70   (2) 140
    (c) 35   (4) 105

42. Solve for x: $\dfrac{3x-2}{5x+7} = \dfrac{1}{12}$
    (a) -4 (b) 3
    (c) 1   (d) -2

43. Solve for y: $3(y-4) - 5(y+5) = -21$
    (a) -4 (b) -7
    (c) -3 (d) -8

44. Two-third of a number is 32 less than three-fifth of the number. Find the number.
    (a) 360   (b) -480
    (c) -360 (d) 480

45. One-third of a number is subtracted from three times the number, the result is 800. Find the number.
    (a) 300   (b) 400
    (c) 200   (d) 600

46. Solve for t, $3t - 8 \leq -t$ in the set of whole numbers.
    (a) 0, 1, 2
    (b) 1, 2, 3
    (c) 0, 1, 2, 3
    (d) 1, 2, 3, 4

47. In a two-digit number, the tens digit is twice the units digit. If the sum of its digits is 9. Find the number.
    (a) 63   (b) 82
    (c) 72   (d) 36

Level 3

48. The present age of a man is seven times the present age of his son. Two years ago, the age of the man was eleven times the age of the son. Find the present age of the man (in years).
    (a) 35   (b) 26
    (c) 47   (d) 45

49. The present age of A is thrice that of B. Five years from now, A's age will be 8 years more than twice B's age. Find the present age of B (in years).
    (a) 10   (b) 13
    (c) 12   (d) 15

50. The sum of the present ages of Ram and Shyam is 75 years. Ten years ago, Ram's age was 4 times the age of Shyam. Find the difference between their present ages (in years).
    (a) 22   (b) 23
    (c) 33   (d) 30

51. A road divider of certain length is painted one-sixth yellow, three-fifth black and the remaining 28 m is painted white. Find the length of the divider.
    (a) 100 m   (b) 120 m
    (c) 150 m   (d) 92 m

52. Mr Sumanth spends two-fifth of his salary on house rent and one-fourth on food. After spending ₹2000 on miscellaneous, if he could save an amount of ₹5000, then find his monthly income (in ₹).
    (a) 20,000   (b) 25,000
    (c) 15,000   (d) 16,000

53. In a two-digit number, the units digit is 3 more than the ten's digit. The sum of the digits is 18 less than the original number. Find the product of the digits.
    (a) 54   (b) 40
    (c) 10   (d) 28

54. A number is added to two-third of itself, 1 is subtracted from the sum and the result is divided by 12. If the final result is 12, then find the number.
    (a) 20   (b) 87
    (c) 84   (d) 74

55. The present age of A is 4 years less than twice the present age of B. B's present age is 6 years more than twice his age 15 years ago. Find the difference of their ages.
    (a) 30 years   (b) 32 years
    (c) 20 years   (d) 22 years

56. A mother said that, her age is one year less than thrice her daughter's age. The daughter is 9 years less than the difference between their present ages. Find the sum of their ages (in years).
   (a) 45   (b) 47
   (c) 39   (d) 35

57. There are two numbers; the difference between them is equal to twice the smaller number. The sum of the two numbers is 68. Find the product of the two numbers.
   (a) 867   (b) 965
   (c) 814   (d) 986

58. A student painted a circular region of certain area such that 4/7th of the area was pink, 1/10th area was green and 2/7th was yellow. The remaining area of 6 m² was white. Find the area of the region which is painted pink (in sq. units).
   (a) 95   (b) 140
   (c) 240   (d) 80

59. In a two-digit number, tens digit is a multiple of the units digit. The sum of the number and the number formed by reversing the digits is 132. Which of the following can be the product of the two digits?
   (a) 16   (b) 27
   (c) 35   (d) 18

60. There are three house-hold articles. The cost of the first article is two-fifth the cost of the third article and the cost of the third article is twice the cost of second article. If the total cost of the three articles is ₹228. Find the cost of the first article (in ₹).
   (a) 40   (b) 48
   (c) 50   (d) 54

61. In an isosceles triangle, the difference between one of the equal sides and the unequal side (longest of the three) is 3/10 of the sum of the equal sides. If the perimeter of the triangle is 90 cm, then find the length of unequal side in centimetres.
   (a) 40   (b) 80
   (c) 25   (d) 50

62. Mr Anthony travelled 4/9 of a certain distance by bus, 1/3 by car and the remaining 6 km by scooter. Find the distance travelled by bus (in km).
   (a) 12   (b) 18
   (c) 9   (d) 27

63. Which of the following is a solution of $\dfrac{2x-5}{3} > \dfrac{3x+3}{4}$?
   (a) $x = -5$
   (b) $x = -2$
   (c) Both (a) and (b)
   (d) Neither (a) nor (b)

64. The unit digit of a two-digit number is 6. If 9 is added to the number, then the number obtained is 5/4th of the number itself. Find the sum of the digits.
   (a) 7   (b) 8
   (c) 9   (d) 10

65. A purse contains a certain number of coins of denominations Re.1 and 50 paise. The total value of the coins (in ₹) is 14 less than the total number of coins. Find the number of 50 paise coins.
   (a) 12   (b) 18
   (c) 22   (d) 28

66. ₹x is divided among A, B and C. The share of A is two-fifth of the total money, the share of B is two-third of the remaining money and the share of C is ₹600. Find the value of x.
   (a) 3000   (b) 4000
   (c) 5000   (d) 6000

67. Ten years ago, Mohan's age was 35 years less than twice his present age. Find Mohan's present age (in years).
   (a) 15   (b) 20
   (c) 25   (d) 10

68. Ram, Shyam and Tarun have a total of ₹600 with them. The amount with Ram is equal to half of the total amount with the others. Find the amount with Ram (in ₹).
   (a) 150   (b) 300
   (c) 120   (d) 200

69. The difference of the digits of a two-digit number is 8. The sum of its digits can be _____.

(a) 8

(b) 10

(c) Either (a) or (b)

(d) Neither (a) or (b)

70. Ramesh and Suresh have a total of ₹200. If Ramesh gives ₹40 to Suresh, the amounts with both would get interchanged. Find the amount with Suresh (in ₹).

(a) 70 (b) 60

(c) 50 (d) 80

Chapter 7

TEST YOUR CONCEPTS

Very Short Answer Type Questions

1. True
2. False
3. False
4. False
5. True
6. One
7. −2
8. 3
9. 10
10. $10p + q$
11. 18
12. 4
13. 18
14. $x \leq 4$
15. 125
16. 8
17. $1\frac{1}{2}$
18. 50
19. 4
20. (b)
21. (a)
22. (b)
23. (c)
24. (c)
25. (b)
26. (d)
27. (a)
28. (c)

Short Answer Type Questions

29. 14, 84
30. 108
31. 7
32. 18
33. 100
34. 2
35. 18 m
36. 640 m
37. 48
38. 25
39. 28
40. 4
41. 1
42. 1/9
43. All rational numbers greater than $\frac{2}{3}$.
44. 12
45. 7
46. 2
47. 6
48. 50
49. 10
50. 1, 2

Equations and their Applications

Essay Type Questions

51. 10 years
52. 22 years
53. 24
54. 75 years
55. 11 years; 55 years
56. 36
57. 27 years
58. 2
59. All rational numbers less than 190/33.
60. All rational numbers less than 5.
61. Any integer ≥ 12
62. $x < -1$
63. 20
64. 8
65. 2400

CONCEPT APPLICATION

Level 1

1. (b)   2. (d)   3. (d)   4. (c)   5. (b)   6. (d)   7. (b)   8. (a)   9. (c)   10. (b)
11. (b) 12. (d) 13. (c) 14. (c) 15. (b) 16. (b) 17. (b) 18. (d) 19. (c) 20. (d)
21. (a) 22. (b) 23. (b) 24. (d) 25. (a) 26. (e) 27. (c) 28. (e) 29. (a) 30. (b)

Level 2

31. (d) 32. (c) 33. (b) 34. (c) 35. (a) 36. (c) 37. (b) 38. (a) 39. (a) 40. (c)
41. (c) 42. (c) 43. (d) 44. (b) 45. (a) 46. (a) 47. (a)

Level 3

48. (a) 49. (b) 50. (c) 51. (b) 52. (a) 53. (c) 54. (b) 55. (c) 56. (c) 57. (a)
58. (d) 59. (b) 60. (b) 61. (a) 62. (a) 63. (d) 64. (c) 65. (d) 66. (a) 67. (c)
68. (d) 69. (c) 70. (d)

CONCEPT APPLICATION

Level 1

1. $\frac{3}{4}x + 5 = 8$

    $\frac{3}{4}x = 3$

    $x = 4$.

    Hence, the correct option is (b)

2. $5p - 10 = 5$

    $5p = 15$

    $p = 3$.

    Hence, the correct option is (d)

3. $\frac{x}{2} + \frac{x}{3} = 5$

    $\frac{3x + 2x}{6} = 5$

    $\frac{5x}{6} = 5$

    $x = 6$.

    Hence, the correct option is (d)

4. $0.2x + 0.2$

    Given, $x = 10$

    $= 0.2 \times 10 + 0.2$

    $= 2 + 0.2$

    $= 2.2$

    Hence, the correct option is (c)

5. $\frac{0.2}{x} + 0.1 \geq 2.1$

    $\frac{0.2}{x} \geq 2.1 - 0.1$

    $\frac{0.2}{x} \geq 2$

    $\frac{x}{0.2} \leq \frac{1}{2}$

    $x \leq 0.1$.

    Hence, the correct option is (b)

6. $\frac{x}{12} + \frac{1}{2} = x - 5$

    $\frac{x + 6}{12} = x - 5$

    $x + 6 = 12x - 60$

    $66 = 11x$

    $x = 6$.

    Hence, the correct option is (d)

7. $1.7y + 2.3y = 2$

    $4y = 2$

    $y = \frac{1}{2}$

    Hence, the correct option is (b)

8. Given, $5\left(\frac{x}{6} + \frac{1}{2}\right) = 5x - 10$

    $\frac{5x}{6} + \frac{5}{2} = 5x - 10$

    $\frac{5x + 15}{6} = 5x - 10$

    $5x + 15 = 30x - 60$

    $75 = 25x$

    $x = 3$.

    Hence, the correct option is (a)

9. Given, $\frac{1}{x} + \frac{3}{x} = \frac{11}{3}$

    $\frac{1 + 3}{x} = \frac{11}{3}$

    $x = \frac{12}{11}$.

    Hence, the correct option is (c)

10. $9.1x + 3x + 1.9x \leq 42$

    $14x \leq 42$

    $x \leq 3$.

    Hence, the correct option is (b)

11. $5z + 1.5(0.5z + 10) \geq 590$

    $5z + 0.75z + 15 \geq 590$

    $5.75z \geq 590 - 15$

    $5.75z \geq 575$

    $z \geq \frac{575}{5.75}$

    $z \geq 100$.

    Hence, the correct option is (b)

Equations and their Applications   7.21

12. $8.3x - 9.8 = x + 26.7$
    $8.3x - x = 9.8 + 26.7$
    $7.3x = 36.5$
    $x = \dfrac{36.5}{7.3}$
    $x = 5.$
    Hence, the correct option is (d)

13. $3(a - 2) - 2(a + 9) = 1$
    $3a - 6 - 2a - 18 = 1$
    $a - 24 = 1$
    $a = 25.$
    Hence, the correct option is (c)

14. $2(p - 5) = \dfrac{p}{2} + 5$
    $2p - 10 = \dfrac{p}{2} + 5$
    $2p - \dfrac{p}{2} = 5 + 10$
    $\dfrac{3p}{2} = 15$
    $p = 15 \times \dfrac{2}{3}$
    $p = 10.$
    Hence, the correct option is (c)

15. Given, $3a - 4 = -16$
    $\Rightarrow 3a = -16 + 4 \Rightarrow 3a = -12$
    $\Rightarrow a = \dfrac{-12}{3} = -4$
    $\therefore a = -4.$
    Hence, the correct option is (b)

16. Given
    $\dfrac{b}{5} - 5 = 7 \Rightarrow 7 + 5$
    $\Rightarrow \dfrac{b}{5} = 12$
    $\Rightarrow b = 12 \times 5 = 60$
    Hence, the correct option is (b)

17. Given,
    $5x - 6 = 8x - 4$
    $\Rightarrow 5x - 8x = -4 + 6$
    $\Rightarrow -3x = 2 \Rightarrow x = -2/3$
    Hence, the correct option is (b)

18. Given $1.5k - 3.7 = 0.8$
    $\Rightarrow 1.5k = 4.5$
    $\Rightarrow k = \dfrac{4.5}{1.5} = 3$
    Hence, the correct option is (d)

19. (C), (A), (B) and (D) is the required sequential order.
    Hence, the correct option is (c)

20. (D), (A), (B) and (C) is the required sequential order.
    Hence, the correct option is (d)

21. (B), (C), (D) and (A) is the required sequential order.
    Hence, the correct option is (a)

22. (A), (C), (D) and (B) is the required sequential order.
    Hence, the correct option is (b)

23. $\rightarrow b : \dfrac{17}{3}x + 20 = 71$
    $\Rightarrow \dfrac{17x}{3} = 71 - 20$
    $\Rightarrow \dfrac{17x}{3} = 51$
    $\Rightarrow \dfrac{x}{3} = 3 \Rightarrow x = 9$

24. $\rightarrow d : \dfrac{4}{5}x + 9 = 2x - 3$
    $\Rightarrow \dfrac{4x}{5} - 2x = -9 - 3$
    $= \dfrac{-6x}{5} = -12 \Rightarrow \dfrac{6x}{5} = 12$
    $\Rightarrow \dfrac{x}{5} = 2 \Rightarrow x = 10$

25. $\rightarrow a : \dfrac{y}{6} + \dfrac{4}{5} < 2 \Rightarrow \dfrac{y}{6} < 2 - \dfrac{4}{5}$
    $\dfrac{y}{6} < \dfrac{10 - 4}{5} \Rightarrow \dfrac{y}{6} < \dfrac{6}{5}$
    $\Rightarrow y < \dfrac{36}{5}$

26. $\rightarrow e$: Let the smaller part be x.
    $\Rightarrow x + 2x = 69 \Rightarrow 3x = 69 \Rightarrow x = 23.$
    $\therefore$ Greater part $= 2x = 2 \times 23 = 46.$

7.22 Chapter 7

27. $\to c : \dfrac{15}{2} y + 10 = -5$

$\dfrac{15}{2} y = -5 \Rightarrow y = -2$

28. $\to e : \dfrac{7x}{2} + 2 = 14 + 3x$

$\dfrac{7x}{2} - 3x = 12 \Rightarrow \dfrac{7x - 6x}{2} = 12$

$\Rightarrow \dfrac{x}{2} = 12 \Rightarrow x = 24.$

29. $\to a : \dfrac{x}{5} + 4 > 5$

$\Rightarrow \dfrac{x}{5} > 5 - 4$

$\Rightarrow \dfrac{x}{5} > 1$

$\Rightarrow x > 5$

30. $\to b$: Let the smaller part be x.

$\Rightarrow 2x + x = 96 \Rightarrow 3x = 96 \Rightarrow x = 32$

$\therefore$ Larger part $= 2x = 2 \times 32 = 64.$

Level 2

31. $\dfrac{9x - 5}{7} + \dfrac{6 - 3x}{2} = 2$

$= \dfrac{18x - 10 + 42 - 21x}{14} = 2$

$\Rightarrow -3x + 32 = 28$

$\Rightarrow -3x = -4 \Rightarrow x = \dfrac{4}{3}.$

Hence, the correct option is (d)

32. $3y + 1\dfrac{1}{2} + 6(4 - 5y) = 12$

$3y + \dfrac{3}{2} + 24 - 30y = 12$

$-27y = 12 - 24 - \dfrac{3}{2}$

$-27y = -12 - \dfrac{3}{2}$

$-27y = -\dfrac{27}{2}$

$y = \dfrac{27}{2 \times 27}$

$y = \dfrac{1}{2}.$

Hence, the correct option is (c)

33. $\dfrac{z + 5}{7} + \dfrac{4(z - 11)}{9} + 3 = 0$

$\dfrac{9(z + 5) + 28(z - 11)}{63} + 3 = 0$

$9z + 45 + 28z - 308 + 189 = 0$

$37z + 234 - 308 = 0$

$37z = 74$

$z = 2.$

Hence, the correct option is (b)

34. $\dfrac{0.3(3x - 4)}{5} + \dfrac{0.4x + 3.6}{2} = 3.5x$

$\dfrac{0.9x - 1.5}{5} + \dfrac{0.4x + 3.6}{2} = 3.5x$

$\dfrac{1.8x - 2.4 + 2.0x + 18}{10} = 3.5x$

$3.8x + 15.6 = 35x$

$15.6 = 31.2x$

$x = \dfrac{15.6}{31.2}$

$x = \dfrac{1}{2}$

$\therefore x = 0.5$

Hence, the correct option is (c)

35. $\dfrac{7t + 13}{15} + \dfrac{7(2t - 1)}{5} = 6$

$\dfrac{7t + 13 + (14t - 7)3}{15} = 6$

$7t + 13 + 42t - 21 = 90$

$49t - 8 = 90$

$49t = 98$

$t = \dfrac{98}{49}$

$t = 2.$

Hence, the correct option is (a)

Equations and their Applications 7.23

36. Let the quantity be x

$\dfrac{2}{3}x - \dfrac{1}{3}x = 10$

$\dfrac{x}{3} = 10$

$x = 30$

Hence, the correct option is (c)

37. Let the number be x.

According to the problem

$2x + \dfrac{x}{2} - 10 = x - 1$

$2x + \dfrac{x}{2} - x = 10 - 1$

$x + \dfrac{x}{2} = 9$

$x = 6$.

Hence, the correct option is (b)

38. Let the number be x.

According to the problem

$5x + 5 = 50$

$x + 1 = 10$

$x = 9$.

Hence, the correct option is (a)

39. Let the number be x.

According to the problem

$\dfrac{5}{6}\left(7x + \dfrac{x}{5}\right) = 30$

$\Rightarrow \dfrac{5}{6}\left(\dfrac{35 + x}{5}\right) = 30$

$\Rightarrow 6x = 30$

$x = 5$.

Hence, the correct option is (a)

40. Let the number be x.

$\dfrac{x}{4} + \dfrac{x}{2} + \dfrac{x}{3} + x = 25$

$\dfrac{3x + 6x + 4x}{12} + x = 25$

$\dfrac{13x + 12x}{12} = 25$

$\dfrac{25x}{12} = 25$

$x = 12$.

Hence, the correct option is (c)

41. Given

$\dfrac{x}{5} + \dfrac{x}{7} = 12$

$\Rightarrow \dfrac{7x + 5x}{35} = 12$

$\Rightarrow 12x = 12 \times 35$

$\Rightarrow x = \dfrac{12 \times 35}{12} = 35$.

Hence, the correct option is (c)

42. Given $\dfrac{3x - 2}{5x + 7} = \dfrac{1}{12}$

$\Rightarrow 36x - 24 = 5x + 7$

$\Rightarrow 31x = 31$

$\Rightarrow x = 1$.

Hence, the correct option is (c)

43. Given $3(y - 4) - 5(y + 5) = -21$

$\Rightarrow 3y - 12 - 5y - 25 = -21$

$\Rightarrow -2y = 16$

$\Rightarrow y = \dfrac{-16}{2} = -8$.

Hence, the correct option is (d)

44. Let the number be x.

$\dfrac{2x}{3} = \dfrac{3x}{5} - 32$

$\dfrac{2x}{3} - \dfrac{3x}{5} = -32$

$\dfrac{10x - 9x}{15} = -32$

$x = -480$.

Hence, the correct option is (b)

45. Let the number be x.

$\therefore 3x - \dfrac{x}{3} = 800$

$\Rightarrow \dfrac{9x - x}{3} = 800$

$\Rightarrow \dfrac{8x}{3} = 800$

7.24 Chapter 7

$\Rightarrow x = 800 \times \dfrac{3}{8}$

$\Rightarrow x = 300.$

Hence, the correct option is (a)

46. Given, $3t - 8 \leq -t$

$\Rightarrow 3t + t \leq 8$

$\Rightarrow 4t \leq 8$

$\Rightarrow t \leq 8/4$

$\Rightarrow t \leq 2,$ but $t \in W.$

$\therefore t = 0, 1, 2.$

Hence, the correct option is (a)

47. Let the number be ab.

$a = 2b$ (given) $\rightarrow$ (1)

$a + b = 9$ (given) $\rightarrow$ (2)

$\Rightarrow$ From (1) and (2), we get $3b = 9$

$\Rightarrow b = 3$

$a = 6.$

$\Rightarrow$ The number is 63.

Hence, the correct option is (a)

Level 3

48. Let the son's present age $= x$ years.

$\Rightarrow$ Present age of the man $= 7x$ years.

$\Rightarrow$ 2 years ago their ages are $7x - 2, x - 2.$

Given that $7x - 2 = 11(x - 2)$

$7x - 2 = 11x - 22$

$22 - 2 = 11x - 7x$

$4x = 20$

$x = 5.$

$\therefore$ Present age of the man $= 7 \times 5 = 35$ years.

Hence, the correct option is (a)

49. Let the present age of $B = x$ years.

$\Rightarrow$ Present age of $A = 3x.$

After 5 years, their ages will be $x + 5$ and $3x + 5$ years.

Given that, $3x + 5 = 2(x + 5) + 8$

$3x + 5 = 2x + 10 + 8$

$3x - 2x = 18 - 5$

$x = 13$ years.

$\therefore$ Present age of B is 13 years.

Hence, the correct option is (b)

50. Let the present age of Shyam be $x.$

$\Rightarrow$ Ram's present age $= 75 - x.$

10 years ago their ages were

$x - 10, 75 - x - 10 = 65 - x.$

Given that

$65 - x = 4(x - 10)$

$65 - x = 4x - 40$

$65 + 40 = 4x + x$

$105 = 5x$

$x = \dfrac{105}{5}$

$x = 21.$

Ram's present age $= 75 - x = 75 - 21 = 54.$

$\therefore$ Difference between their present ages $= 54 - 21$ $= 33.$

Hence, the correct option is (c)

51. Let the length of road divider $= x$ m.

$\dfrac{x}{6} + \dfrac{3x}{5} + 28 = x$

$\dfrac{5x + 18x + 840}{30} = x$

$23x + 840 = 30x$

$840 = 7x$

$x = \dfrac{840}{7}$

$x = 120.$

$\therefore$ Length of the divider $= 120$ m.

Hence, the correct option is (b)

Equations and their Applications | **7.25**

52. Let the one month salary of Mr Sumanth be ₹x.

    According to the problem

    $\dfrac{2x}{5} + \dfrac{x}{4} + 2000 + 5000 = x$

    $\dfrac{8x + 5x}{20} + 7000 = x$

    $\dfrac{13x}{20} - x = -7000$

    $\Rightarrow \dfrac{13x - 20x}{20} = -7000$

    $\Rightarrow \dfrac{-7x}{20} = -7000$

    $\dfrac{x}{20} = 1000$

    $x = 20000.$

    Hence, the correct option is (a)

53. Let the ten's digit of the number be x.

    $\Rightarrow$ Units digit $= x + 3$.

    $\therefore$ The number is $10x + (x + 3)$.

    $x + (x + 3) = 10x + (x + 3) - 18$

    $2x + 3 = 11x + 3 - 18$

    $2x = 11x - 18$

    $18 = 9x$

    $x = 2.$

    $\therefore$ The number is $10 \times 2 + (2 + 3) = 25$ and the digits are 2, 5.

    $\therefore$ Product of the digits $= 2 \times 5 = 10.$

    Hence, the correct option is (c)

54. Let the number be x.

    $\dfrac{x + \dfrac{2x}{3} - 1}{12} = 12$

    $\dfrac{3x + 2x - 3}{36} = 12$

    $5x - 3 = 432$

    $5x = 435$

    $x = 87.$

    $\therefore$ The number is 87.

    Hence, the correct option is (b)

55. Let the present age of B be x years.

    $\Rightarrow$ Present age of A is $2x - 4$ years

    $x = 2(x - 15) + 6$

    $x = 2x - 30 + 6$

    $24 = x$

    $\therefore$ A's present age $= 2 \times 24 - 4 = 48 - 4 = 44$ years.

    $\therefore$ Difference between their ages $= 44 - 24 = 20$ years.

    Hence, the correct option is (c)

56. Let the daughter's present age be x years.

    $\therefore$ Mother's present age $= (3x - 1)$ years

    $x = (3x - 1) - x - 9$

    $x = 3x - 1 - x - 9$

    $x = 2x - 10$

    $x = 10$

    $\therefore$ Mother's age $= 3(10) - 1 = 29$ years.

    $\therefore$ Sum of their ages $= 29 + 10 = 39$ years.

    Hence, the correct option is (c)

57. Let the smaller number be x.

    The other number is $68 - x$.

    $68 - x - x = 2x$

    $68 = 4x$

    $\therefore x = 17.$

    $\therefore$ Bigger number $= 68 - 17 = 51.$

    $\therefore$ Product of the number $= 867.$

    Hence, the correct option is (a)

58. Let the area of the circle be x sq. units.

    $\dfrac{4x}{7} + \dfrac{x}{10} + \dfrac{2x}{7} + 6 = x$

    $\dfrac{6x}{7} + \dfrac{x}{10} + 6 = x$

    $\dfrac{60x + 7x}{70} + 6 = x$

    $\dfrac{67x}{70} + 6 = x$

    $6 = x - \dfrac{67x}{70}$

7.26 Chapter 7

$6 = \dfrac{70x - 67x}{70}$

$420 = 3x$

$\therefore x = 140$

$\dfrac{4}{7}x = \dfrac{4}{7} \times 140 = 80$ sq. units.

Hence, the correct option is (d)

59. Let the units digit of the number be x and the ten's digit be nx the number formed is $10 \times nx$ $+ x$ and the number by reversing the digits is $10x + nx$.

Given that $10nx + x + 10x + nx = 132$.

$11nx + 11x = 132$

$11(nx + x) = 132$

$nx + x = 12$

$\therefore$ Sum of the digits = 12.

This is possible with the digits 9, 3 or 8, 4 or 6, 6.

Sum of the Numbers	Product of the Digits
93 + 39 = 132	27
84 + 48 = 132	32
66 + 66 = 132	36

Hence, the correct option is (b)

60. Let the cost of the third article be ₹x.

$\therefore$ Cost of the first article is ₹$\dfrac{2}{5}x$.

Cost of the second article is ₹$\dfrac{x}{2}$.

Total cost of the three articles is ₹228.

i.e., $x + \dfrac{2x}{5} + \dfrac{x}{2} = 228$

$\dfrac{10x + 4x + 5x}{10} = 228$

$19x = 2280$

$x = 120$.

$\therefore$ Cost of the first article $= \dfrac{2}{5} \times 120$

$= 2 \times 24 = ₹48$

Hence, the correct option is (b)

61. Let the equal side of the triangle be x cm.

Given perimeter of the triangle = 90 cm then the unequal side = $90 - 2x$

Given that $(90 - 2x) - x = \dfrac{3}{10}(2x)$

$90 - 3x = \dfrac{3x}{5}$

$90 = \dfrac{3x}{5} + 3x$

$90 = \dfrac{18x}{5}$

$\therefore x = 25$ cm.

$\therefore$ Length of the unequal side $= 90 - 2x = 90 - 50 = 40$ cm.

Hence, the correct option is (a)

62. Let the total distance travelled = x km.

Given that $\dfrac{4x}{9} + \dfrac{x}{3} + 6 = x$

$\dfrac{4x + 3x + 54}{9} = x$

$7x + 54 = 9x$

$54 = 2x$

$x = 27$ km.

$\therefore$ Distance travelled by bus $= \dfrac{4}{9}x = \dfrac{4}{9} \times 27$

$= 12$ km.

Hence, the correct option is (a)

63. $\dfrac{2x - 5}{3} > \dfrac{3x + 3}{4}$

Put $x = -5 \Rightarrow -5 > -3$ which is false.

Put $x = -2$.

$\Rightarrow -3 > \dfrac{-3}{4}$ which is false.

Hence, the correct option is (d)

64. Let the tens digit be x.

Given, units digit be 6.

$\therefore (10x + 6) + 9 = \dfrac{5}{4}(10x + 6)$

$\Rightarrow 10x + 15 = \dfrac{5}{4}(10x + 6)$

$\Rightarrow 2x + 3 = \dfrac{1}{2}(5x + 3)$

Equations and their Applications

$\Rightarrow 4x + 6 = 5x + 3$

$\Rightarrow x = 3$

$\therefore$ The sum of the digits = 6 + 3 = 9.

Hence, the correct option is (c)

65. Let the number of ₹1 coins and 50 paise coins be x and y respectively.

$x(1) + y(1/2) = x + y - 14$

$\Rightarrow y - \dfrac{y}{2} = 14$

$\Rightarrow y = 28.$

Hence, the correct option is (d)

66. Total money = ₹x.

The share of A = ₹$\dfrac{2}{5}x$.

The share of B = $\dfrac{2}{3}$ (₹x – ₹$\dfrac{2x}{5}$) = ₹$\dfrac{2x}{5}$.

The share of C = ₹600.

$\therefore \dfrac{2x}{5} + \dfrac{2x}{5} + 600 = x$

$\Rightarrow x - \dfrac{4x}{5} = 600 \Rightarrow \dfrac{x}{5} = 600$

$\Rightarrow x = 600(5)$

$\Rightarrow x = 3000.$

Hence, the correct option is (a)

67. Let the present age of Mohan be M.

$(M - 10) = 2M - 35$

$\Rightarrow M = 25.$

Hence, the correct option is (c)

68. Let the amounts with Ram Shyam and Tarun be ₹R, ₹S and ₹T respectively.

$R + S + T = 600$ (1)

$R = \dfrac{1}{2}(S + T)$

$2R = S + T$

(1) $\Rightarrow 3R = 600 \Rightarrow R = 200$.

Hence, the correct option is (d)

69. Let the number be $x\,y$.

Difference of x and y = 8 (given)

$\therefore x > y$ or $x < y$

If $x > y$, $x - y = 8$

If $x < y$, $y - x = 8$

$x - y = 8 \Rightarrow (x, y) = (8, 0)$ or $(9, 1)$

$y - x = 8 \Rightarrow (x, y) = (1, 9)$

$x\,y$ = 80 or 91 or 19. (Only three numbers are possible)

$\therefore x + y$ is either 8 or 10.

Hence, the correct option is (c)

70. Let the amounts with Ramesh and Suresh be ₹R and ₹S respectively.

$R + S = 200$ (1)

If Ramesh gives ₹40 to Suresh, Ramesh would have ₹$(R - 40)$ and Suresh would have ₹$S+40$.

The amounts with both would then get interchanged.

$\therefore R - 40 = S$ and $S + 40 = R$.

These equations are the same. Solving one of these along with (1), $R = 120$ and $S = 80$.

Hence, the correct option is (d)

Chapter 8

Formulae

a) $5(3x) - 3(x-4) = 0$
$15x - 3x + 12 = 0$
$12x + 12 = 0$
$12x = -12$
$x = -12/12$
$x = -1 \qquad x^2 = 1$

REMEMBER

Before beginning this chapter, you should be able to:
- Review the geometric figures, triangles, squares, rectangles, etc.
- Know the mathematical expressions

KEY IDEAS

After completing this chapter, you should be able to:
- Use formula in different cases of mathematical problems
- Understand the subject of a formula, change of subject (auxiliary formulae)
- Evaluate the subject of a formula and study characteristics of subject in a formula
- Know how to frame a formula

Chapter 8

INTRODUCTION

You have already learnt in your earlier classes about geometrical figures like triangle, rectangle, square, etc.

Let us recall some aspects of what you have learnt. Consider one of the figures, i.e., rectangle. If a rectangle is given and you are asked to find its perimeter, you add all the sides of the rectangle and give the obtained result, i.e., the sum of its two lengths and two breadths. It can be generalised that the perimeter of a rectangle = 2(length + breadth).

If length is l units and breadth is b units, the symbolical representation of perimeter of the rectangle is $P = 2(l + b)$ units.

The above statement can be used in finding the perimeter effortlessly. This statement is called a formula.

Definition: A formula is an equation based on a rule, concept or a principle and is used frequently to solve given problems.

It shows the relation between two or more variables or unknowns. For example, when perimeter $P = 2(l + b)$, P, l and b are called the variables.

Note   The plural form of formula is formulae.

Subject of a Formula

The subject of a formula is the variable which is expressed in terms of all the other variables.

For example,

1. In the formula $A = \dfrac{1}{2}(a + b)$, the subject is A.

2. In the formula $S = ut + \dfrac{1}{2}at^2$, the subject is S.

3. In the formula $A = \sqrt{S(S-a)(S-b)(S-c)}$, where $S = \dfrac{a+b+c}{2}$, the subject is A.

Change of Subject

The desired variable in a given formula may be expressed in terms of the other variables by rewriting the formula using the rules applicable in solving a simple equation. Changing a term from one side of an equation to the other side is called '**transposition**'.

Thus, in a given formula, we may change the subject as per our requirement. The new formula deduced from the original formula is called the **auxiliary formula or derived formula**.

For example,

1. Consider the formula for evaluating perimeter of a square of side 's' is $A = 4s$.

    Rewriting the formula by expressing s in terms of A we get, $s = \dfrac{A}{4}$.

    Hence, changing the subject as s we get, $s = \dfrac{A}{4}$.

    Thus, $s = \dfrac{A}{4}$ is an auxiliary formula.

2. Consider the formula for evaluating the curved surface area (A) of a cylinder with base radius r and height h is $A = 2\pi rh$.

Rewriting this formula by expressing r in terms of a and h we get, $r = \dfrac{A}{2\pi h}$. (1)

Rewriting $A = 2\pi rh$ by expressing h in terms of A and r we get, $h = \dfrac{A}{2\pi r}$. (2)

Hence, changing the subject of the given formula as r we get (1) and as h we get (2).

∴ (1) and (2) are the auxiliary formulae.

Evaluation of the Subject of a Formula

The value of the subject in a formula is found by substituting the values of the other variables in the formula.

EXAMPLE 8.1

Using the formula $A = \pi r^2$, find A when $r = 14$ cm (take $\pi = 22/7$).

SOLUTION

$A = \pi r^2$, here $r = 14 \Rightarrow A = \dfrac{22}{7} \times 14 \times 14 = 616$

∴ $A = 616$ cm^2.

EXAMPLE 8.2

Using the formula $S = 2(lb + bh + lh)$, find S when $l = 12$ cm, $b = 8$ cm and $h = 4$ cm.

SOLUTION

Given $S = 2(lb + bh + lh)$ and $l = 12$, $b = 8$ and $h = 4$. Substituting the given values, we get, $S = 2(12 \times 8 + 8 \times 4 + 12 \times 4) = 2(96 + 32 + 48) = 352$.
∴ $S = 352$ cm^2.

Characteristics of Subject in a Formula

1. The subject symbol alone always occurs on the left side of the equality.
2. Co-efficient of a subject is always one.
3. In changing the subject of a formula, all the properties used in simple equations are used.

EXAMPLE 8.3

Write the formula for finding the circumference (C) of a circle with radius r units. What is the subject in this formula?

SOLUTION

We know that the circumference (C) of a circle of radius r is given by $C = 2\pi r$.
In this formula, the subject is C.

Chapter 8

EXAMPLE 8.4

Make h the subject of the formula, $V = lbh$ and find h, when $V = 960$ cm^3, $l = 20$ cm and $b = 12$ cm.

SOLUTION

$V = lbh$ on dividing both the sides with lb, we get $\dfrac{V}{lb} = h$. Hence, $h = \dfrac{V}{l \times b}$.

$h = \dfrac{960}{20 \times 12} = 4$

$\therefore h = 4$ cm.

EXAMPLE 8.5

(a) In the formula $S_n = \dfrac{n}{2}\{2a + (n-1)d\}$, make d as the subject.

(b) Find the value of d, when $S_n = 240$, $n = 10$ and $a = 6$.

SOLUTION

(a) Given formula is, $S_n = \dfrac{n}{2}\{2a + (n-1)d\}$

$\Rightarrow \dfrac{n}{2}\{2a + (n-1)d\} = S_n$

$\Rightarrow 2a + (n-1)d = \dfrac{2S_n}{n}$

$\Rightarrow (n-1)d = \dfrac{2S_n}{n} - 2a$

$\Rightarrow d = \dfrac{2}{n-1}\left[\dfrac{S_n}{n} - a\right]$

(b) Substituting the values $S_n = 240$, $n = 10$ and $a = 6$.

We get, $d = \dfrac{2}{(10-1)}\left[\dfrac{240}{10} - 6\right] = \dfrac{2}{9} \times 18 = 4$.

$\therefore d = 4$.

The focal length of a lens is given by the formula, $\dfrac{1}{f} = \dfrac{1}{u} + \dfrac{1}{v}$.

EXAMPLE 8.6

Make f as the subject of the formula.

SOLUTION

Given $\dfrac{1}{f} = \dfrac{1}{u} + \dfrac{1}{v} \Rightarrow \dfrac{1}{f} = \dfrac{v+u}{uv}$

$\Rightarrow f = \dfrac{uv}{v+u}$ $(\because$ Reciprocal form$)$

EXAMPLE 8.7

If $u = 15$ cm and $f = 6$ cm then find v.

SOLUTION

$$\frac{1}{f} = \frac{1}{u} + \frac{1}{v}$$

$$\frac{1}{v} = \frac{1}{f} - \frac{1}{u}$$

$$\frac{1}{v} = \frac{u - f}{fu} \Rightarrow v = \frac{fu}{u - f} = \frac{6 \times 15}{15 - 6} = 10 \text{ cm.}$$

EXAMPLE 8.8

If $f = 3$ cm and $v = 4$ cm, then find u.

SOLUTION

$$\frac{1}{f} = \frac{1}{u} + \frac{1}{v} \Rightarrow u = \frac{fv}{v - f}$$

$$= \frac{3 \times 4}{4 - 3} = 12 \text{ cm.}$$

Framing a Formula

A formula can be framed based on the relation between the given variables subject to a certain given condition or conditions.

For example,

1. We know that the perimeter (P) of a sector is equal to the sum of two times the radius (r) and the length of the arc (l). ∴ The formula for the perimeter of a sector is $P = 2r + l$.

2. The volume (V) of sphere is equal to $\frac{4}{3}\pi$ times of cube of its radius (r).

   The formula for the volume of a sphere is $V = \frac{4}{3}\pi r^3$.

EXAMPLE 8.9

Frame the formula: Hypotenuse (h) of right triangle is the square root of sum of the squares of perpendicular sides a, b.

SOLUTION

$h = \sqrt{a^2 + b^2}$.

EXAMPLE 8.10

The slant height (l) of a cone is the square root of the sum of the squares of its radius (r) and its vertical height (h). If for a cone, $l = 17$ cm and $r = 15$ cm, then find h.

SOLUTION

$l = \sqrt{r^2 + h^2}$

Squaring on both sides, $l^2 = r^2 + h^2$

$h^2 = l^2 - r^2 = 17^2 - 15^2 = 289 - 225 = 64$

$h > 0$

$\therefore h = 8$ cm.

Points to Remember

- A formula is an equation based on a rule, concept or a principle and is used frequently to solve the given problems.
- The plural form of formula is formulae.
- Changing a term from one side of an equation to the other side is called 'transposition'.
- The new formula deduced from the original formula is called the auxiliary formula or derived formula.
- The subject symbol always occurs on the left side of the equality.
- The co-efficient of a subject is always one.

TEST YOUR CONCEPTS

Very Short Answer Type Questions

Directions for questions 1 to 5: State whether the following statements are true or false.

1. An equation which is used frequently to solve problems is called a formula.
2. The number of auxiliary formulae which can be derived from $A = S^2$ is two.
3. Auxiliary formula of $A = \pi r^2$ is $r = \dfrac{A}{\pi}$.
4. If $h^2 = a^2 + b^2$, then $b = \sqrt{h^2 - a^2}$.
5. Thirty six is divided into two parts such that one of the parts is twice the other. The two parts are 12 and 24.

Directions for questions 6 to 13: Fill in the blanks.

6. There are b boys and g girls in a class and the average of number of boys and the number of girls is 18. Then $b + g =$ _____.
7. The number of variables present in RHS of $A = 2(lb + bh + lh)$ is _____.
8. If the cost of two pens is ₹x, then the cost of three pens is _____.
9. In $A = \dfrac{x}{360°} \times \pi r^2$, $x = 60°$ and $r = 6$ cm. Then $A =$ _____.
10. The formulae obtained by transforming the subject in the given formula are called _____.
11. The symbolic form of "total surface area (A) of a cube is six times the square of its side (s)" is _____.
12. The symbolic form of "area (A) of a rhombus is half of the product of its diagonals (d_1, d_2)" is _____.
13. The symbolic form of "simple interest (I) is one hundredths of the product principle (P), time period (T) in years and rate of interest (R)" is _____.

Directions for questions 14 and 15: State the following in words.

14. $P = S - C$ where P is the profit, S is the selling price and C is the cost price.
15. $P = \dfrac{36}{7} r$ where P is the perimeter of the semi-circular region and r is the radius of the semicircle.

Directions for questions 16 to 20: Select the correct alternative from the given choices.

16. The number of auxiliary formulae that can be derived from $S = \dfrac{(100 - l)}{100} c$ is _____.
    (a) 4
    (b) 3
    (c) 2
    (d) 1

17. If $V = lbh$, then $b =$ _____.
    (a) $\dfrac{v}{lh}$
    (b) $\dfrac{l}{vh}$
    (c) $\dfrac{h}{vl}$
    (d) lvh

18. The symbolic form of "five times b is added to six times a to get the result c" is _____.
    (a) $5b + 6a = c$
    (b) $5b - 6a = c$
    (c) $6b + 5a = c$
    (d) $6b - 5a = c$

19. The subject of the formula, $A = 2h(l + b)$ is _____.
    (a) l
    (b) b
    (c) h
    (d) A

20. A variable standing alone on the left side of an equation is called _____.
    (a) the formula
    (b) the subject of the formula
    (c) the transposition
    (d) None of these

Short Answer Type Questions

21. If $V = s^3$ and $V = 216$ cm^3, then find the value of s in cm.

22. In the formula $\angle P + \angle Q + \angle R + \angle S = 360°$, if $\angle P = 100°$, $\angle Q = 100°$ and $\angle R = 100°$, then find $\angle S$.

23. If $P = \pi r + 2r$ and $P = 36$ cm, then find the value of r (in cm).

24. If $A = c(a - b)$, then make 'a' the subject of the formula.

25. If $h = \dfrac{V}{A}$, $h = 5$ cm and $V = 60$ cm^3, then find the value of A (in cm^2).

26. The sum of the interior angles in a 6-sided figure is 720° and the six angles are x, y, z, z, y and x. Express the relation among x, y and z by making z as the subject.

27. If $k = a + bc$, then make 'c' as the subject of the formula.

28. The compound interest on a sum of ₹P, for T years at $R\%$ per annum is given by $I = P\left[\left(1+\dfrac{R}{100}\right)^T - 1\right]$. Make R as the subject of the formula.

29. In the previous question, if $I = 662$, $P = 2000$, and $T = 3$, then find the value of R.

30. Let C denote the temperature of a body in degree Celsius. Let F denote its temperature in degree Fahrenheit. The relation between C and F is given by $\dfrac{C}{100} = \dfrac{F - 32}{180}$. If $F = 2.2C$, then find the value of C.

Essay Type Questions

31. Write all the possible related auxiliary formulae from $A = \pi(R^2 - r^2)$.

32. Make g the subject of the formula, $C = \dfrac{100S}{100 + g}$. Find g(in %), when C = ₹400 and S = ₹450.

33. Make r the subject of the formula, $V = \dfrac{\pi r^2 h}{3}$. Find r, when $V = 27\pi$ cm^3 and $h = 4$ cm.

34. Make 'a' the subject of the formula, $A = \dfrac{\sqrt{3}a^2}{4}$. Find a, when $A = 64\sqrt{3}$ cm^2.

35. Make y the subject in $\dfrac{x + y}{x - y} = \dfrac{a + b}{a - b}$.

36. Frame the formula from the following table. Make Y the subject of the formula.

X	1	2	5	7	8
Y	1	4	25	49	64

37. Frame the formula from the following table. Make a the subject of the formula.

a	25°	34°	75°	4°	89°	85°
b	65°	56°	15°	86°	1°	5°

38. Frame the formula from the following table. Make Z the subject of the formula.

X	1	2	1	2	3
Y	1	2	2	1	4
Z	$2\sqrt{2}$	$2\sqrt{2}$	$\sqrt{5}$	$\sqrt{5}$	5

39. If $(x + a)^2 = x^2 + 1 + \dfrac{1}{4x^2}$, then find a.

40. A number, x is divided by 10 and 7 is added to the quotient and then the sum is multiplied by 3 to give the result N. Frame the formula by making x as the subject.

41. If $A = c(a^2 + b^2)$, then make a the subject of the formula.

42. The following table shows the relation between a and b.

a	1	2	3	4	5
b	2	6	12	20	30

Express the relation between a and b with b as the subject.

Formulae 8.9

43. If $\dfrac{x+y}{z} = \dfrac{a+b}{c}$, then make y the subject of the formula.

44. If $d = ut + \dfrac{1}{2}at^2$, $u = 20$, $a = 10$ and $d = 50t$, then find t.

45. Find the relation between x and y from the data given in the following table.

x	2	3	4	5
y	5	10	17	26

CONCEPT APPLICATION

Level 1

Directions for questions 1 to 14: Select the correct alternative from the given choices.

1. The symbolic form of "the sum of four angles in a quadrilateral PQRS is 360°" is ____.
   (a) $\angle P + \angle Q + \angle R + \angle S = 180°$
   (b) $\angle P + \angle Q + \angle R + \angle S = 360°$
   (c) $\angle P + \angle Q + \angle R + \angle S = 90°$
   (d) None of these.

2. The symbolic form of "time taken (t) for a journey is the quotient of distance covered (d) and average speed (s)" is ____.
   (a) $t = \dfrac{d}{s}$
   (b) $d = \dfrac{t}{s}$
   (c) $\dfrac{s}{d}$
   (d) $t = d + s$

3. The symbolic form of "area of a sector (A) is half of the product length of the arc (l) and radius (r)" is ____.
   (a) $A = lr$
   (b) $A = \dfrac{r}{l}$
   (c) $A = \dfrac{lr}{2}$
   (d) $A = \dfrac{l+r}{2}$

4. The symbolic form of "volume (v) of a cube is cube of its length (s)" is ____.
   (a) $V = 3s$
   (b) $V = \dfrac{s}{3}$
   (c) $V = s$
   (d) $V = s^3$

5. The symbolic form of "area (A) of a trapezium is half of the product of distance between the parallel sides (h) and sum of the lengths of parallel sides (a, b)" is ____.
   (a) $A = \dfrac{h}{2}(a+b)$
   (b) $A = \dfrac{h}{2}(a-b)$
   (c) $A = \dfrac{hab}{2}$
   (d) None of these.

6. The number of auxiliary formulae that can be derived from $P = \dfrac{x}{360°}(2\pi r)$ is ____.
   (a) 1
   (b) 2
   (c) 3
   (d) 4

7. If $A = \dfrac{S}{N}$, then $N =$ ____.
   (a) $\dfrac{S}{A}$
   (b) $\dfrac{A}{S}$
   (c) SA
   (d) $S + A$

8. The symbolic form of "6 less than twice p is equal to 3 more than q" is ____.
   (a) $2p + 6 = q - 3$
   (b) $6 < 2p$
   (c) $2p - 6 = q + 3$
   (d) $3 > q$

9. The subject of the formula, $s = ut + \dfrac{ut^2}{2}$ is ____.
   (a) s
   (b) u
   (c) t
   (d) a

10. The formula obtained by transforming the subject in a given formula is called ____.
    (a) the subject of the formula
    (b) the transposition
    (c) an auxiliary formula
    (d) None of these

11. The cost price C is given by the formula. $C = \dfrac{100S}{100+g}$, where S = selling price and g = gain in %. Make S the subject of the formula. Find S, if C = ₹800 and $g = 20$.

The following are the steps involved in solving the above problem. Arrange them in sequential order.

(A) $\Rightarrow 100S = C(100 + g)$

(B) $S = \dfrac{(100 + g)C}{100}$

(C) Given $C = \dfrac{100S}{100 + g}$

(D) $\therefore S = \dfrac{(100 + 20) \times 800}{100} = ₹960$

(a) ABCD (b) BCAD
(c) CADB (d) CABD

12. In the formula $S_n = \dfrac{n}{2}\{2a + (n-1)d\}$, make d as the subject.

The following are the steps involved in solving the above problem. Arrange them in sequential order.

(A) $(n-1)d = \dfrac{2S_n}{n} - 2a$.

(B) Given, $S_n = \dfrac{n}{2}[2a + (n-1)d] \Rightarrow n[2a+(n-1)d] = 2S_n$.

(C) $\Rightarrow d = \dfrac{2}{n-1}\left[\dfrac{S_n}{n} - a\right]$

(D) $2a + (n-1)d = \dfrac{2S_n}{n}$.

(a) DBAC (b) BDAC
(c) ABDC (d) BDCA

13. The compound interest on a certain sum is given by C.I. $= P\left(1 + \dfrac{R}{100}\right)^n - P$. Find C.I. when $P = ₹1000$, $R = 10\%$ p.a. and $n = 2$.

The following are the steps involved in solving the above problem. Arrange them in sequential order.

(A) $\therefore$ C.I. $= ₹210$.

(B) $1000\left(\dfrac{11}{10}\right)\left(\dfrac{11}{10}\right) - 1000 = 1210 - 100$.

(C) Given $CI = P\left(1 + \dfrac{R}{100}\right)^n - P$, $P = ₹1000$, $R = 10\%$ p.a and $n = 2$.

(D) C.I. $= 1000 + \left(1 + \dfrac{10}{100}\right)^2 - 1000$.

(a) BCDA (b) DCBA
(c) CDBA (d) BDCA

14. The focal length of a lens is given by the formula, $\dfrac{1}{f} = \dfrac{1}{u} + \dfrac{1}{v}$. Make f as the subject of the formula. If $u = 20$ cm and $v = 30$ cm then find f.

The following are the steps involved in solving the above problem. Arrange them in sequential order.

(A) Given $\dfrac{1}{f} = \dfrac{1}{u} + \dfrac{1}{v}$

(B) $\Rightarrow f = \dfrac{uv}{u+v}$

(C) $f = \dfrac{20 \times 30}{20 + 30} = \dfrac{600}{50} = 12$ cm

(D) $\Rightarrow \dfrac{1}{f} = \dfrac{v + u}{uv}$

(a) ADBC (b) BADC
(c) ACDB (d) DBAC

Directions for questions 15 to 18: Match the column A with column B.

Column A	Column B
15. Changing a term from one side of an equation to the other side ( )	(a) Subject
16. Coefficient of the subject of a formula ( )	(b) Transposition
17. In a formula, a variable which is expressed in terms of other variables ( )	(c) $C = \pi d$
18. The circumference (C) of a circle is π times its diameter (d). ( )	(d) 1

Directions for questions 19 to 22: Match the column A with column B.

Column A	Column B
19. A symbol that occurs alone on LHS of the equality ( )	(a) Formula
20. The symbolic form of "the sum of the angles of $\triangle ABC$ is $180°$." ( )	(b) Subject
21. The symbolic form of "perimeter (P) of $\triangle ABC$ is the sum of its sides." ( )	(c) $P = AB + BC + AC$
22. An equation based on a rule ( )	(d) $\angle A + \angle B + \angle C = 180°$

Formulae **8.11**

Level 2

Directions for questions 23 to 25: Select the correct alternative from the given choices.

23. If $A = 2\pi r$, then $r = $ _____.
    (a) $2\pi A$ (b) $\dfrac{2\pi}{A}$
    (c) $\dfrac{2A}{\pi}$ (d) $\dfrac{A}{2\pi}$

24. In $A = 2h(l + b)$, if $A = 54$ m², $l = 5$ m and $b = 4$ m, then find h.
    (a) 6 m (b) 4 m
    (c) 3 m (d) 2 m

25. If $A = 2(lb + bh + lh)$, then which of the following is/are true?
    (a) $l = \dfrac{A - 2bh}{2(b + h)}$
    (b) $b = \dfrac{A - 2lh}{2(l + h)}$
    (c) $h = \dfrac{A - 2lb}{2(l + b)}$
    (d) All of these

Directions for questions 26 to 28: These questions are based on the following information.

The length of an arc of a circle is given by the formula, $l = \dfrac{x}{360°} \times 2\pi r$.

26. Make r as the subject of the formula.
    (a) $r = \dfrac{720° \pi x}{l}$
    (b) $r = \dfrac{180° l}{\pi x}$
    (c) $r = \dfrac{720° l}{\pi x}$
    (d) None of these.

27. If $x = 60°$ and $r = 3$ cm, then find l.
    (a) 22 cm (b) 2π cm
    (c) π cm (d) 11 cm

28. If $l = 4\pi$ cm and $r = 18$ cm, then find x.
    (a) 60° (b) 90°
    (c) 80° (d) 40°

Directions for questions 29 to 43: Select the correct alternative from the given choices.

29. The number of variables in the formula, $S = ut + \dfrac{at^2}{2}$ is _____.
    (a) 4 (b) 3
    (c) 2 (d) 1

30. The number of all possible squares in $n \times n$ network is equal to $1^2 + 2^2 + 3^2 + ….. + n^2$. Find the number of possible squares in 7×7 network.
    (a) 103 (b) 91
    (c) 120 (d) 140

31. Write the relation between H and m from the given table.

Number of hours (H)	1	3	5	6	10
Number of minutes (m)	60	180	300	360	600

    (a) $H = 60$ m (b) $m = 60$ H
    (c) $H + m = 1$ (d) None of these

32. Simple interest on a certain sum is given by $I = \dfrac{PTR}{100}$. Make T as the subject of the formula. Find T when $P = ₹1000$, $R = 10\%$ p.a and $I = ₹300$.
    (a) $T = \dfrac{100I}{PR}$; 6 years
    (b) $T = \dfrac{100I}{PR}$; 3 years
    (c) $T = \dfrac{100P}{IR}$; 4 years
    (d) $T = \dfrac{100R}{PI}$; 5 years

33. If $A = S^2$ and $A = 324$ cm², then find the value of S (in cm).
    (a) 162 (b) $(324)^2$
    (c) 18 (d) 16

34. In the formula, $\angle A + \angle B + \angle C = 180°$, if $\angle A = 90°$ and $\angle B = 55°$, then $\angle C = $ _____.
    (a) 45° (b) 45°
    (c) 25° (d) 35°

8.12 Chapter 8

35. If $A = \dfrac{d_1 d_2}{2}$, $d_1 = 6$ cm and $d_2 = 8$ cm, then find the value of A (in cm²).
    (a) 12 (b) 18
    (c) 36 (d) 24

36. If $A = 2h(l + b)$, then $b = $ _____.
    (a) $2Ah - l$ (b) $\left(\dfrac{A}{2h}\right) - l$
    (c) $2A\, l - h$ (d) $\dfrac{A}{2l} - h$

37. If $S = (n - 2)180°$ and $S = 540°$, then find n.
    (a) 4 (b) 3
    (c) 5 (d) 7

38. The sum of four angles of a quadrilateral is 360°. From the following figure, express the relation between x and y by making x as the subject.

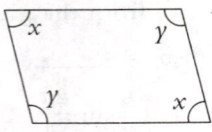

    (a) $x = 180° - y$ (b) $y = 180° - x$
    (c) $x = 180° + y$ (d) $y = 180° + x$

39. If $M = a + \dfrac{b}{c}$, then $b = $ _____.
    (a) $Mc - a$ (b) $M - ac$
    (c) $\dfrac{(M - a)}{c}$ (d) $(M - a)c$

Level 3

40. If $V = lbh$, then $b = $ _____.
    (a) $\dfrac{v}{lb}$ (b) $\dfrac{v}{bh}$
    (c) $\dfrac{v}{lh}$ (d) None of these

41. In $A = s^2 - (s - 2w)^2$, if $w = 1$ m and $s = 6$ m, then find A (in m²).
    (a) 20 (b) 10
    (c) 15 (d) 16

42. Frame the formula: volume (v) of a cuboid is the product of its length (l), breadth (b) and height (h).
    (a) $v = lbh$
    (b) $v = l + b + h$
    (c) $v = \dfrac{lb}{h}$
    (d) None of these

43. If $S = \dfrac{(100 + g)C}{100}$, then which of the following is/are true?
    (a) $C = \dfrac{100S}{(100 + g)}$
    (b) $g = \dfrac{100(S - C)}{C}$
    (c) Both (a) and (b)
    (d) None of these

Directions for questions 44 to 50: Select the correct alternative from the given choices.

44. The number of auxiliary formulae that can be derived from the formula, $D = \left[\dfrac{n(n - 1)}{2}\right] - n$.
    (a) 1 (b) 2
    (c) 3 (d) None of these

45. The number of diagonals of a convex polygon of sides n is equal to $\dfrac{n(n - 3)}{2}$. Find the number of diagonals in hexagon.
    (a) 9 (b) 6
    (c) 10 (d) 8

46. Write the relation between x and y from the given table.

    | x | 26° | 34° | 75° | 30° | 82° | 10° |
    | y | 64° | 56° | 15° | 60° | 8° | 80° |

    (a) $x = y - 45°$
    (b) $x = y + 54°$
    (c) $x = 90° - y$
    (d) None of these

47. The area of four walls of room is given by $A = 2h(l + b)$. Make l as the subject of the formula. Find l when $A = 100$ m², $h = 5$ m and $b = 4$ m.

(a) $l = \dfrac{A}{2h} - b;\ 6\ m$

(b) $l = \dfrac{A}{2h} - b;\ 5\ m$

(c) $l = \dfrac{A}{2b} - h;\ 4\ m$

(d) $l = \dfrac{A}{2b} - h;\ 8\ m$

48. If $A = \dfrac{d}{2}(a + b)$, then which of the following is/are true?

    (a) $d = \dfrac{2A}{a+b}$
    (b) $a = \left(\dfrac{2A}{d}\right) - b$
    (c) Both (a) and (b)
    (d) None of these

49. The following table shows the relation between the angles x and y.

X	90°	100°	110°	120°	130°	140°
Y	90°	80°	70°	60°	50°	40°

Express the relation between x and y with x as the subject.

(a) $y = 180° - x$

(b) $x = 180° - y$

(c) $y = 180° + x$

(d) $x = 180° + y$

50. If $\dfrac{a+b}{c+d} = \dfrac{x}{y}$, then $y = $ _____.

(a) $y = x(a+b)(c+d)$

(b) $y = \dfrac{x}{(a+b)(c+d)}$

(c) $y = \dfrac{x(c+d)}{a+b}$

(d) $y = \dfrac{x(a+b)}{c+d}$

ASSESSMENT TESTS

Test 1

Directions for questions 1 to 11: Select the correct alternative from the given choices.

1. Make v as the subject of the formula, $\dfrac{1}{f} = \dfrac{1}{u} + \dfrac{1}{v}$.

   The following are the steps involved in solving the above problem. Arrange them in sequential order.

   (A) $\dfrac{1}{v} = \dfrac{1}{f} - \dfrac{1}{u}$

   (B) $\dfrac{1}{v} = \dfrac{u-f}{fu}$

   (C) $v = \dfrac{fu}{u-f}$

   (D) $\dfrac{1}{u} + \dfrac{1}{v} = \dfrac{1}{f}$

   (a) DBAC
   (b) DACB
   (c) DABC
   (d) DCBA

2. The sum of the digits of a two-digit number is 11. If 9 is subtracted from the number, the digits interchange their places. Find the number.

   The following are the steps involved in solving the above problem. Arrange them in sequential order.

   (A) Let the units digit be x. Therefore the tens digit is $(11 - x)$.

   ∴ The number is $10(11-x) + x = 110 - 9x$.

   (B) Given that $110 - 9x - 9 = 9x + 11 \Rightarrow x = 5$.

   (C) Units digit is 5 and tens digit is 6 and the required number is 65.

   (D) The number formed by interchanging the digits is $10x + (11-x) = 9x + 11$.

   (a) ADBC
   (b) ABDC
   (c) ABCD
   (d) BADC

3. If $A = c(a - b)$, then $a = $ _____.
   (a) $\left(\dfrac{A}{c}\right) - b$
   (b) $\left(\dfrac{A}{c}\right) + b$
   (c) $\left(\dfrac{A}{b}\right) - c$
   (d) $\left(\dfrac{A}{b}\right) + c$

4. The sum of the interior angles in a 6-sided polygon is 720° and the six angles are x, y, z, z, y and x. Express the relation among x, y and z by making z as the subject.
   (a) $z = 360° - x + y$
   (b) $z = 360° - (x + y)$
   (c) $z = 360° + x - y$
   (d) $z = 360° + x + y$

5. If $k = a + bc$, then $c = $ _____.
   (a) $\dfrac{k+a}{b}$
   (b) $\dfrac{k+b}{a}$
   (c) $\dfrac{k-a}{b}$
   (d) $\dfrac{k-b}{a}$

6. If $A = c(a^2 + b^2)$, then which of the following is/are true?
   (a) $c = \dfrac{A}{a^2 - b^2}$
   (b) $a = \sqrt{\dfrac{A}{c} + b^2}$
   (c) Both (a) and (b)
   (d) None of these

7. The following table shows the relation between a and b.

a	1	2	3	4	5
b	2	6	12	20	30

   Express the relation between a and b with b as the subject.
   (a) $b = 2a$
   (b) $b = 3a$
   (c) $b = (a + 1)^2$
   (d) $b = a(a + 1)$

8. If $\dfrac{x+y}{z} = \dfrac{a+b}{c}$, then $y = $ _____.
   (a) $\dfrac{(a+b)z}{c} - x$
   (b) $\dfrac{(a+b)z - x}{c}$
   (c) $\dfrac{(a+b)z}{c} + x$
   (d) $\dfrac{(a+b)z + x}{c}$

9. One-fifth of a number is 5 more than one-tenth of the number. Find the number.
   (a) 50
   (b) 75
   (c) 25
   (d) 100

10. Twice a number is added to half the number, the result is 250. Find one-tenth of the number.
    (a) 20
    (b) 10
    (c) 50
    (d) 25

11. Which of the following is a solution of $2x - 5 > 4x - 3$?
    (a) 0
    (b) 1
    (c) −1
    (d) −2

Directions for questions 12 to 15: Match the Column A with Column B.

Column A		Column B
12. If $\dfrac{t}{5} - \dfrac{t}{10} = 11 - t$, then $t = $	( )	(a) 5
13. If $6.7t + 9.2t + 10.7t - 0.6t = 100 + 6t$, then $t = $	( )	(b) 10
14. If three-fifths of a certain number exceeds its one-fourth by 7, then the number is	( )	(c) 15
15. The solution of $\dfrac{3x}{4} - \dfrac{x}{4} \leq 4$ is _____.	( )	(d) 20
		(e) 25
		(f) 30

Test 2

Directions for questions 16 to 26: Select the correct alternative from the given choices.

16. Make l as the subject of the formula, $A = 2(lb + bh + hl)$.

    The following are the steps involved in solving the above problem. Arrange them in sequential order.

    (A) $lb + bh + hl = \dfrac{A}{2}$

    (B) $l = \dfrac{A - 2bh}{2(b + h)}$

    (C) $2(lb + bh + hl) = A$

    (D) $l\,(b + h) = \dfrac{A}{2} - bh$

    (a) CBAD   (b) CABD
    (c) CADB   (d) ACDB

17. The sum of the digits of a two-digit number is 12. If 18 is subtracted from the number, the digits interchange their places. Find the number.

    The following are the steps involved in solving the above problem. Arrange them in sequential order.

    (A) Units digit is 5, tens digit is 7 and the number is 75.

    (B) Given that $120 - 9x - 18 = 9x + 12 \Rightarrow 90 = 18x \Rightarrow x = 5$.

    (C) The number formed by interchanging the digits is $10x + (12 - x) = 9x + 12$.

    (D) Let the digit in the units place be x. Then the digit in the tens place be $(12 - x)$. ∴ The number is $10\,(12 - x) + x = 120 - 10x + x = 120 - 9x$.

    (a) ABCD   (b) DCBA
    (c) DBCA   (d) DABC

18. If $A = 2h(l + b)$, then $b =$ _____.

    (a) $2Ah - l$

    (b) $\left(\dfrac{A}{2h}\right) - l$

    (c) $2Al - h$

    (d) $\dfrac{A}{2l} - h$

19. The sum of all the four angles of a quadrilateral is 360°. From the following figure, express the relation between x and y by making x as the subject.

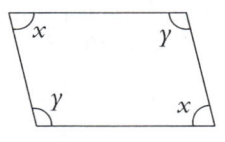

    (a) $x = 180° - y$ (b) $y = 180° - x$
    (c) $x = 180° + y$ (d) $y = 180° + x$

20. If $M = a + \dfrac{b}{c}$, then $b =$ _____.

    (a) $Mc - a$ (b) $M - ac$
    (c) $\dfrac{(M - a)}{c}$ (d) $(M - a)c$

21. If $A = \dfrac{d}{2}(a + b)$, then which of the following is/are true?

    (a) $d = \dfrac{2A}{a + b}$

    (b) $a = \left(\dfrac{2A}{d}\right) - b$

    (c) Both (a) and (b)

    (d) None of these

22. The following table shows the relation between the angles x and y.

x	90°	100°	110°	120°	130°	140°
y	90°	80°	70°	60°	50°	40°

    Then which of the following is true?

    (a) $x = y$ (b) $x + y = 180°$
    (c) $x - y = 20°$ (d) $x = 2y$

23. If $\dfrac{a + b}{c + d} = \dfrac{x}{y}$, then $y =$ _____.

    (a) $y = x(a + b)(c + d)$

    (b) $y = \dfrac{x}{(a + b)(c + d)}$

    (c) $y = \dfrac{x(c + d)}{a + b}$

    (d) $y = \dfrac{x(a + b)}{c + d}$

24. Two-thirds of a number is 32 less than three-fifths of the number. Find the number.
    (a) 360 (b) −480
    (c) −360 (d) 480

25. If one-third of a number is subtracted from three times the number, the result is 800. Find the number.
    (a) 300 (b) 400
    (c) 200 (d) 600

26. Which of the following is a solution of $\dfrac{2x-5}{3} > \dfrac{3x+3}{4}$?
    (a) $x = -5$
    (b) $x = -2$
    (c) Both (a) and (b)
    (d) Neither (a) nor (b)

Directions for questions 27 to 30: Match the Column A with Column B.

Column A		Column B
27. If $\dfrac{x}{10}+\dfrac{x}{15}+\dfrac{x}{30}=3$, then $x=$	( )	(a) 10
28. If $1.5t + 2.5t + 3.5t = 70 + 0.5t$, then $t =$	( )	(b) 20
29. If 2/3 of a certain number exceeds its one-sixth by 10, then the number is	( )	(c) −5
30. If $3x + 5 > 25 - x$, $x \in Q$ then $x >$ ___	( )	(d) 5
		(e) 15
		(f) 25

Formulae 8.17

TEST YOUR CONCEPTS

Very Short Answer Type Questions

1. True
2. False
3. False
4. True
5. True
6. 36
7. 3
8. ₹(3x/2)
9. 6π cm²
10. Auxiliary formulae
11. $A = 6s^2$
12. $A = \dfrac{d_1 d_2}{2}$
13. $I = \dfrac{PTR}{100}$
14. Profit is equal to the difference of selling price and cost price.
15. Perimeter of semi-circular region is equal to $\left(\dfrac{36}{7}\right)$ times its radius.
16. (c)
17. (a)
18. (a)
19. (d)
20. (b)

Short Answer Type Questions

21. 6 cm
22. 60°
23. 7 cm
24. $a = \dfrac{A}{c} + b$
25. 12 cm²
26. $z = 360° - (x + y)$
27. $c = \dfrac{k-a}{b}$
28. $R = \left[\left(\dfrac{1}{P}+1\right)^{\frac{1}{T}} - 1\right]100$
29. 10
30. 80

Essay Type Questions

31. $R = \sqrt{\dfrac{A}{\pi} + r^2}$; $r = \sqrt{R^2 - \dfrac{A}{\pi}}$
32. $g = 100\left(\dfrac{S-C}{C}\right)$; $g = 12\dfrac{1}{2}\%$
33. $r = \sqrt{\dfrac{3V}{\pi h}}$; $r = 4.5$ cm
34. $a = \sqrt{\dfrac{4A}{\sqrt{3}}}$; $a = 16$ cm
35. $y = \dfrac{bx}{a}$
36. $Y = X^2$
37. $a = 90° - b$
38. $Z = \sqrt{X^2 + Y^2}$
39. $a = \dfrac{1}{2x}$
40. $x = \left(\dfrac{N}{3} - 7\right)10$
41. $a = \sqrt{\dfrac{A}{c} - b^2}$
42. $b = a(a+1)$
43. $y = \dfrac{(a+b)z}{c} - x$
44. 6
45. $y = x^2 + 1$

ANSWER KEYS

CONCEPT APPLICATION

Level 1

1. (b) 2. (a) 3. (c) 4. (d) 5. (a) 6. (b) 7. (a) 8. (c) 9. (a) 10. (c)
11. (d) 12. (b) 13. (c) 14. (a) 15. (b) 16. (d) 17. (a) 18. (c) 19. (b) 20. (d)
21. (c) 22. (a)

Level 2

23. (d) 24. (c) 25. (d) 26. (b) 27. (c) 28. (d) 29. (a) 30. (d) 31. (b) 32. (b)
33. (c) 34. (d) 35. (d) 36. (b) 37. (c) 38. (a) 39. (d)

Level 3

40. (a) 41. (a) 42. (a) 43. (c) 44. (a) 45. (a) 46. (c) 47. (a) 48. (c) 49. (b)
50. (c)

ASSESSMENT TESTS

Test 1

1. (c) 2. (a) 3. (b) 4. (b) 5. (c) 6. (d) 7. (d) 8. (a) 9. (a) 10. (b)
11. (d) 12. (b) 13. (a) 14. (d) 15. (f)

Test 2

16. (c) 17. (b) 18. (b) 19. (a) 20. (d) 21. (c) 22. (b) 23. (c) 24. (b) 25. (a)
26. (d) 27. (e) 28. (a) 29. (b) 30. (d)

Formulae   8.19

CONCEPT APPLICATION

Level 1

1. $\angle P + \angle Q + \angle R + \angle S = 360°$
   Hence, the correct option is (b)

2. $t = d/s$
   Hence, the correct option is (a)

3. $A = (lr)/2$
   Hence, the correct option is (c)

4. $V = s^3$
   Hence, the correct option is (d)

5. $A = \dfrac{h}{2}(a + b)$
   Hence, the correct option is (a)

6. Given $P = \dfrac{x}{360°}(2\pi r)$
   As there are two variables (x and r) on the RHS, two auxiliary formulae can be derived.
   Hence, the correct option is (b)

7. Given $A = \dfrac{S}{N}$
   Multiplying with N on both the sides, we get
   $A \times N = S$.
   Dividing with A on both the sides, we get
   $N = \dfrac{S}{A}$.
   Hence, the correct option is (a)

8. Twice p is $2p$, 6 less than twice p is $2p - 6$, 3 more than q is $q + 3$.
   ∴ The symbolic form of "6 less than twice p is equal to 3 more than q" is $2p - 6 = q + 3$.
   Hence, the correct option is (c)

9. A variable standing alone on the LHS of a formula is called the subject of the formula.

   Given $s = ut + \dfrac{at^2}{2}$.
   ∴ The subject is s.
   Hence, the correct option is (a)

10. The formula obtained by transforming the subject in a given formula is called an auxiliary formula.
    Hence, the correct option is (c)

11. (C), (A), (B) and (D) is the required sequential order.
    Hence, the correct option is (d)

12. (B), (D), (A) and (C) is the required sequential order.
    Hence, the correct option is (b)

13. (C), (D), (B) and (A) is the required sequential order.
    Hence, the correct option is (c)

14. (A), (D), (B) and (C) is the required sequential order.
    Hence, the correct option is (a)

15. → b: Changing a term from one side of an equation to the other side is called transposition.

16. → d: Coefficient of the subject of a formula is always one.

17. → a: A variable which is expressed in terms of other variables is called subject.

18. → c: The symbolic form of circumference (C) of a circle is π times it diameter is $C = \pi d$.

19. → b: A symbol occurs alone on LHS of the equality is called the subject of the formula.

20. → d: The symbolic form of "sum of the angles of DABC is 180°", is $\angle A + \angle B + \angle C = 180$.

21. → c: The symbolic form of "the perimeter of DABC is the sum of its sides" is $P = AB + BC + AC$.

22. → a: An equation based on a rule is called formula.

Level 2

23. $A = 2\pi r$. (Dividing both the sides with 2π).
    $\dfrac{A}{2\pi} = \dfrac{2\pi r}{2\pi} \Rightarrow r = \dfrac{A}{2\pi}$.
    Hence, the correct option is (d)

24. Given $A = 2h(l + b) \Rightarrow h = \dfrac{A}{2(l+b)}$
    $\Rightarrow h = \dfrac{54}{2(5+4)} \Rightarrow h = 3\,\text{m}.$
    Hence, the correct option is (c)

25. Given $a = 2(lb + bh + lh)$.

Making l as the subject of the formula:

$A = 2(lb + bh + 2lh)$

$\Rightarrow A = 2(lb + lh) + 2bh$.

Subtracting $2bh$ from both the sides, we get

$A - 2bh = 2l(b + h)$

Dividing both the sides with $2(b + h)$, we get

$\dfrac{A - 2bh}{2(b+h)} = l \Rightarrow l = \dfrac{A - 2bh}{2(b+h)}$

Similarly, we get

$b = \dfrac{A - 2lh}{2(l+h)}$ and $h = \dfrac{A - 2lb}{2(l+b)}$

Hence, the correct option is (d)

26. Given, $l = \dfrac{x}{360°} \times 2\pi r$

$\Rightarrow l \times 360° = x \times 2\pi r$

$\Rightarrow \dfrac{360° l}{2\pi x} = r$

$\Rightarrow r = \dfrac{180° l}{\pi x}$.

Hence, the correct option is (b)

27. $l = \dfrac{x}{360°} \times 2\pi r$

$\Rightarrow l = \dfrac{60°}{360°} \times 2\pi (3)$

$\Rightarrow l = \pi$ cm.

Hence, the correct option is (c)

28. Given $l = \dfrac{x}{360°} \times 2\pi r$

$\Rightarrow x = \dfrac{360 l}{2\pi r} = \dfrac{360° \times 4\pi}{2\pi (18)} = 40°$

Hence, the correct option is (d)

29. $S = ut + \dfrac{at^2}{2}$

There are four different variables.

Hence, the correct option is (a)

30. Number of all possible squares in $n \times n$ network.

$= (1^2 + 2^2 + 3^2 + \ldots + n^2)$

$= 1^2 + 2^2 + 3^2 + \ldots + 7^2)$ ($\because n = 7$)

$= 140$.

Hence, the correct option is (d)

31. We have, 1 hour = 60 minutes $\Rightarrow m = 60$ H.

Hence, the correct option is (b)

32. $I = \dfrac{PTR}{100}$

$\Rightarrow 100 I = PTR \Rightarrow \dfrac{100I}{PR} = T \Rightarrow T = \dfrac{100I}{PR}$

Given, $P = ₹1000$, $R = 10\%$ p.a, $I = ₹300$.

$T = \dfrac{100 \times 300}{1000 \times 10} = 3$ years.

Hence, the correct option is (b)

33. Given $A = S^2$ and $A = 324$

$A = S^2 \Rightarrow \sqrt{A} = \sqrt{S^2}$

$\Rightarrow S = \sqrt{A}$

$\Rightarrow S = \sqrt{324}$

$\Rightarrow S = 18$ cm.

Hence, the correct option is (c)

34. Given $\angle A + \angle B + \angle C = 180°$.

$\Rightarrow \angle C = 180° - (\angle A + \angle B)$

$\Rightarrow \angle C = 180° - (90° + 55°)$ (given)

$\Rightarrow \angle C = 35°$.

Hence, the correct option is (d)

35. Given $A = \dfrac{d_1 d_2}{2}$,

$d_1 = 6$ cm and $d_2 = 8$ cm.

$\Rightarrow A = \dfrac{6(8)}{2} = 24$ cm^2.

Hence, the correct option is (d)

36. Given $A = 2h(l + b)$.

$\Rightarrow \dfrac{A}{2h} = l + b$

$\Rightarrow \dfrac{A}{2h} - l = b$

$\Rightarrow b = \left(\dfrac{A}{2h}\right) - l$.

Hence, the correct option is (b)

37. Given $S = (n - 2)180°$.

And $S = 540°$.

$\Rightarrow 540° = (n - 2)180°$

$\Rightarrow \dfrac{540°}{180°} = n - 2 \Rightarrow 3 = n - 2$

$\Rightarrow n = 3 + 2 = 5$.

Hence, the correct option is (c)

38. From the given data,

$x + y + x + y = 360°$.

$\Rightarrow 2(x + y) = 360°$

$\Rightarrow x + y = \dfrac{360°}{2} \Rightarrow x = 180° - y$.

Hence, the correct option is (a)

39. Given $M = a + b/c$.

$\Rightarrow M - a = b/c$

$\Rightarrow (M - a)c = b$

$\Rightarrow b = (M - a)c$.

Hence, the correct option is (d)

Level 3

40. Given, $V = lbh$

Dividing on both the sides with lb.

$\Rightarrow \dfrac{V}{lb} = \dfrac{lbh}{lb} \Rightarrow h = \dfrac{V}{lb}$.

Hence, the correct option is (a)

41. $A = s^2 - (s - 2w)^2$

$\Rightarrow A = (6)^2 - (6 - 2)^2 = 36 - 16 = 20 \ m^2$.

Hence, the correct option is (a)

42. $V = lbh$.

Hence, the correct option is (a)

43. Given, $S = \dfrac{(100s + g)C}{100}$

Making C as the subject of the formula:

$S = \dfrac{(100 + g)C}{100}$

Multiplying on both the sides with 100.

$\Rightarrow 100S = (100 + g)C$

Dividing on both the sides with $(100 + g)$

$\Rightarrow \dfrac{100S}{100 + g} = C$

$\Rightarrow C = \dfrac{100S}{(100 + g)}$

Making g as the subject of the formula:

$S = \dfrac{(100 + g)C}{100}$

$\Rightarrow \dfrac{100S}{C} = 100 + g$

$\Rightarrow g = \dfrac{100S}{C} - 100$

$\Rightarrow g = \dfrac{100(S - C)}{C}$.

Hence, the correct option is (c)

44. As there is only one variable on RHS of the formula, only one auxiliary formula can be derived.

Hence, the correct option is (a)

45. The number of diagonals $= \dfrac{n(n - 3)}{2}$.

$= \dfrac{6(6 - 3)}{2}$ (given $n = 6$) $= 9$

Hence, the correct option is (a)

46. From the given data,

$x + y = 90° \Rightarrow x = 90° - y$.

Hence, the correct option is (c)

47. $A = 2h(l + b)$

$\Rightarrow \dfrac{A}{2h} = l + b \Rightarrow \dfrac{A}{2h} - b = l$

$\Rightarrow l = \dfrac{A}{2h} - b$

Given $A = 100 \ m^2$, $h = 5$ m and $b = 4$ m.

$\Rightarrow l = \dfrac{100}{2 \times 5} - 4$

$\Rightarrow l = 6$ m.

Hence, the correct option is (a)

48. Given $A = \dfrac{d}{2}(a + b)$

Option (a):

$2A = d(a + b)$

$\Rightarrow \dfrac{2A}{a + b} = d$

$\Rightarrow d = \dfrac{2A}{a + b}$

8.22 Chapter 8

∴ Choice (a) is true.

Option (b) :

$2A = d(a + b)$

$\Rightarrow \dfrac{2A}{d} = a + b \Rightarrow \left(\dfrac{2A}{d}\right) - b = a$

$\therefore a = \left(\dfrac{2A}{d}\right) - b.$

∴ Choice (b) is also true.

Hence, the correct option is (c)

49. From the given table,

$x + y = 180°$

$\Rightarrow x = 180° - y$

Hence, the correct option is (b)

50. Given $\dfrac{a+b}{c+d} = \dfrac{x}{y}$

$\Rightarrow (a + b)y = x(c + d)$

$\Rightarrow y = \dfrac{x(c+d)}{a+b}$

Hence, the correct option is (c)

ASSESSMENT TESTS

Test 1

Solutions for questions 1 to 11:

1. (D), (A), (B) and (C) is the required sequential order.
   Hence, the correct option is (c)

2. (A), (D), (B) and (C) is the required sequential order.
   Hence, the correct option is (a)

3. Given $A = c(a - b)$

   $\Rightarrow \dfrac{A}{c} = a - b \Rightarrow \left(\dfrac{A}{c}\right) + b = a$

   $\Rightarrow a = \left(\dfrac{A}{c}\right) + b$

   Hence, the correct option is (b)

4. From the given data, $x + y + z + z + y + x = 720$

   $2(x + y + z) = 720°$

   $\Rightarrow x + y + z = 360°$

   $\Rightarrow z = 360° - (x + y)$

   Hence, the correct option is (b)

5. Given $k = a + bc$

   $\Rightarrow k - a = bc$

   $\Rightarrow \dfrac{k-a}{b} = c$

   $\Rightarrow c = \dfrac{k-a}{b}$

   Hence, the correct option is (c)

6. Given

   $A = c(a^2 + b^2)$

   Option (a):

   $\Rightarrow \dfrac{A}{a^2 + b^2} = c$

   $\Rightarrow c = \dfrac{A}{a^2 + b^2}$

   ∴ Choice (a) is false.

   Option (b):

   $\dfrac{A}{c} = a^2 + b^2$

   $\Rightarrow \dfrac{A}{c} - b^2 = a^2$

   $\Rightarrow a = \sqrt{\dfrac{A}{c} - b^2}$

   ∴ Choice (b) is false.

   Hence, the correct option is (d)

7. From the given table,

   $b \Rightarrow 2 = 1(1 + 1)$

   $\Rightarrow 6 = 2(2 + 1)$

   $\Rightarrow 12 = 3(3 + 1)$ ………

   $b = a(a + 1)$

   Hence, the correct option is (d)

Formulae 8.23

8. Given $\dfrac{x+y}{z} = \dfrac{a+b}{c}$

$\Rightarrow x + y = \dfrac{(a+b)z}{c}$

$\Rightarrow y = \dfrac{(a+b)z}{c} - x$

Hence, the correct option is (a).

9. Let the number be x.

$\therefore \dfrac{x}{5} = \dfrac{x}{10} + 5$

$\Rightarrow \dfrac{x}{5} - \dfrac{x}{10} = 5$

$\dfrac{2x - x}{10} = 5 \Rightarrow x = 50$

Hence, the correct option is (a).

10. Let the number be x.

$\therefore 2x + \dfrac{x}{2} = 250$

$\Rightarrow \dfrac{4x + x}{2} = 250$

$\Rightarrow 5x = 500 \Rightarrow x = 100$

$\therefore$ One-tenth of the number $= \dfrac{x}{10} = \dfrac{100}{10} = 10$.

Hence, the correct option is (b).

11. $2x - 5 > 4x - 3$

$\Rightarrow 2x - 4x > -3 + 5$

$\Rightarrow -2x > 2 \Rightarrow -x > 1$

$\Rightarrow x < -1$

$\therefore$ Choice (d) follows

($\because x = -2$ and $-2 < -1$)

Hence, the correct option is (d).

Solutions for questions 12 to 15:

12. $\rightarrow b$: $\dfrac{t}{5} - \dfrac{t}{10} = 11 - t$

$\dfrac{2t - t}{10} = 11 - t$

$11t = 11 \times 10$

$\therefore t = 10$.

13. $\rightarrow a$: $6.7t + 9.2t + 10.7t - 0.6t = 100 + 6t$

$\Rightarrow 26.6t - 0.6t = 100 + 6t$

$\Rightarrow 26t - 6t = 100 \Rightarrow 20t = 100$

$\Rightarrow t = \dfrac{100}{20} = 5$

$\therefore t = 5$.

14. $\rightarrow d$: Let the number be x.

$\dfrac{3x}{5} - \dfrac{x}{4} = 7$

$\Rightarrow \dfrac{12x - 5x}{20} = 7$

$\therefore x = 20$

15. $\rightarrow f$: $\dfrac{3x}{4} + \dfrac{x}{4} \geq 15$

$\dfrac{2x}{4} \geq 15$

$\dfrac{x}{2} \geq 15$

$x \geq 30$

Test 2

Solutions for questions 16 to 26:

16. (C), (A), (D) and (B) is the required sequential order.

Hence, the correct option is (c).

17. (D), (C), (B) and (A) is the required sequential order.

Hence, the correct option is (b).

18. Given $A = 2h(l + b)$.

$\Rightarrow \dfrac{A}{2h} = l + b$

$\Rightarrow \dfrac{A}{2h} - l = b$

$\Rightarrow b = \left(\dfrac{A}{2h}\right) - l$

Hence, the correct option is (b)

19. From the given data,

$x + y + x + y = 360°$.

$\Rightarrow 2(x + y) = 360°$

$\Rightarrow x + y = \dfrac{360°}{2}$

$\Rightarrow x = 180° - y$

Hence, the correct option is (a)

20. Given $M = a + b/c$

    $\Rightarrow M - a = b/c$

    $\Rightarrow (M - a)c = b$

    $\Rightarrow b = (M - a)c$

    Hence, the correct option is (d)

21. Given

    $A = \dfrac{d}{2}(a + b)$

    Option (a):

    $2A = d(a + b)$

    $\Rightarrow \dfrac{2A}{a+b} = d \Rightarrow d = \dfrac{2A}{a+b}$

    $\therefore$ Choice (a) is true.

    Option (b):

    $2A = d(a + b)$

    $\Rightarrow \dfrac{2A}{d} = a + b \Rightarrow \left(\dfrac{2A}{d}\right) - b = a$

    $\therefore a = \left(\dfrac{2A}{d}\right) - b$

    $\therefore$ Choice (b) is also true.

    Hence, the correct option is (c)

22. From the given table,

    $x + y = 180°$.

    Hence, the correct option is (b)

23. Given $\dfrac{a+b}{c+d} = \dfrac{x}{y}$

    $\Rightarrow (a + b)y = x(c + d)$

    $\Rightarrow y = \dfrac{x(c+d)}{a+b}$

    Hence, the correct option is (c)

24. Let the number be x.

    $\dfrac{2x}{3} = \dfrac{3x}{5} - 32$

    $\dfrac{2x}{3} - \dfrac{3x}{5} = -32$

    $\dfrac{10x - 9x}{15} = -32$

    $x = -480$.

    Hence, the correct option is (b)

25. Let the number be x.

    $\therefore 3x - \dfrac{x}{3} = 800$

    $\Rightarrow \dfrac{9x - x}{3} = 800 \Rightarrow \dfrac{8x}{3} = 800$

    $\Rightarrow x = 800 \times \dfrac{3}{8} \Rightarrow x = 300$

    Hence, the correct option is (a)

26. $\dfrac{2x-5}{3} > \dfrac{3x+3}{4}$

    Put $x = -5 \Rightarrow -5 > -3$ which is false.

    Put $x = -2 \Rightarrow -3 > \dfrac{-3}{4}$ which is false.

    Hence, the correct option is (d)

Solutions for questions 27 to 30:

27. $\rightarrow$ e: $\dfrac{x}{10} + \dfrac{x}{15} + \dfrac{x}{30} = 3$

    $\Rightarrow \dfrac{3x + 2x + x}{30} = 3$

    $\therefore x = 15$

28. $\rightarrow$ a: $1.5t + 2.5t + 3.5t = 70 + 0.5t$

    $\Rightarrow 7.5t - 0.5t = 70$

    $\Rightarrow 7t = 70 \Rightarrow t = \dfrac{70}{7} = 10$

29. $\rightarrow$ b: Let the number be x.

    Given that, $\dfrac{2x}{3} - \dfrac{x}{6} = 10$

    $\Rightarrow \dfrac{4x - x}{6} = 10$

    $x = 20$

30. $\rightarrow$ d: $3x + 5 > 25 - x$

    $\Rightarrow 3x + x > 25 - 5$

    $\Rightarrow 4x > 20 \Rightarrow x > \dfrac{20}{4} \Rightarrow x > 5$

Chapter 9

Statistics

a) $5(3x) - 3(x-4) = 0$
$15x - 3x + 12 = 0$
$12x + 12 = 0$
$12x = -12$
$x = -12/12$
$x = -1$ $x^2 = 1$

REMEMBER
Before beginning this chapter, you should be able to:
- Understand the meaning of data and representation of data
- Know the process of collection of data

KEY IDEAS
After completing this chapter, you should be able to:
- Know the forms of data and range of data
- Represent data using statistical graphs like pictographs, bar graphs, double bar graphs, pie graphs, line graphs
- Find out arithmetic mean, median, mode of given data
- Understand the term probability

Chapter 9

INTRODUCTION

The word '**statistics**' is derived from the Latin word '**status**' or the Italian word 'status' or the Greek word 'Statistic' which means **political state**. Political states had to collect information about their citizens to facilitate governance and to plan for development. Then, in course of time, statistics evolved as a branch of mathematics which deals with the collection, classification and analysis of numerical data.

In this chapter, we shall learn about statistical graphs and measures of central tendency.

Data

The word data means information in the form of numerical figures or a set of given facts.

For example, the percentage of marks scored by 10 pupils of a class in a test is: 72, 84, 82, 96, 94, 98, 99, 67, 92 and 93.

The set of these figures is the data related to the marks obtained by 10 pupils in a class test.

> **Note**
> 1. Data obtained from direct observation is called raw data. The marks obtained by 10 students in a monthly test are an example of raw data or ungrouped data.
> 2. The difference between the highest and the lowest values of the data is called 'Range'.
> For example, the range of the data of the first 25 natural numbers is $25 - 1 = 24$.

Some Basic Definitions

Before getting into the details of tabular representation of data, let us review some basic definitions.

1. *Observation:* Each numerical figure in a data is called an observation.
2. *Frequency:* The number of times a particular observation occurs is called its frequency.

Tabulation or Presentation of Data

A systematic arrangement of data in a tabular form is called tabulation or presentation of the data. This grouping, results in a table called the frequency table which indicates the number of scores within each group. Many conclusions about the characteristics of the data, the behaviour of variables, etc., can be drawn from this table.

The quantitative data that is to be analysed statistically can be divided into two categories:

1. Individual series
2. Discrete series

1. *Individual series:* Any raw data that is collected forms an individual series.

   For example,

   (a) The weight of 8 students: 35, 40, 36, 42, 32, 30, 50 and 46 (in kg).

   (b) Marks obtained by 12 students in a test in Mathematics: 45, 72, 76, 57, 49, 64, 75, 88, 92, 90, 60 and 55.

2. *Discrete series:* A discrete series is formulated from raw data by taking the frequency of the observations into consideration.

EXAMPLE 9.1

Given below is the data showing the number of children in a family, in a locality of 15 families. Prepare the frequency table. 1, 3, 5, 4, 3, 2, 1, 1, 3, 2, 2, 1, 4, 2 and 1.

SOLUTION

Arranging the data in ascending order we have:

1, 1, 1, 1, 1, 2, 2, 2, 2, 3, 3, 3, 4, 4, 5

To count, we can use tally marks. We record tally marks in bunches of five, the fifth one crossing the other diagonally, i.e., 𝍢

Thus we may prepare a frequency table as shown below.

Number of Children	Tally Marks	Number of Families (frequency)
1	𝍢	5
2	IIII	4
3	III	3
4	II	2
5	I	1

Statistical Graphs

The information provided by a numerical frequency distribution is easily understood when represented by diagrams or graphs. The diagrams act as visual aids and leave a lasting impression on the mind. This enables the investigator to make quick conclusions about the distribution.

There are different types of graphs or diagrams to represent statistical data. Some of them are:

1. Pictograph
2. Bar Graph
3. Double bar graph
4. Pie chart
5. Line graph

1. **Pictograph:** It is a pictorial representation of data using symbols, by observing this pictograph one can easily understand the given data.

2. **Bar graph:** The important features of bar graphs are:

   (a) Bar graphs are used to represent unclassified frequency distributions.

   (b) The frequency of a value of a variable is represented by a bar (rectangle) whose length (i.e., height) is equal (or proportional) to the frequency.

   (c) The breadth of the bar is arbitrary and the breadth of all the bars is equal.

   (d) Uniform spaces should be left between any two consecutive bars.

   (e) All the bars should rest on the same line called base line.

EXAMPLE 9.2

Represent the following frequency distribution as a bar graph.

Value of Variable (x)	1	3	4	6	7
Frequency (F)	4	3	2	5	1

SOLUTION

Either of the following bar graphs (fig (1) or fig (2)) may be used to represent the above frequency distribution. The first graph takes value of the variable along the X-axis and the frequency along the Y-axis, whereas the second one takes the frequency along the X-axis and the value of the variable on the Y-axis.

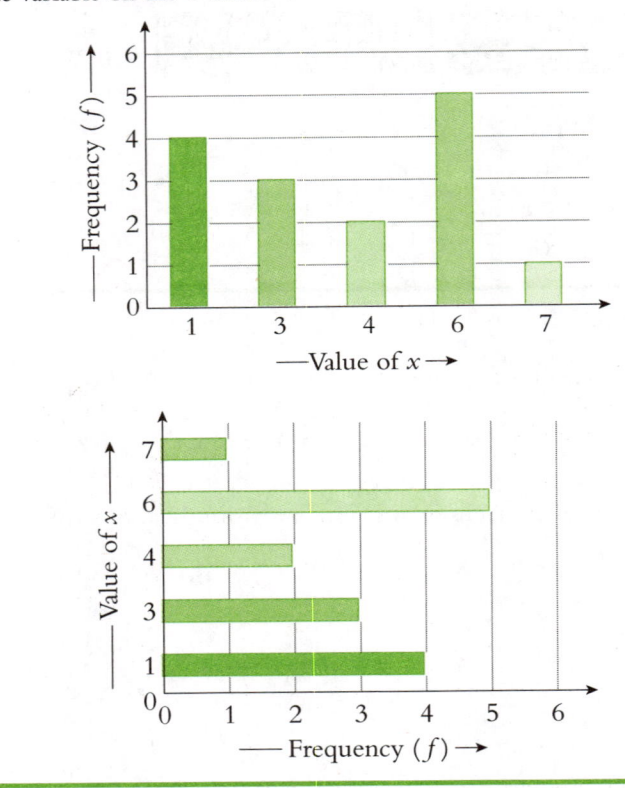

3. **Double bar graph:** In a bar graph if two sets of data are to be presented simultaneously, then the graph is called a double bar graph. Double bar graph is very much useful to compare the data.

EXAMPLE 9.3

Two different models of TV sets are produced in a factory and are given below as a double bar graph: Read the data and answer the following questions.

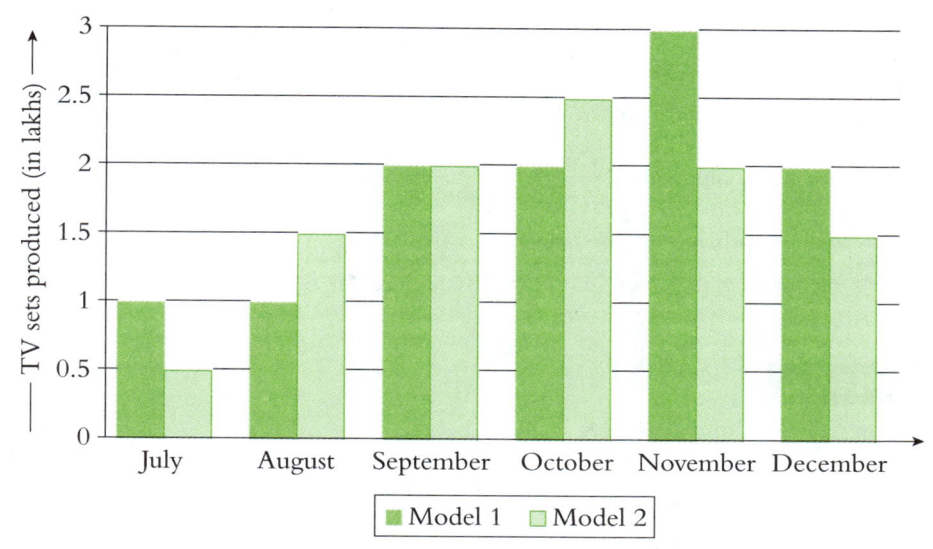

(a) What information does the graph represent?
(b) In which month, both the models of TV sets produced are equal?
(c) In which month, both the models of TV sets produced are maximum?
(d) In which month, both the models of TV sets produced are minimum and how many are they?

SOLUTION

(a) Two different types of TV sets produced each month in a factory.
(b) September
(c) November
(d) July a total of 1.00 + 0.5 = 1.5 lakhs TV sets.

4. **Pie graph or pie chart:** Statistical data can also be presented in the form of a pie graph. In the pie graph, a circle is divided into various sectors, in proportion with the various component parts of the total.

The adjoining figure shows the result of 7th class students.

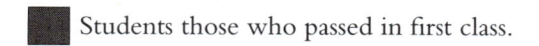

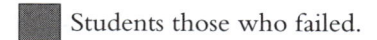

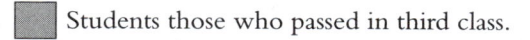

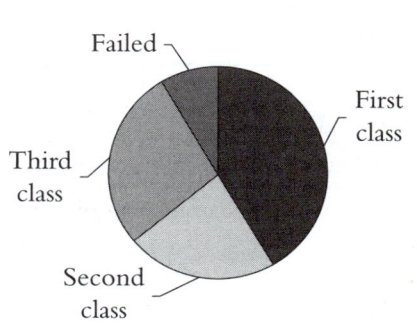

■ Students those who passed in first class.

■ Students those who failed.

■ Students those who passed in third class.

□ Students those who passed in second class.

Steps for the construction of a pie diagram:

Steps:

1. Take the total value of all the items equal to 360°.
2. Convert each component of the data into degrees using the following formula.

$$\text{Degree of any component} = \frac{\text{Component value}}{\text{Total value}} \times 360°.$$

3. Draw a circle with appropriate radius.
4. Mark the angles at the centre of the circle and draw the sectors.

EXAMPLE 9.4

A person spends his time every day as shown below.

Activity	Study	Games	Yoga and Dance	Sleeping	Miscellaneous
Number of hours	9	3	3	7	2

SOLUTION

Activity	Number of Hours	Angle of the Sector
Study	9	$\frac{9}{24} \times 360° = 45°$
Games	3	$\frac{3}{24} \times 360° = 45°$
Yoga and dance	3	$\frac{3}{24} \times 360° = 45°$
Sleeping	7	$\frac{7}{24} \times 360° = 105°$
Miscellaneous	2	$\frac{2}{24} \times 360° = 35°$
Total	24	360°

Steps:

1. Draw a circle with a convenient radius.
2. Convert each component of the data into degrees.
3. Mark the angles at the centre of the circle and draw the sectors.
4. To distinguish different sectors from one another, different shades can be used.

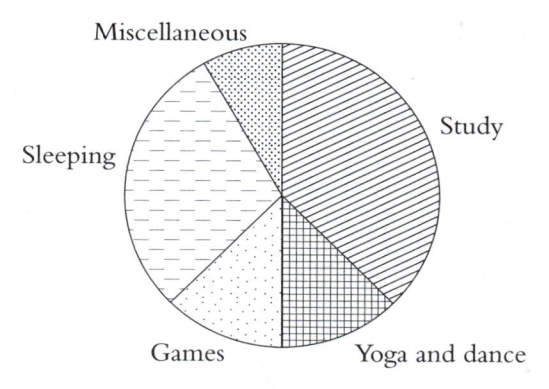

Pie diagram showing the
time spent on different activities

5. **Line graph:** When a table related to two variables x and y are given, take every x value and the corresponding y value as different points. By plotting the obtained points on a graph paper and joining these points by means of lines, the graph obtained is called a line graph.

EXAMPLE 9.5

Draw a line graph to represent the runs made by different batsmen in a match whose details are given in the following table:

Player	P_1	P_2	P_3	P_4	P_5	P_6	P_7
Runs made	30	45	55	20	40	35	50

SOLUTION

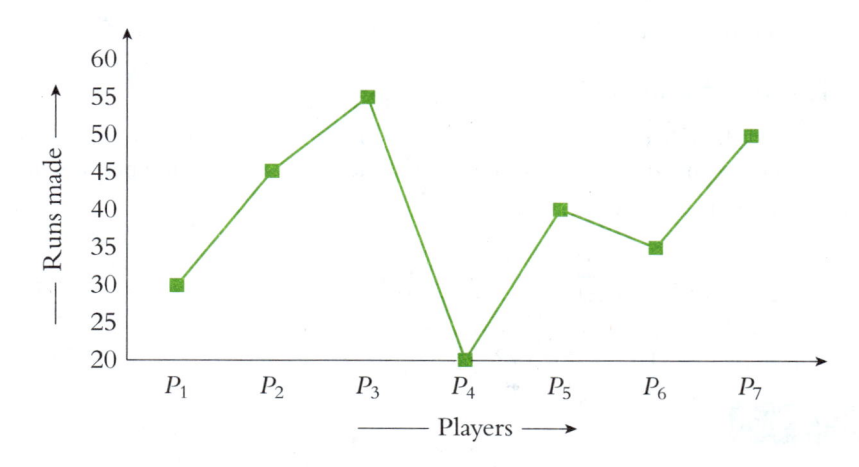

EXAMPLE 9.6

The line graph given below shows the production of a particular model of bike produced in different months. Read the line graph and answer the following questions:

(a) In which month is the production maximum?

(b) In which month is the production least?

(c) In which two successive months has the production increased?
(d) In which two successive months has the production decreased?

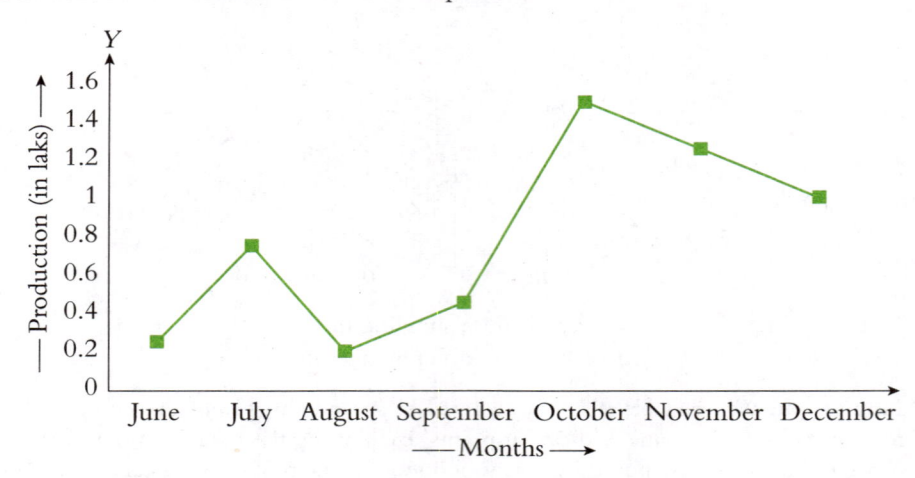

SOLUTION

(a) In the month of October
(b) August
(c) September and October
(d) November and December

Arithmetic Mean or Mean (AM)

The arithmetic mean (or simply the mean) is the most commonly used measure of central tendency.

Arithmetic Mean for Raw Data

Definition: The arithmetic mean of a statistical data is defined as the quotient obtained when the sum of all the observations or entries is divided by the total number of items.

If $x_1, x_2, \ldots x_n$ are the n items given, then

$$A.\,M. = \frac{x_1 + x_2 + \ldots + x_n}{n} = \frac{\sum_{i=1}^{n} x_i}{n} \text{ or briefly } \frac{\sum x}{n}$$

A.M. is usually denoted by $\bar{x}$.

EXAMPLE 9.7

Find the mean of the first 10 prime numbers.

SOLUTION

The first 10 prime numbers are 2, 3, 5, 7, 11, 13, 17, 19, 23, 29.

Arithmetic mean (A. M) = $\dfrac{\text{Sum of observations}}{\text{Total number of observations}}$

$= \dfrac{2 + 3 + 5 + 7 + 11 + 13 + 17 + 19 + 23 + 29}{10} = \dfrac{129}{10} = 12.9$

EXAMPLE 9.8

If the mean of the data 9, 17, 18, 14, x, 16, 15, 11 and 12 is x, then find the value of x.

SOLUTION

Given, mean = x

$$\frac{9+17+18+14+x+16+15+11+12}{9} = x$$

$$\frac{112+x}{9} = x$$

$9x = 112 + x$

$8x = 112$

$x = \frac{112}{8}$

$x = 14$.

EXAMPLE 9.9

There are 7 observations in the data and their mean is 11. If each observation is multiplied by 2, then find the new mean.

SOLUTION

There are 7 observations in the data.

Mean = 11

$\therefore \dfrac{\sum x}{7} = 77$

$x_1 + x_2 + \ldots + x_7 = 77$

Now, $2x_1 + 2x_2 + \ldots + 2x_7$

$= 2(x_1 + x_2 + \ldots + x_y) = 77 \times 2 = 154$

$\therefore$ Mean $= \dfrac{154}{7} = 22$.

Mean of Discrete Series

Let $x_1, x_2, x_3 \ldots x_n$ be n observations with respective frequencies $f_1, f_2, \ldots f_n$.

This can be considered as a special case of raw data where the observation x_1 occurs f_1 times; x_2 occurs f_2 times, and so on.

$\therefore$ The mean of the above data $= \dfrac{f_1x_1 + f_2x_2 + \ldots + f_nx_n}{f_1 + f_2 + \ldots + f_n}$.

It can also be represented by $\bar{x} = \dfrac{\sum\limits_{i=1}^{n} f_i x_i}{\sum\limits_{i=1}^{n} f_i}$.

EXAMPLE 9.10

The marks (maximum marks 100) obtained by 20 students in a test are given below.

Marks Obtained (x)	Number of Students (f)
40	3
55	4
60	2
70	5
75	4
85	1
95	1
	N = 20

Find the mean marks of the 20 students.

SOLUTION

The mean

$$(\bar{x}) = \frac{\sum f_i x_i}{N (\text{or}) \sum f} = \frac{(40 \times 3) + (55 \times 4) + (60 \times 2) + (70 \times 5)(75 \times 4)(85 \times 1) + (95 \times 1)}{20}$$

$$= \frac{1290}{20} = 6.45$$

∴ The mean marks of the 20 students = 64.5.

Median

Another measure of central tendency of a given data is the median.

Definition: If the values x_i in the raw data is arranged either in the increasing or decreasing order of magnitude, then the middle-most value in this arrangement is called the median.

Thus, for the raw (un-grouped) data, the median is computed as follows:

1. The value of the observations are arranged in the order of magnitude.
2. The middle-most value is taken as the median.

   (a) When the number of observations (*n*) is odd, then the median is the value of the $\left(\frac{n+1}{2}\right)$th observation.

   (b) If the number of observations (*n*) is even, then the median is the mean of the $\left(\frac{n}{2}\right)$th observation and the $\left(\frac{n}{2}+1\right)$th observation.

EXAMPLE 9.11

Find the median of the following observations 12, 18, 11, 21, 32, 16 and 22.

SOLUTION

Arranging the given observations in the ascending order we have 11, 12, 16, 18, 21, 22, 32.

The middle term of the above data is $\left(\dfrac{n+1}{2}\right)$ th observation (here $n = 7$)

i.e., 4th observation which is 18.

∴ Median = 18

EXAMPLE 9.12

Find the median of the data 21, 33, 15, 19, 26, 32, 14 and 24.

SOLUTION

Arranging the given values in the ascending order we have 14, 15, 19, 21, 24, 26, 32, 33.
As the given number of observations is even ($n = 8$), then the average of the two middle terms is the median, i.e., $\left(\dfrac{n}{2}\right)$ th and $\left(\dfrac{n}{2}+1\right)$ th observations average.

i.e., median is the average of the 4th and the 5th terms.

∴ Median = $\dfrac{21+24}{2}$ = 22.5.

EXAMPLE 9.13

The median of 11 observations is 10. Find the maximum number of possible observations in the data which are less than 10.

SOLUTION

Median of 11 observations is 10 ⇒ Middle most value is 10

⇒ There are 5 observations before 10 and after 10.

Mode

The most frequently found value in the data is called the mode. This is the measure which can be identified in the simplest way.

EXAMPLE 9.14

Find the mode of the observations 8, 2, 3, 4, 6, 3, 2, 4, 5, 4, 1, 1, 4, 6 and 8.

SOLUTION

Among the given observations the most frequently occurred observation is 4 which occurred four times.

∴ Mode = 4.

Notes
1. For a given data, the mode may or may not exist. In a series of observations, if no item occurs more than once, then the mode is said to be ill-defined.
2. If the mode exists for a given data, it may or may not be unique.

EXAMPLE 9.15

The mode of the unimodal data 7, 8, 9, 8, 9, 10, 9, 10, 11, 10, 11, 12 and x is 10. Find the value of x.

SOLUTION

The given observations are 7, 8, 9, 8, 9, 10, 9, 10, 11, 10, 11, 12 and x.

Observations	Frequency
7	1
8	2
9	3
10	3
11	2
12	1
x	1

And also given that the mode is 10 and the given data is unimodal.
∴ $x = 10$.

Chance and Probability

In our daily life, we come across a lot of situations, where you take a chance and it does not go the way you want it. When you roll a dice and you want 4 to come up, it may or may not come up. In this case, the chances of getting 4 and not getting 4 are not equal. When you toss a coin the chances of getting heads or tails are equal. Such experiments are called random experiments and each of the results that come up is called outcomes. The outcomes which have equal chances of occurring are called equally likely outcomes.

For example,

1. When we toss a coin, the chances of occurrence of heads or tails are equal. Hence, there are equally likely outcomes.
2. When we roll a dice, the occurrence of any number from 1 to 6 are equally likely outcomes.

Probability

Consider the experiment of rolling a dice marked on the faces with numbers 1, 2, 3, 4, 5, 6 (only one number on each face). The outcomes of getting a number are equally likely. The likelihood of getting 4 is one out of six outcomes, i.e., 1/6. In other words, we say that the probability of getting 4 = 1/6.

Similarly the probability of any other number (i.e., 1, 2, 3, 5, 6) is also 1/6.

Event

Every outcome or a collection of outcomes of a random experiment is an event.

For example,

1. When we toss a coin and get a head it is an event.
2. When a dice is rolled and we get an even number it is an event.
3. From a pack of 52 cards, picking up a red card is an event.

EXAMPLE 9.16

There are 10 marbles in a box which are marked with the distinct numbers from 1 to 10. If a marble is drawn, then what is the probability of getting prime numbered marble?

SOLUTION

Number of prime numbered marbles = 4 (i.e., 2, 3, 5, 7)

Total marbles = 10.

∴ The probability of getting a prime numbered marble.

$$= \frac{4}{10} = \frac{2}{5}$$

EXAMPLE 9.17

There are 10 marbles in a box which are marked with the distinct numbers from 1 to 10. If a marble is drawn, then what is the probability of getting a marble being a number multiple of 3?

SOLUTION

Multiples of 3 between 1 and 10 are 3, 6 and 9.

∴ The required probability = 3/10.

Points to Remember

- Each numerical figure in a data is called an observation.
- The number of times a particular observation occurs is called its frequency.
- A systematic arrangement of data in a tabular form is called tabulation or presentation of the data.
- The frequency of a value of a variable is represented by a bar (rectangle) whose length (i.e., height) is equal (or proportional) to the frequency.
- In a bar diagram, all the bars should rest on the same line called base line.
- In a given data, an observation occurring more frequently is called mode of the data.
- If there are n observations (n is odd), then the median of the data is $\left(\dfrac{n+1}{2}\right)$th observation.
- If there are n observations (n is even), then the median of the data is the average of $\left(\dfrac{n}{2}\right)$th and $\left(\dfrac{n}{2}+1\right)$th observations.

Formulae

- In a pie chart, the degree of a component = $\dfrac{\text{The component value}}{\text{The total value}} \times 360°$.

- Arithmetic mean (AM) = $\dfrac{\text{The sum of the observations}}{\text{The total number of the observations}}$.

TEST YOUR CONCEPTS

Very Short Answer Type Questions

Directions for questions 1 to 5: State whether the following statements are true or false.

1. If the range of a data is 9 and its highest value is 81, then its least value is 73.
2. The mean of the observations 7, 8, 9, 11 and 15 is 10.
3. The median of the data 13, 12, 10, 8, 4, 6, 15 is 10.
4. The mode of the data 8, 4, 5, 3, 6, 7, 7, 8, 4, 6, 8, 9, 9 is 7.
5. For an unbiased coin, the probability of getting tail is 1/2.

Directions for questions 6 to 24: Fill in the blanks.

6. The mean of 20, 40, 35, 42 and 45 is _____.
7. In a pie chart, the central angle for a component value of 240, when the total value is 720, is _____.
8. In a bar graph, the height of a bar is 5.2 cm. If 1 cm = 80 units, then the value represented by the bar is _____ units.
9. The range of the data 13, 18, 20, 15, 12, 17, 9, 14, 11 and 16 is _____.
10. If the sum of 10 observations is 95, then their mean is _____.
11. In a pictograph, if 1 picture = 80 cars, then 240 cars is equal to _____ pictures.
12. The median of the data 13, 12, 14, 13, 15, 16, 18 is _____.
13. If a dice is rolled, then the probability of getting a prime number is _____.
14. When an unbiased coin is tossed, the probability of getting a head is _____.
15. The maximum value of the probability of an event is _____.
16. The measures of central tendency are _____.
17. The middle value of a data, when it is arranged in ascending or descending, is called _____.
18. The definition of frequency of an observation is _____.
19. The definition of equally likely events is _____.
20. If a dice is thrown, then the probability of getting an even number is _____.
21. The mode of a data is _____.
22. The central angle of a component in a pie chart is _____.
23. The graphs drawn generally for representing a data are _____.
24. When a dice is thrown, the outcomes are _____.

Directions for questions 25 to 29: These questions are based on the following data. Read the following bar graph and answer the following.

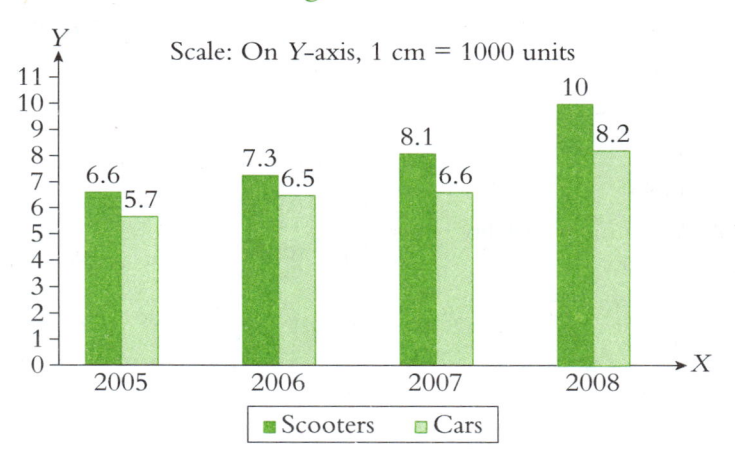

Scale: On Y-axis, 1 cm = 1000 units

Year	Scooters	Cars
2005	6.6	5.7
2006	7.3	6.5
2007	8.1	6.6
2008	10	8.2

9.16 Chapter 9

25. In which year is the difference between the sales of the scooters and the sales of cars the least?
    (a) 2005 (b) 2006
    (c) 2007 (d) 2008

26. Total number of vehicles (scooters and cars) sold in the years 2005 and 2006 is _____.
    (a) 26100 (b) 28500
    (c) 25100 (d) 27500

27. Find the maximum difference between sales of scooters and that of cars, in any year, in the given period.
    (a) 1500 (b) 1700
    (c) 1800 (d) 2000

28. Find the total number of scooters sold in the four years.
    (a) 26000 (b) 27000
    (c) 31000 (d) 32000

29. Find the ratio between the total number of vehicles sold (scooters and cars) in the year 2006 and that in the year 2008.
    (a) 41 : 46 (b) 69 : 91
    (c) 147 : 182 (d) 46 : 49

Directions for questions 30 to 39: Select the correct alternate from the given choices.

30. There are 10 cards numbered from 1 to 10 in a box. If a card is drawn randomly, then find the probability of getting an even numbered card.
    (a) $\dfrac{1}{10}$ (b) $\dfrac{1}{5}$
    (c) $\dfrac{2}{5}$ (d) $\dfrac{1}{2}$

31. The mode of the data 48, 45, 49, 51, 46, 48, 47, 49, 48 and 51 is _____.
    (a) 47 (b) 49
    (c) 51 (d) 48

32. When a dice is rolled, find the probability of getting an even prime number.
    (a) $\dfrac{1}{6}$ (b) $\dfrac{1}{3}$
    (c) $\dfrac{1}{2}$ (d) $\dfrac{5}{6}$

33. The range of 12 observations is 60. If the least observation is 90, then find the greatest observation.
    (a) 130 (b) 140
    (c) 150 (d) 160

34. Find the median of the data 35, 28, 49, 56, 79, 81, 73 and 80.
    (a) 64.5 (b) 65
    (c) 65.5 (d) 66

35. There are 20 marbles in a box which are marked with distinct numbers from 1 to 20. If a marble is drawn, then find the probability that the marble being numbered as a multiple of 5.
    (a) 3/5 (b) 2/5
    (c) 1/5 (d) None of these

36. The mode of the data 35, 25, 26, 33, 31, x and 30 is 26, then x is _____.
    (a) 25 (b) 26
    (c) 31 (d) 33

37. The median of the data 18, 42, 31, 25, 26, 38 and 43 is _____.
    (a) 31 (b) 26
    (c) 25 (d) 18

38. The sum of 20 observations is 500, then the mean is _____.
    (a) 15 (b) 20
    (c) 25 (d) 30

39. There are 100 cards numbered from 1 to 100 in a box. If a card is drawn from the box and the probability of an event is 1/2, then the number of favourable cases to the event is _____.
    (a) 20 (b) 25
    (c) 40 (d) 50

Short Answer Type Questions

40. Find the mean of 4, 6, 7, 9 and 4.
41. Find the median of the data 9, 12, 11, 10, 8, 9, 11.
42. Find the mode of the data 7, 8, 9, 9, 10, 7, 11, 10, 7, 6.
43. Find the range of the data: 14, 15, 16, 18, 19, 25, 30, 41, 26, 16, 13, 18, 20 and 26.
44. When a coin is flipped once, what is the probability of getting HEAD?
45. What is a pie chart?
46. The marks obtained by 15 students in an examination are given below:

    40, 20, 24, 19, 20, 35, 12, 48, 29, 40, 45, 48, 42, 23, 35. Find the average mark of the students.

47. The heights of 7 students are given below (in cm):

    120, 126, 132.41, 121.52, 120.35, 132, 125. Find the median height of the students.

48. The mode of the data; 20, 23, 22, 23, 22, 20, x, 21, is 22. Find the value of x.
49. The arithmetic mean of 10 observations is 45. If one of the observations, 54, is deleted, then find the mean of the remaining observations.
50. If the mean of 4, x and y is 6, then find the mean of x, y and 10.
51. A bar graph is drawn to the scale, 1 cm = $4x$ units. The length of the bar representing a quantity 1000 units is 1.25 cm. Find x.
52. In a pie graph, a component is represented as a sector with sector angle 108°. Find the percentage of the component value in total.
53. If the mean of 2, 3, x, 7, 8 is x, then find the value of x.
54. Anand's income and expenditure are in the ratio 9 : 5. Find the central angle of the sector, which represents Anand's savings.
55. In a bar graph, a bar of length 8.6 cm is represented by 430 units. Find the length of another bar which is represented by 340 units.
56. Find the mode of the data 2, 4, 6, 4, 6, 7, 6, 7 and 8.
57. The mean height of a group of 30 students is 150 cm. If a 150 cm tall student is included in the group, then find the mean height of the new group.
58. The mean of observations 2, 3, 4, 5, 8, 14 and x is x. Find x.
59. In a box, there are 20 marbles. Each marble is marked with a distinct number from 1 to 20. Find the probability of drawing a marble from the box which is marked with a number that is a perfect square.
60. Find the mean of the data 7, 8, 10, 13, 17, 23, 30, 38, 47 and 57.

Essay Type Questions

61. Find the central angles of the components of a data given below, when represented in a pie chart?

Component	Food	Education	Rent	Clothing	Savings	Total
Expenses (in ₹)	3600	2400	2100	1200	1500	10800

62. For the information given in the above question, find the heights of the bars, when the data represented in a bar graph: Scale of the graph:1 cm = ₹500.

63. The marks (out of 100) obtained by a student in his quarterly examination in various subjects are given below:

Subject	Telugu	Hindi	English	Maths	Science	Social
Marks obtained	85	74	68	90	75	50

Represent the data in a bar graph.

9.18 Chapter 9

64. The percentage of marks obtained by a student in various exams is given below. Represent the data by a line graph.

Name of the exam	UT – 1	UT – 1	Quarterly	UT – 3	Half yearly	UT – 4	Annual
Percentage of marks	50	85	60	80	68	75	70

65. The following table shows the favourite games of 240 students of a school. Draw a pie chart for the following data.

Game	Football	Badminton	Cricket	Volleyball	Hockey
Number of students	35	30	80	40	55

66. The following table gives the average temperatures, during the year 2006 and 2007. Draw a double bar graph for the given data.

Month	Jan	Feb	Mar	Apr	May	Jun	July	Aug	Sep	Oct	Nov	Dec
Temp (in °C) 2006	20	32	36	38	42	35	32	30	28	25	20	15
Temp (in °C) 2007	20	30	40	40	40	35	30	30	25	20	18	15

67. In a pie chart, the central angles of two components A and B are 108° and 81° respectively. Find the difference in the percentages that A forms of the total value of all components and that B forms of the total value of all components. (in %)

CONCEPT APPLICATION

Level 1

Directions for questions 1 to 5: Select the correct alternative from the given choices.

1. If the mean of n observations is 12 and the sum of the observations is 132, then find the value of n.
   (a) 9
   (b) 10
   (c) 11
   (d) 12

2. The range of the data 14, 15, 18, 25, 11, 40, 36, 30 is _____.
   (a) 29
   (b) 27
   (c) 24
   (d) 26

3. The median of 5, 7, 9, 10, 11 is _____.
   (a) 7
   (b) 9
   (c) 11
   (d) 10

4. In a pie chart, the sum of the angles of all its components is _____.
   (a) 90°
   (b) 180°
   (c) 240°
   (d) 360°

5. In a bar graph, the height of a bar is proportional to the _____.
   (a) Width of the bar
   (b) Range of the data
   (c) Value of the component
   (d) Number of observations in the data

Directions for the questions from 6 to 10: These questions are based on the following data. Read the following bar graph and answer the questions.

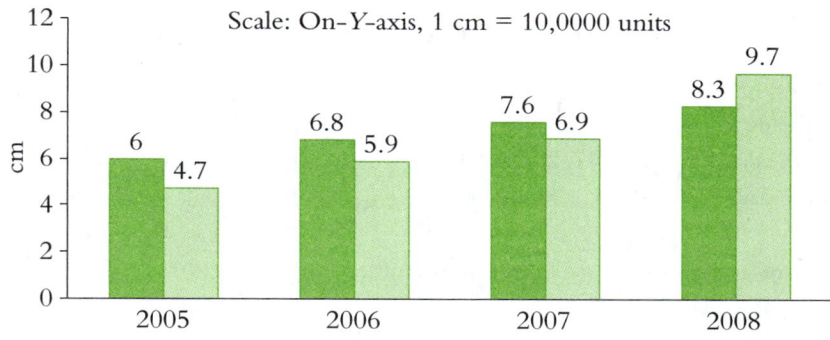

Scale: On-Y-axis, 1 cm = 10,0000 units

Bargraph reprsents the sales of cooldrinks of two companies A and B from 2005 to 2008

■ Company A   □ Company B

6. In which year is the difference between the sales of A and the sales of B the highest?
   (a) 2008   (b) 2005
   (c) 2006   (d) 2007

7. Total sales of A and B in the year 2006 is _____.
   (a) 1160000   (b) 1270000
   (c) 1380000   (d) 1490000

8. Find the minimum difference between the sales of A and sales of B in any year in the given period.
   (a) 90000   (b) 70000
   (c) 50000   (d) 30000

9. In which year is the sales of B more than the sales of A?
   (a) 2005   (b) 2006
   (c) 2007   (d) 2008

10. Find the ratio of the total sales in the year 2007 and that in 2008.
    (a) 107 : 145 (b) 127 : 145
    (c) 29 : 36   (d) 107 : 127

11. When a dice is thrown, the total number of possible outcomes is _____.
    (a) 6   (b) 1
    (c) 3   (d) 4

Directions for questions 12 and 13: Select the correct alternative from the given choices.

12. The arithmetic mean of 13 observations is 60. If one of the observation 50 is deleted and another observation 63 is included. Find the new arithmetic mean. The following are the steps involved in solving the above problem. Arrange them in sequential order.

    (A) New sum of observations = 780 − 50 + 63 = 793.
    (B) The sum of the observations = 13 × 60 = 780.
    (C) ∴ New arithmetic mean = $\frac{793}{13}$.
    (D) The required mean = 61.

    (a) BACD   (b) BCAD
    (c) BDCA   (d) ABCD

13. If a dice is rolled, find the probability of getting a composite number.

    The following are the steps involved in solving the above problem. Arrange them in sequential order.

    (A) The possible composite numbers when a dice is rolled are 4 and 6. When a dice is rolled, the possible outcomes are 1, 2, 3, 4, 5, 6.
    (B) The required probability = $\frac{1}{3}$.
    (C) The probability of getting a composite number = $\frac{\text{Number of favourable outcomes}}{\text{Total number of outcomes}} = \frac{2}{6}$.

(a) ABC (b) CBA
(c) CAB (d) ACB

Directions for questions 14 to 17: Match the Column A with Column B.

Column A		Column B
14. The mode of the data 3, 4, 9, 7, 8, 9, 3, 9, 8 is _____. ( )		(a) 17
15. The median of the data 15, 13, 12, 17, 20, 24, 11 is _____. ( )		(b) 15
16. The mean of the data 22, 16, 15, 23, 13, 12, 18 is _____. ( )		(c) $\frac{1}{26}$
17. The probability of getting a red card from a well shuffled pack of cards is _____. ( )		(d) 9
		(e) $\frac{1}{2}$
		(f) 3

Level 2

Directions for questions 18 to 39: Select the correct alternative from the given choices.

18. In a pie chart, the central angle of a component is 72° and its value is 24. Find the total value of all the components of the data.
    (a) 240 (b) 120
    (c) 360 (d) 420

19. In a bar graph, the height of a bar is 5 cm and it represents 40 units. Find the height of a bar representing 56 units (in cm).
    (a) 11.2 (b) 5.6
    (c) 7   (d) 8

20. In a pictograph, each picture represents 1000 units of certain production in a certain year. There are 4 full pictures and (3/4) th of a picture in a row. Find the number of units produced in that year.
    (a) 4340 (b) 4750
    (c) 4250 (d) 4725

21. The mean of a data is 15 and the sum of the observations is 195. Find the number of observations in the data.
    (a) 13 (b) 19
    (c) 16 (d) 17

22. If the mode of 22, 21, 23, 24, 21, 20, 23, 26, x and 26 is 23, then x is _____.
    (a) 20 (b) 21
    (c) 23 (d) 24

23. If the range of 10 observations is 15 and its highest score is 28, then the least score of the data is _____.
    (a) 3   (b) 13
    (c) 14 (d) 5

24. The mean of the data 15, 20, 20, 16, 22, 17, 23, 18, 24, 25; is _____.
    (a) 26 (b) 22
    (c) 20 (d) 18

25. If an unbiased dice is rolled, then the probability of getting an even number is _____ than/to that of getting an odd number.
    (a) More   (b) Less
    (c) Equal (d) Cannot say

26. In a class of 15 students, the total marks obtained by all the students in a test is 600. Find the average mark of the class.
    (a) 35 (b) 30
    (c) 45 (d) 40

27. If the mean of 5, 7, x, 10, 5 and 7 is 7, then find the value of x.
    (a) 6   (b) 7
    (c) 8   (d) 9

28. There are 20 students in a class. Of them, the height of ten students is 150 cm each, the height of 6 students is 142 cm each and the height of 4 students is 132 cm each. Find the average height of all the students.
    (a) 144   (b) 140
    (c) 138   (d) 146

29. Find the mean weight (in kg) of 50 boys of a class whose total weight is 1550 kg.
    (a) 30   (b) 35
    (c) 32   (d) 31

30. In a pie chart, find the central angle of a component which is 30% in the total value of all the components.
    (a) 108°   (b) 30°
    (c) 70°   (d) 120°

31. If the central angle of a component is 72°, then find the percentage of the value of the component in the total value of all the components.
    (a) 70%   (b) 18%
    (c) 24%   (d) 20%

32. The mean of p, q and r is same as the mean of q, $2r$ and s. Then which of the following is correct?
    (a) $p = q = r$ (b) $q = r = s$
    (c) $q = r$ (d) $p = r + s$

33. A bar graph is drawn to the scale of 1 cm = $2m$ units. The length of the bar representing a quantity of 875 units is 1.75 cm. Find m.
    (a) 125   (b) 225
    (c) 250   (d) 375

34. In a pie graph, a component is represented as a sector with sector angle 72°. Find the percentage of the component value in the total.
    (a) 21%   (b) 27.5%
    (c) 22.5%   (d) 20%

35. If mean of 2, 4, p, 8 and 10 is p, then find the value of p?
    (a) 4   (b) 5
    (c) 8   (d) 6

36. A person contributed his total salary of a month for three social welfare societies A, B and C. The ratio of total salary, share of A and share of B is 20 : 5 : 8. Find the central angle of sector, which represents the share of C.
    (a) $72\frac{1°}{2}$ (b) 126°
    (c) 105°   (d) $67\frac{1°}{2}$

37. In a pictograph, a picture is represented by 375 units. If the data is represented by 30 pictures, then find the total numbers of units in the data.
    (a) 12.5   (b) 11250
    (c) 37.5   (d) 12750

38. The mean weight of 21 students is 21 kg. If a student weighing 21 kg is removed from the group, then what is the mean weight of the remaining students?
    (a) 20 kg   (b) 21 kg
    (c) 19 kg   (d) None of these

39. The mean of 10, 15, 19, 30, 43, 69 and x is x. Find the median of the data.
    (a) 19   (b) 43
    (c) 30   (d) Cannot say

Level 3

Directions for questions 40 and 41: Select the correct alternative from the given choices.

40. In a factory, the daily wages (in ₹) of 12 workers are 75, 90, 100, 120, 110, 114, 50, 60, 70, 105, 108 and 102. Find the number of workers whose daily wage is less than the mean daily wage.
    (a) 5   (b) 6
    (c) 4   (d) 7

Chapter 9

41. If the mean of 9, 10, 15, x, 6, 8 and 12 is 11. Find the median of the scores.
    (a) 4
    (b) 10
    (c) 13
    (d) 5

Directions for questions 42 to 50: Select the correct alternative from the given choices.

42. The mode of the data 9, x, 6, 3, 4, 9, 8, 6, 4, 6 is 6. Which of the following cannot be the value of x?
    (a) 8
    (b) 7
    (c) 6
    (d) 9

43. The median, mode and mean of a data is 9. If there are five integers in the data and the range of the data is 4. If the least value of the data is 7, then find the number of different observations in the data.
    (a) 3
    (b) 4
    (c) 2
    (d) Cannot say

44. The median of 10, 12, x, 6, 18 is 10. Then which of the following is true about the value of x?
    (a) $6 \leq x \leq 10$
    (b) $x < 6$
    (c) $x > 18$
    (d) Either (a) or (b)

45. The mean of 10 observations is 15. If one observation 15 is added, then find the new mean.
    (a) 16
    (b) 11
    (c) 15
    (d) 10

46. The range of the data 15, 18, 17, 16, 14, x, 12, 10, 9, 15 is 9. Which of the following is true about x?
    (a) $x > 18$
    (b) $x < 9$
    (c) $9 \leq x \leq 18$
    (d) Cannot say

47. The range of a data is x, the median and the mode of the data is 7 each. If the number of observations is odd and all observations are integers, then find the least value of x (Range $\neq$ 0).
    (a) 1
    (b) 2
    (c) 3
    (d) 4

48. For a pie chart, a student calculates the central angle of a component as 72° by taking the total value as 220 instead of 240. Find the correct central angle of the component.
    (a) 60°
    (b) 66°
    (c) 74°
    (d) 80°

49. In a bar graph, length of a bar is 6.4 cm and it represents 256 units. Find the number of units represented by a bar of length 5.3 cm.
    (a) 228
    (b) 196
    (c) 212
    (d) 224

50. A bag contains 4 green balls, 4 red balls and 2 blue balls. If a ball is drawn from the bag then, what is the probability of getting neither green nor red ball?
    (a) 2/5
    (b) 1/2
    (c) 4/5
    (d) 1/5

TEST YOUR CONCEPTS

Very Short Answer Type Questions

1. False
2. True
3. True
4. False
5. True
6. 36.4
7. 120°
8. 416
9. 11
10. 9.5
11. 3
12. 14
13. 1/2
14. 1/2
15. 1
16. mean, median and mode
17. median
20. $\dfrac{1}{2}$

22. $\dfrac{\text{Component value}}{\text{Total value of the components}} \times 360°$
23. Pictograph, bar graph, double bar graph, ….
24. 1, 2, 3, 4, 5, 6
25. (b)
26. (a)
27. (c)
28. (d)
29. (b)
30. (d)
31. (d)
32. (a)
33. (c)
34. (a)
35. (c)
36. (b)
37. (a)
38. (c)
39. (d)

Short Answer Type Questions

40. 6
41. 10
42. 7
43. 28
44. $\dfrac{1}{2}$
46. 32
47. 125
48. 22
49. 44
50. 8

51. 200
52. 30%
53. 5
54. 160°
55. 6.8 cm
56. 6
57. 150 cm
58. 6
59. 1/5
60. 25

Chapter 9

Essay Type Questions

61. 120°, 80°, 70°, 40°, 50°
62. 7.2 cm, 4.8 cm, 4.2 cm, 2.4 cm, 3 cm
67. 7.5%

CONCEPT APPLICATION

Level 1

1. (c)	2. (a)	3. (b)	4. (d)	5. (c)	6. (a)	7. (b)	8. (b)	9. (d)
10. (c)	11. (a)	12. (a)	13. (d)	14. (d)	15. (b)	16. (a)	17. (e)	

Level 2

18. (b)	19. (c)	20. (b)	21. (a)	22. (c)	23. (b)	24. (c)	25. (c)	26. (d)
27. (c)	28. (a)	29. (d)	30. (a)	31. (d)	32. (d)	33. (c)	34. (d)	35. (d)
36. (b)	37. (b)	38. (b)	39. (c)					

Level 3

40. (a)	41. (b)	42. (d)	43. (a)	44. (d)	45. (c)	46. (c)	47. (a)	48. (b)
49. (c)	50. (d)							

Statistics 9.25

CONCEPT APPLICATION

Level 1

1. Mean =

    $\dfrac{\text{Sum of the observations}}{\text{Total number of observations}}$

    $\Rightarrow n = \dfrac{132}{12} = 11.$

2. Range = Highest value − Least value

    Range = 40 − 11 = 29.

    Hence, the correct option is (a)

3. 9 is the middle most value.

    ∴ Median = 9.

    Hence, the correct option is (b)

4. In a pie chart, the sum of angles of all its components is 360°.

    Hence, the correct option is (d)

5. The value of a component is directly proportional to its height.

    Hence, the correct option is (c)

6. For 2005:

    Difference = (6 − 4.7) × 100000 = 130000

    For 2006:

    Difference = (6.8 − 5.9) × 100000 = 90000

    For 2007:

    Difference = (7.6 − 6.9) × 100000 = 70000

    For 2008:

    Difference = (9.7 − 8.3) × 100000 = 140000

    Hence, the correct option is (a)

7. Total sales of A and B in the year 2006

    = (6.8 + 5.9) × 100000 = 1270000.

    Hence, the correct option is (b)

8. From solution of Q.1., minimum difference is appeared in the year 2007, i.e., 70000.

    Hence, the correct option is (b)

9. Clearly, in the year 2008, the sales of cool drinks of company B is more than that of sales of cool drinks of company A.

    Hence, the correct option is (d)

10. Required ratio = (7.6 + 6.9) : (8.3 + 9.7) = 14.5 : 18

    = 29 : 36.

    Hence, the correct option is (c)

11. When a dice is thrown, the possible outcomes are 1, 2, 3, 4, 5 and 6.

    ∴ The total number of outcomes = 6.

    Hence, the correct option is (a)

12. (B), (A), (C) and (D) is the required sequential order.

    Hence, the correct option is (a)

13. (A), (C) and (B) is the required sequential order

    Hence, the correct option is (d)

14. d: The mode of the data is 9.

15. b: The median of the data 11, 12, 13, 15, 17, 20, 24 is 15.

16. a: Mean = $\dfrac{22 + 16 + 15 + 23 + 13 + 12 + 18}{7}$

    $= \dfrac{119}{7} = 17.$

17. e: Required probability = $\dfrac{26}{52} = \dfrac{1}{2}$

Level 2

18. Let the total value = x

    ∴ $\dfrac{24}{x} \times 360° = 72°$

    $x = 120.$

    Hence, the correct option is (b)

19. 5 cm = 40 units.

    1 cm = 8 units.

    56 units = $\dfrac{56}{8} = 7$ cm.

    Hence, the correct option is (c)

HINTS AND EXPLANATION

20. $4 \times 1000 + \dfrac{3}{4} \times 1000 = 4000 + 3 \times 250$

    $= 4000 + 750 = 4750$ units.

    Hence, the correct option is (b)

21. The number of observations =

    $\dfrac{\text{The sum of the observations}}{\text{The mean of the observations}}$

    $= \dfrac{195}{15} = 13.$

    Hence, the correct option is (a)

22.

Observations	Frequency
20	1
21	2
22	1
23	2
24	1
26	1

Given Mode = 23

∴ $x = 23$. (∵ Frequency of 3 must be more than the others)

Hence, the correct option is (c)

23. Lowest score = 28 − 15 = 13.

    Hence, the correct option is (b)

24. Mean =

    $\dfrac{(15 + 20 + 20 + 16 + 22 + 17 + 23 + 18 + 24 + 25)}{10}$

    $= \dfrac{200}{10} = 20.$

    Hence, the correct option is (c)

25. The probability of getting an even number is equal to that of getting an odd number.

    Hence, the correct option is (c)

26. Average mark = $\dfrac{600}{15} = 40$

    Hence, the correct option is (d)

27. Mean = $\dfrac{5 + 7 + x + 10 + 5 + 7}{6} = 7$

    $34 + x = 42$

    $x = 42 − 34$

    $x = 8.$

    Hence, the correct option is (c)

28. The average height =

    $\dfrac{10 \times 150 + 6 \times 142 + 4 \times 132}{20}$

    $= \dfrac{1500 + 852 + 528}{20} = \dfrac{2880}{20} = 144$

    Hence, the correct option is (a)

29. $\dfrac{1550}{50} = 31$ kg.

    Hence, the correct option is (d)

30. The required central angle =

    $\dfrac{30\%}{100\%} \times 360° = 108°$

    Hence, the correct option is (a)

31. $\dfrac{x}{100\%} \times 360° = 72° \Rightarrow x = 20\%.$

    Hence, the correct option is (d)

32. Given, $\dfrac{p + q + r}{3} = \dfrac{q + 2r + s}{3}$

    ∴ $p = r + s$

    Hence, the correct option is (d)

33. Given, 1 cm = 2 m units → (1)

    And 1.75 cm = 875 units

    ⇒ 1 cm = 500 units → (2)

    From (1) and (2), we get,

    2 m = 500

    m = 250

    Hence, the correct option is (c)

34. Required percentage =

    $\dfrac{72°}{360°} \times 100\% = 20\%.$

    Hence, the correct option is (d)

35. Given, mean = 6

    $\dfrac{2 + 4 + p + 8 + 10}{5} = 5$

    $\Rightarrow 4p = 24 \Rightarrow p = 6.$

    Hence, the correct option is (d)

36. Given, the ratio of total salary, share of A and share of B is 20 : 5 : 8.

∴ Share of C is 7 parts out of 20 parts.

Required central angle = $\frac{7}{20} \times 360° = 126°$.

Hence, the correct option is (b)

37. One picture is represented by 375 units.

    ∴ 30 pictures are represented by (375 × 30)

    i.e., 11250 units.

    Hence, the correct option is (b)

38. Given that the mean weight of 21 students = 21 kg.

    ∴ The total weight of 21 students = 21 (21 kg).

    If a student weighing 21 kg is removed from the group, then the total weight of 20 students = 21(21 kg) 21 kg

    = 420 kg.

∴ The mean weight of 20 students = $\frac{420}{20}$ = 21 kg.

Hence, the correct option is (b)

39. Given that $\frac{10 + 15 + 19 + 30 + 43 + 69 + x}{7} = x$

    ⇒ 7x = 186 + x

    ⇒ 6x = 186

    ⇒ x = 31

    ∴ Given observations in ascending order are 10, 15, 19, 30, 31, 43, 69.

    ∴ The median = 30.

    Hence, the correct option is (c)

Level 3

40. Mean =

    $\frac{(75 + 90 + 100 + 120 + 110 + 114 + 50 + 60 + 70 + 105 + 108 + 102)}{12}$

    = $\frac{1104}{12}$ = 92

    ∴ Mean daily usage = ₹92.

    From the data, there are 5 workers whose daily wage is less than the average daily wage.

    Hence, the correct option is (a)

41. $\frac{9 + 10 + 15 + x + 6 + 8 + 12}{7} = 11$ (given)

    $\frac{60 + x}{7} = 11$

    x = 77 − 60

    x = 17.

    ∴ Descending order:

    6, 8, 9, 10, 12, 15, 17

    ∴ Median = 10.

    Hence, the correct option is (b)

42. The given data is 9, x, 6, 3, 4, 9, 8, 6, 4, 6

    D.O: 9, 9, 8, 6, 6, 6, 4, 4, 3 (except x)

Given mode = 6.

∴ x should not be 9 or 4.

Hence, the correct option is (d)

43. There are five integers in the data.

    Given mode = 9.

    ∴ There must be at least two 9's.

    And also given that the least value is 7 and range is 4.

    ∴ The greatest value is 11.

    ∴ 7, 9, 9, 11 are the four observations of the data.

    Let the 5th observation be x.

    Since mean is 9, the sum of the observations must be 45.

    ∴ 45 − (7 + 9 + 9 + 11) = x

    45 − (36) = x

    ∴ x = 9

    ∴ Number of different observations in the data is 3.

    Hence, the correct option is (a)

44. Given, median of 10, 12, 7, 6, 18 is 10.

    This can be arranged either as x, 6, 10, 12, 18 or 6, x, 10, 12, 18.

    ∴ x can be either 6 ≤ x ≤ 10 or x < 6.

    Hence, the correct option is (d)

45. Mean of 10 observations is $\dfrac{\Sigma x}{10} = 15$.

$\Sigma x = 150$.

If one observation 15 is added, then the sum of new observations = 150 + 15 = 165.

∴ New Mean = $\dfrac{165}{10+1} = 15$.

Hence, the correct option is (c)

46. 15, 18, 17, 16, 14, x, 12, 10, 9, 15.

Range = Maximum value − Minimum value

18 − 9 = 9.

Given, Range is also '9'.

∴ x value should not be greater than 18 or less than 9.

∴ 9 < x < 18.

Hence, the correct option is (c)

47. Median of the data and mode are 7 each and the number of observations is odd.

Since 7 is the mode, 7 should appear at least two times and median is 7, it must be the middle one.

6, 7, 7 can be taken as observations.

∴ Range = 7 − 6 = 1.

Hence, the correct option is (a)

48. Let the value of the component be x.

$\dfrac{x}{220} \times 360° = 72°$

$x = 44$

∴ The correct central angle of the component

$= \dfrac{44}{240} \times 360° = 60°$.

49. 6.4 cm = 256 units

1 cm = $\dfrac{256}{6.4} = \dfrac{2560}{64}$

1 cm = 40 units

5.3 cm = 5.3 × 40 = 212 units.

Hence, the correct option is (c)

50. From the given data, the ball must be blue in colour.

Probability of getting blue ball = $\dfrac{2}{10} = \dfrac{1}{5}$.

Hence, the correct option is (d)

Chapter 10

Set Theory

REMEMBER

Before beginning this chapter, you should be able to:
- Make groups of objects on a given pattern
- Identify the similarities and differences in a group

KEY IDEAS

After completing this chapter, you should be able to:
- Learn about set, elements of a set, representation of sets and cardinal number of a set
- Explain the definitions of empty set, singleton set, disjoint sets, subset, universal set and complement of a set
- Study operations on sets like union, intersection, difference of sets
- Understand and draw Venn diagrams
- Apply some formulae on cardinality of sets

INTRODUCTION

In everyday life we come across different collections of objects. For example, a herd of sheep, a cluster of stars, a posse of policemen, etc. In mathematics, we call such collections as 'Sets'. The objects are referred to as the elements of the sets.

Set

A set is a well-defined collection of objects. Let us understand what we mean by a well-defined collection of objects.

We say that a collection of objects is well-defined if there is some reason or rule by which we can say whether a given object of the universe belongs to or does not belong to the collection.

To understand this better, let us look at some examples.

1. Let us consider the collection of odd natural numbers less than 10. In this example, we can definitely say what the collection is. The collection comprises of numbers such as 1, 3, 5, 7 and 9.

2. Let us consider the collection of intelligent boys in a class. In this example, we cannot say precisely which boys of the class belong to our collection. So, this collection is not well-defined.

Hence, the first collection is a set whereas the second collection is not a set.

In the first example given above, the set of the first 5 odd natural numbers can be represented as

$A = \{1, 3, 5, 7, 9\}$.

Elements of a Set

The objects in a set are called its elements or members. We usually denote the sets by capital letters A, B, C or X, Y, Z, etc.

If a is an element of a set A, then we say that a belongs to A and we write, $a \in A$.

If a is not an element of A, then we say that a does not belong to A and we write, $a \notin A$.

Some sets of numbers and their notations:
N = The set of all natural numbers = $\{1, 2, 3, 4, 5, \ldots\}$
W = The set of all whole numbers = $\{0, 1, 2, 3, 4, 5, \ldots\}$
Z = The set of all integers = $\{0, \pm 1, \pm 2, \pm 3, \ldots\}$
Q = The set of all rational numbers = $\left\{ x \mid x = \dfrac{p}{q} \text{ where } p, q \in Z \text{ and } q \neq 0 \right\}$.

Cardinal Number of a Set

The number of elements in a set A is called its cardinal number. It is denoted by $n(A)$. A set which has finite number of elements is a finite set and a set which has infinite number of elements is an infinite set.

For example,

1. The set of vowels in English alphabet is a finite set.

2. The set of all persons living on the earth is a finite set.

3. The set of all natural numbers is an infinite set.

Representation of Sets

We represent sets by the following methods:

1. **Roster method:** In this method, a set is described by listing out all the elements in the set.

   For example,

   (a) Let V be the set of all vowels in English alphabet. Then we represent V as, $V = \{a, e, i, o, u\}$.

   (b) Let E be the set of all even natural numbers less than 12. Then we represent the set E as, $E = \{2, 4, 6, 8, 10\}$.

2. **Set-builder method:** In this method, a set is described by using some property common to all its elements.

   For example,

   (a) Let V be the set of all vowels in English alphabet. Then we represent the set V as, $V = \{x \,/\, x \text{ is a vowel in English alphabet}\}$.

   (b) Let P be the set of all even numbers less than 100.

   Then we represent the set P as, $P = \{x/x \leq 100 \text{ and } x \text{ is an even number}\}$.

Some Simple Definitions of Sets

1. **Empty set or null set:** A set with no elements in it is called an empty set (or) void set (or) null set. It is denoted by $\{\ \}$ or ϕ. (read as phi)

   For example,

   (a) Set of all natural numbers less than 1.

   (b) Set of all woman Chief Ministers of Andhra Pradesh.

2. **Singleton set:** A set consisting of only one element is called a singleton set.

   For example,

   (a) The set of all even prime numbers is a singleton set as 2 is the only even prime number.

   (b) The set of all positive integers which are factors of every natural number is a singleton set as 1 is the only integer which is a factor of all natural numbers.

3. **Disjoint sets:** Two sets A and B are said to be disjoint, if they have no elements in common.

   For example,

   (a) Sets $A = \{2, 5, 6, 7, 10\}$ and $B = \{1, 3, 4, 8, 9\}$ are disjoint as they have no element in common.

   (b) Sets $A = \{1, 2, 3, 4, 5, 6\}$ and $B = \{2, 5, 7\}$ are not disjoint as they have some common elements 2 and 5.

4. **Subset:** Let A and B be two sets. If every element of set A is also an element of set B, then A is said to be a subset of B. We write this symbolically as $A \subseteq B$.

   For example,

   (a) Set $A = \{1, 2, 5, 7\}$ is a subset of set $B = \{0, 1, 2, 3, 4, 5, 6, 7, 8\}$.

   (b) Set of all primes is a subset of the set of all natural numbers.

Notes (i) Empty set is a subset of every set.
(ii) Every set is a subset of itself.
(iii) If a set A has n elements, then the number of subsets of A is 2^n.

5. **Proper subset:** If A is a subset of B and $A \neq B$, then A is called a proper subset of B and is denoted by $A \subset B$. If $A \subset B$, then $n(A) < n(B)$.

Notes (i) If a set A has n elements, then the number of non-empty subsets of A is $2^n - 1$.
(ii) If a set A has n elements, then the number of proper subsets of A is $2^n - 1$.

EXAMPLE 10.1

If $N = \{\alpha, \beta, \gamma\}$, then find the number of all possible proper subsets of N

SOLUTION

If a set has n elements, then the number of proper subsets $= 2^n - 1$
$= 2^3 - 1 = 8 - 1 = 7$.

6. **Universal set:** A set which consists of all the sets under consideration or discussion is called the universal set. It is usually denoted by 'U' or 'μ'.

For example:

Let $A = \{1, 2, 3, 4\}$, $B = \{2, 5, 7\}$ and $C = \{3, 4, 8, 10\}$.

Here μ can be $\{1, 2, 3, 4, 5, 6, 7, 8, 9, 10\}$.

7. **Complement of a set:** Let μ be the universal set and $A \subset \mu$. Then, the set of all those elements of μ which are not in set A is called the complement of the set A. It is denoted by A' or A^c.

$A' = \{x/x \in \mu \text{ and } x \notin A\}$.

For example,

(a) Let $\mu = \{1, 2, 3, 4, 5, 6, 7, 8, 9, 10\}$ and $A = \{2, 3, 5, 7\}$.

Then, $A' = \{1, 4, 6, 8, 9, 10\}$.

(b) Let $\mu = N$ (natural numbers set) and E be the set of all even natural numbers.

Then, E' will be the set of all odd natural numbers, i.e., $E' = \{1, 3, 5, 7, 9, 11, ...\}$.

Notes (i) A and A' are disjoint sets.
(ii) $\mu' = \phi$ and $\phi' = \mu$.

Operations on Sets

1. **Union of sets:** Let A and B be two sets. Then, the union of A and B, denoted by $A \cup B$, is the set of all those elements which are either in A or in B or in both A and B, i.e., $A \cup B = \{x/x \in A \text{ or } x \in B\}$.

For example,

Let $A = \{1, 2, 3, 4, 5, 6\}$ and $B = \{2, 4, 6, 8, 9, 10\}$.

Then, $A \cup B = \{1, 2, 3, 4, 5, 6, 8, 9, 10\}$.

Notes 1. If $A \subseteq B$, then $A \cup B = B$.
2. $A \cup \mu = \mu$ and $A \cup \phi = A$.

2. **Intersection of sets:** Let A and B be two sets. Then the intersection of A and B, denoted by $A \cap B$, is the set of all those elements which are common to both A and B.

i.e., $A \cap B = \{x/x \in A \text{ and } x \in B\}$.

For example,

Let $A = \{1, 2, 3, 5, 7\}$ and $B = \{2, 4, 7, 10\}$. Then, $A \cap B = \{2, 7\}$.

Notes 1. If A and B are disjoint sets, then $A \cap B = \phi$.
2. If $A \subseteq B$, then $A \cap B = A$.
3. $A \cap \mu = A$ and $A \cap \phi = \phi$.

EXAMPLE 10.2

Given that $\mu = \{$Whole numbers up to $36\}$, $A = \{3, 6, 9 \ldots, 36\}$ and $B = \{4, 8, 12 \ldots 36\}$. Find $n(A \cap B)'$.

SOLUTION

$\mu = \{0, 1, 2, 3, \ldots, 36\}$
$A = \{$Multiples of 3 from 3 to 36$\}$
$B = \{$Multiples of 4 from 4 to 36$\}$
$A \cap B = \{$common multiples of 3 and 4$\} = \{12, 24, 36\}$
$n(A \cap B)' = n(\mu) - n(A \cap B) = 37 - 3 = 34$.

3. **Difference of sets:** Let A and B be two sets, then the difference $A - B$ is the set of all those elements which are in A but not in B, i.e., $A - B = \{x/x \in A \text{ and } x \notin B\}$.

For example:

Let $A = \{1, 2, 4, 5, 7, 8, 10\}$ and $B = \{2, 3, 4, 5, 6, 10\}$. Then, $A - B = \{1, 7, 8\}$ and $B - A = \{3, 6\}$.

Notes 1. $A - B \neq B - A$ unless $A = B$.
2. For any set A, $A' = \mu - A$.

Venn Diagrams

We also represent sets pictorially by means of diagrams called Venn diagrams. In Venn diagrams, the universal set is usually represented by a rectangular region and its subsets by closed regions inside the rectangular region. The elements of the set are written in the closed regions.

For example:

Let $\mu = \{1, 2, 3, 4, 5, 6, 7, 8, 9\}$, $A = \{1, 2, 4, 6\}$ and $B = \{2, 3, 4, 5\}$.

We represent these sets in the form of Venn diagram as follows:

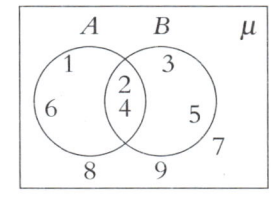

We can also represent the sets in Venn diagrams by shaded regions:

For example,

(i) Venn diagram of $A \cup B$, where A and B are two overlapping sets, is

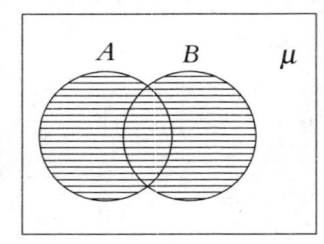

(ii) Let A and B be two overlapping sets. Then, the Venn diagram of $A \cap B$ is

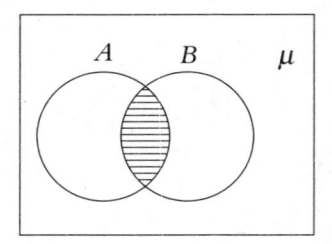

(iii) For a non-empty set A, Venn diagram of A' is

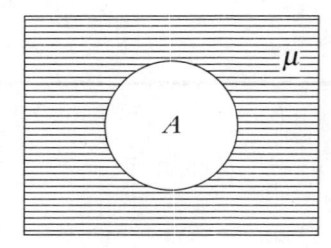

(iv) Let A and B be two overlapping sets. Then, the Venn diagram of $A - B$ is

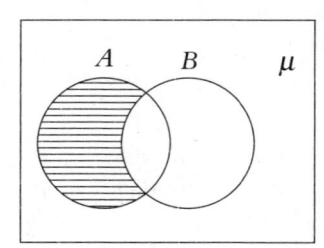

(v) Let A and B be two sets such $A \subset B$. We can represent this relation using Venn diagram as follows:

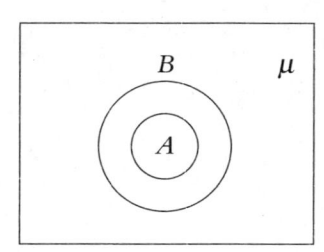

Some Formulae on the Cardinality of Sets

Set $A = \{1, 2, 3, 4, 5, 6, 7\}$ and $B = \{2, 4, 8, 16\}$.

Then, $A \cup B = \{1, 2, 3, 4, 5, 6, 7, 8, 16\}$ and $A \cap B = \{2, 4\}$.

In terms of the cardinal numbers, $n(A) = 7$, $n(B) = 4$, $n(A \cap B) = 2$ and $n(A \cup B) = 9$.

So, $n(A) + n(B) - n(A \cap B) = 7 + 4 - 2 = 9 = n(A \cup B)$.

We have the following formula:

For any two sets A and B,

$n(A \cup B) = n(A) + n(B) - n(A \cap B)$.

EXAMPLE 10.3

If $n(A) = 4$, $n(B) = 6$ and $n(A \cup B) = 8$, then find $n(A \cap B)$.

SOLUTION

Given, $n(A) = 4$, $n(B) = 6$ and $n(A \cup B) = 8$.
We know that, $n(A \cup B) = n(A) + n(B) - n(A \cap B)$.
So, $8 = 4 + 6 - n(A \cap B) \Rightarrow n(A \cap B) = 10 - 8 = 2$.

EXAMPLE 10.4

If $n(A) = 8$, $n(B) = 6$ and the sets A and B are disjoint, then find $n(A \cup B)$.

SOLUTION

Given, $n(A) = 8$ and $n(B) = 6$.
A and B are disjoint $\Rightarrow A \cap B = \phi \Rightarrow n(A \cap B) = 0$.
$\therefore n(A \cup B) = n(A) + n(B) - n(A \cap B) = 8 + 6 - 0 = 14$.

EXAMPLE 10.5

There are 40 persons in a group; four of them can speak neither English nor Hindi. The sum of the number of persons who can speak English and that of those who can speak Hindi is 44. Find the number of those who can speak both English and Hindi.

SOLUTION

$n(E \cup H) + n(E' \cap H') = 40$
$n(E' \cap H') = 4 \Rightarrow n(E \cup H) = 36$
$\therefore n(E) + n(H) = 44$
$\Rightarrow n(E \cup H) + n(E \cap H) = 44$
$\therefore n(E \cap H) = 44 - 36 = 8$.

EXAMPLE 10.6

There are 100 children in a colony. Of them 70 watch Disney channel, 50 watch both Cartoon network and Disney channels and 20 watch none of these. Find the number of children who watch only cartoon network.

SOLUTION

Given $n(D) = 70$.

$n(D \cup C) = 100 - 20 = 80$

$n(D \cap C) = 50$

$n(D \cup C) = n(D) + n(C) - n(D \cap C)$

$80 = 70 + n(C) - 50$

$\Rightarrow n(C) = 60$

$\therefore$ The number of children who watch only cartoon network
$= 60 - 50 = 10$.

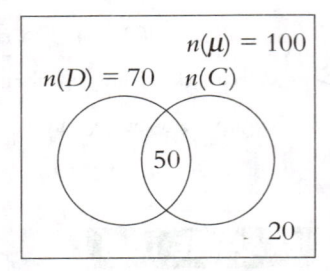

EXAMPLE 10.7

If $n(\mu) = 40$, $n(A' \cap B') = 6$, $n(A \cap B') = 10$ and $n(B \cap A') = 16$, then find $n(A \cap B)$.

SOLUTION

$n(\text{only } A) = n(A \cap B') = 10$.

$n(\text{only } B) = n(B \cap A') = 16$.

$n(\text{neither } A \text{ nor } B) = n(A' \cap B') = 6$.

$\therefore a = 10$, $c = 16$ and $d = 6$.

$a + b + c + d = 40$

$b = 40 - 32 = 8$.

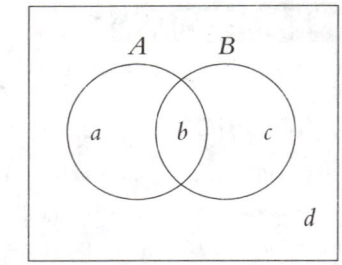

EXAMPLE 10.8

X and Y are disjoint sets. If $n(X) = 40$ and $n(Y) = 28$, then find $n(X - Y) + n(Y - X)$.

SOLUTION

$n(X - Y) = n(X) - n(X \cap Y)$

$n(Y - X) = n(Y) - n(X \cap Y)$

X and Y are disjoint sets i.e., $n(X \cap Y) = 0$.

$\therefore n(X - Y) = n(X)$ and $n(Y - X) = n(Y)$

$\therefore n(X - Y) + n(Y - X) = n(X) + n(Y)$

$\quad\quad\quad\quad\quad\quad\quad\quad = 40 + 28 = 68$.

Points to Remember

- A set is a well-defined collection of objects.
- Empty set is a subset of every set.
- Every set is a subset of itself and it is called improper subset.
- If a set A has n elements, then the number of subsets of A is 2^n.
- If a set A has n elements, then the number of proper subsets of A is $2^n - 1$.
- If $A \subseteq B$, then $A \cup B = B$.
- If A and B are disjoint sets, then $A \cap B = \phi$.
- If $A \subseteq B$, then $A \cap B = A$.
- $A - B \neq B - A$ unless $A = B$.
- $n(A \cup B) = n(A) + n(B) - n(A \cap B)$.

10.10 Chapter 10

TEST YOUR CONCEPTS

Very Short Answer Type Questions

Directions for questions 1 to 5: State whether the following statements are true or false.

1. If $A = \{2, 3, 4\}$ and $B = \{3, 5\}$, then $A \cap B$ has only one element.
2. If $A = \{1, 2, 3, 4, 5\}$, then $2 \in A$.
3. If $A = \{1, 2, 3\}$ and $B = \{2, 4\}$, then $A \cup B$ has 5 elements.
4. If $A \subset B$, then $B \cap A = B$.
5. If $A \subseteq B$ and $B \subseteq A$, then $A = B$.

Directions for questions 6 to 10: Fill in the blanks.

6. Two sets having no element in common are called _____ sets.
7. The set of whole numbers is a/an _____ set. (finite/infinite)
8. If $X = \{1, 3, 5, 7, 9\}$, then the cardinal number of X is _____.
9. If $Y = \{2, 4, 6, 8\}$, then the number of proper subsets of X is _____.
10. If $Z = \{a, b, c\}$, then the number of non-empty subsets of Z is _____.

Directions for questions 11 to 14: Which of the following collections is/are sets?

11. All rich people in your city.
12. All intelligent students in your school.
13. All fat boys in your colony.
14. All the boys of your class whose height exceeds 150 cm.

Directions for questions 15 to 28: Select the correct alternative from the given choices.

15. If a set has 2 elements, then how many proper subsets in all are there for the given set?
    (a) 4 (b) 3
    (c) 2 (d) 1

16. Which of the following is a null set?
    (a) {1}
    (b) {ϕ}
    (c) {x/x is a composite number less than 5}
    (d) {x/x is an even prime number more than 2}

17. If $A = \{a, e, i, o, u, a, e, i\}$, then $n(A) =$ _____.
    (a) 4 (b) 8
    (c) 16 (d) 5

18. If $A = \{x/x + 10 = 10\}$, then $n(A) =$ _____.
    (a) 0 (b) 1
    (c) 4 (d) None of these

19. If $A = \{a, b, c\}$ and $X = \{x, y, z\}$, then $A \cap X =$ _____.
    (a) A (b) X
    (c) ϕ (d) None of these

20. If $X = \{a, e, i, o, u\}$, then which of the following is a correct statement?
    (a) $e \in X$ (b) $e \subset X$
    (c) $e \notin X$ (d) $e \not\subset X$

21. If $Y = \{\{a, e\}, \{i, o\}, u\}$, then which of the following is a correct statement?
    (a) $\{a, e\} \subset Y$
    (b) $\{a, e\} \in Y$
    (c) $\{\{a, e\}\} \in Y$
    (d) $\{a, e\} \notin Y$

22. If $E = \{x/x$ is a factor of 8$\}$, $F = \{x/x$ is a factor of 16$\}$ and $G = \{x/x$ is a factor of 13$\}$, then which of the following statements is true?
    (a) $E \subset G$ (b) $G \subset E$
    (c) $E \subset F$ (d) $F \subset E$

23. Write the difference of the sets containing the letters of the words 'MATHEMATICS' and 'SOCIALMAN'.
    (a) {M, A, T}
    (b) {L, M, N}
    (c) {T, H, E}
    (d) {I, C, S}

24. If $n(A) = 20$, $n(B) = 30$ and $n(A \cup B) = 45$, then find $n(A \cap B)$.
    (a) 5 (b) 10
    (c) 15 (d) 20

25. If $A = \{a, e, i, o, u\}$ and $B = \{a, i, e, c, d\}$, then $n(A - B)$ is
    (a) 0 (b) 1
    (c) 2 (d) 3

26. If $A = \{1, 2, 3\}$, then $n(P(A))$
    (a) 3 (b) 8
    (c) 4 (d) 6

27. The cardinal number of the set containing letters of the word 'MOONROCK'.
    (a) 8 (b) 6
    (c) 4 (d) 2

28. If A and B are disjoint, then $n(A \cap B) =$
    (a) 4 (b) 2
    (c) 1 (d) 0

Short Answer Type Questions

29. Find the cardinal number of a set containing woman prime ministers of India.

30. Write the set builder form for the given set in the above question.

31. Write the cardinal number of the set containing the letters of the word 'MATHEMATICS'.

32. Write the difference of the sets containing the letters of the words 'STATISTICS' and 'ARITHMETIC'.

33. If $A \cap B = \phi$, then name what type of sets are A and B and also give example.

34. Suggest a universal set for the set given below
    $A = \{$Even numbers less than 100$\}$
    $B = \{$Odd natural numbers less than 100 in natural numbers$\}$

35. Write $A \cup B$ in the set builder form for the above question.

36. Let the set of all natural numbers N be the universal set and $K = \{x{:}x$ is an even prime number$\}$. Find K'.

37. Write down all the possible subsets of $\{x, y, z\}$.

38. Write down all the possible subsets of $\{p, q\}$.

39. Write all the possible proper subsets of $\{0, -1, 1\}$.

40. Write all the possible proper subsets of $\left\{-\dfrac{1}{2}, \dfrac{1}{2}\right\}$.

41. $A = \{x{:}x$ is a multiple of 3, $x < 10\}$ and $B = \{x{:}x$ is multiple of 5, $x < 15\}$. Find $n(A) - n(B)$.

42. If $n(\mu) = 200$ and $n(A' \cup B') = 120$, then find $n(A \cap B)$.

43. Write the roster form of $X = \{x : x = n^2 + 2n + 1, n \in N$ and $n < 10\}$.

44. A set X has 255 proper subsets. Find its cardinal number.

45. $A = \{0, 1^2, 2^2, 3^2, 4^2, 5^2\}$ and $B = \{0, 1, 4, 9, 16, 25\}$. Find $A \cap B$.

46. Find $A \cup B$ in the above question.

47. If $A = \{2, 3, 5, 7, 11\}$ and $B = \{1, 3, 5, 7, 9\}$, then find $A - B$ and $B - A$.

48. If $U = \{4, 5, 6, 7, 8, 9, 10, 11, 12\}$ and $C = \{4, 6, 8, 10, 12\}$, then find C'.

49. If $O = \{1\}$, then find the number of all possible proper subsets of O.

50. If $N = \{a, b, c\dots\dots z\}$, then find the number of all possible subsets of N.

51. If $n(A) = 20$, $n(A \cap B) = 10$ and $n(A \cup B) = 70$, then find $n(B)$.

52. If $X = \{0, 1, 2, 3, 4, 5, 6, 7, 8, 9, 10\}$ and $Y = \{2, 4, 6, 8, 10\}$, then find $X - Y$.

10.12 Chapter 10

Essay Type Questions

53. From the given Venn diagram, find (a) $P \cup Q$ (b) $P \cap Q$ (c) $(P \cap Q)'$.

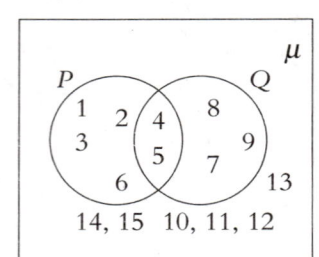

54. From the given Venn diagram, find (a) $A' \cup B'$ (b) $B' \cap C'$ (c) $C' - A'$.

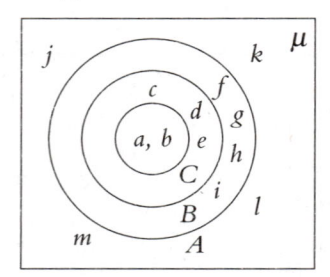

55. In a group of 36 students, 18 like volleyball, 12 like hockey and 14 like neither of the games. How many like both games?

56. In the question above, how many do not like volley ball and how many do not like hockey?

57. In a group of persons, the number of persons who like only tea is half that of those who like only coffee or one-third of that of those who like neither tea nor coffee or one-fourth of that of those who like both tea and coffee. If there are 100 persons in the group, then find the number of persons who like both tea and coffee.

58. In a class of 60 students, 30 passed in Physics and 24 passed in Biology. Find the maximum number of students who could have failed in both the subjects.

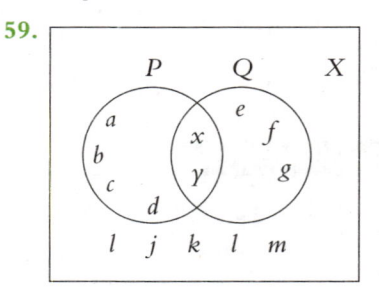

59.

From the above Venn diagram, find $n(P - Q) + n(Q - P)$.

60. In a class of 70 students, each student passed Hindi or English. Of them 15 students passed both Hindi and English, and 40 students passed Hindi. Find the number of students who passed English.

61. In a school, there are 150 students. If 90 of them play chess, 70 of them play both chess and carrom and 40 play none of these games, then find the number of children who play only carrom.

CONCEPT APPLICATION

Level 1

Directions for questions 1 to 15: Select the correct alternative from the given choices.

1. If $\mu = \{1, 2, 3, 4, 5, 6, 7, 8\}$ and $A = \{2, 5, 8\}$, then find $n(A')$.
   (a) 3 (b) 5
   (c) 4 (d) 6

2. If $X = \{$Non-prime numbers$\}$ and $Y = \{$Non-composite numbers$\}$, then $n(X \cap Y) =$ _____.
   (a) 0 (b) 2
   (c) 1 (d) 3

3. If $P = \{$Factors of 6$\}$ and $Q = \{$Factors of 12$\}$, then $n(P \cup Q) =$ _____.
   (a) 4 (b) 8
   (c) 10 (d) 6

4. Which of the following is/are true?
   (a) If $M = N$, then $M' = N'$
   (b) If $M' = N'$, then $M = N$.
   (c) Both (a) and (b)
   (d) Neither (a) nor (b)

5. If $n(A) = 10$, $n(B) = 20$ and $n(A \cup B) = 26$, then $n(A \cap B) = $ _____.

   (a) 4
   (b) 2
   (c) 6
   (d) 8

Directions for questions 6 to 9: These questions are based on the following data.

$A = \{1, 2, 3, 4, 5, 6, 7, 8, 9\}$
$B = \{2, 4, 6, 8\}$
$C = \{1, 3, 5, 7, 9\}$

6. $(A \cup (B \cup C)) = $ _____.

   (a) A
   (b) B
   (c) $B \cup C$
   (d) Both (a) and (c)

7. $A \cap B = $ _____.

   (a) B
   (b) C
   (c) $B \cap C$
   (d) $A \cap C$

8. $A \cap C = $ _____.

   (a) B
   (b) C
   (c) $B \cap C$
   (d) $A \cap B$

9. $(A \cap B) \cup (A \cap C) = $ _____.

   (a) A
   (b) $A \cup B$
   (c) $A \cup C$
   (d) All of these

Directions for question 10 to 15: Select the correct alternative from the given choices.

10. Which of the following is a singleton set?

    (a) $\{0\}$
    (b) $\{\phi\}$
    (c) $\{x : x$ is an even prime number$\}$
    (d) All of the above

11. If $A = \{1, 3, 5, 2, 4, 1, 3, 5, 7, 8, 9, 6, 10\}$, then $n(A) = $ _____.

    (a) 13
    (b) 8
    (c) 10
    (d) None of these

12. If $A = \{x : x + 5 = 5\}$, then $n(A) = $ _____.

    (a) 0
    (b) 1
    (c) 5
    (d) None of these

13. If $X = \{1, 2, 3, 4\}$, then which of the following is a correct statement?

    (a) $4 \in X$
    (b) $4 \subset X$
    (c) $4 \notin X$
    (d) $4 \not\subset X$

14. $P = \{x:x$ is a multiple of 4, $x < 20\}$ and $Q = \{x:x$ is a multiple of 6, $x < 30\}$. Find $n(P) + n(Q)$.

    The following are the steps involved in solving the above problem. Arrange them in sequential order.

    (A) $P = \{4, 8, 12, 16\}$ and $Q = \{6, 12, 18, 24\}$
    (B) $\Rightarrow n(P) = 4$ and $n(Q) = 4$
    (C) $n(P) + n(Q) = 4 + 4 = 8$

    (a) CBA
    (b) ACB
    (c) BAC
    (d) ABC

15. In a group of 50 students, 30 like Basketball, 20 like football and 10 like neither of the games. How many like both the games?

    The following are the steps involved in solving the above problem. Arrange them in sequential order.

    (A) $n(B \cap F) = 50 - 40 = 10$.
    (B) $\Rightarrow 40 = 30 + 20 - n(B \cap F)$
      $\Rightarrow 40 = 50 - n(B \cap F)$
    (C) Let $n(B) = 30$, $n(F) = 20$ and $n(B \cup F) = 50 - 10 = 40$.
    (D) We know that $n(B \cup F) = n(B) + n(F) - n(B \cap F)$.

    (a) CDBA
    (b) CBDA
    (c) ADBC
    (d) BCDA

Directions for questions 16 to 19: Match the Column A with Column B.

Column A		Column B
16. If $A = \{x:x<11, x \in W\}$ and $B = \{x:x<11, x \in N\}$ ( )		(a) A
17. If $x = \{\}$, then $n(Px))$ ( )		(b) B
18. The cardinal number of the set containing the letter of the word 'GOOGLE' ( )		(c) 1
19. If $A \subset B$, then $A \cap B$ ( )		(d) 2
		(e) 3
		(f) 4

Chapter 10

Directions for questions 20 to 23: State whether the following statements are true or false.

20. If $A = \{T, I, G, E, R\}$ and $B = \{G, I, N, T, E, R\}$, then $A \cup B$ has 6 elements.

21. The cardinal number of the set containing the letters of the word 'GINGERCOOK' is 8.

22. If $A \subset B$, then $A \cup B = A$.

23. If $X = \{1, 2, 3, \{4, 5\}, 6, \{7, 8, 9\}, 10\}$, then $\{4, 5\} \subset X$.

Level 2

Directions for questions 24 to 27: Select the correct alternative from the given choices.

24. If $A = \{$Positive perfect squares less than 100$\}$ and $B = \{$Positive perfect cubes less than 100$\}$, then find $n(A \cap B)$.
    (a) 1
    (b) 2
    (c) 3
    (d) 4

25. If $\mu = \{$All prime numbers$\}$ and $O = \{$All odd prime numbers$\}$, then find $n(O')$.
    (a) 1
    (b) 2
    (c) 3
    (d) More than 3

26. $\mu = \{$Two digit perfect squares$\}$
    $X = \{$Two digit perfect squares for which sum of digits is a perfect square$\}$
    $Y = \{$Two digit perfect squares for which sum of digits is a prime number$\}$
    Find $(X \cup Y)'$.
    (a) $\{49\}$
    (b) $\{25\}$
    (c) $\{36\}$
    (d) $\{64\}$

27. If $\mu = \{0, 1, 2, 3, 4, 5, 6, 7, 8, 9\}$, $X = \{2, 3, 5, 7\}$ and $Y = \{2, 5, 8\}$. Find $n(X' \cup Y')$.
    (a) 8
    (b) 7
    (c) 6
    (d) 9

Directions for questions 28 to 31: These questions are based on the following data.
$A = \{1, 2, 3, 4, 5, 6, 7, 8, 9\}$
$B = \{2, 4, 6, 8\}$
$C = \{1, 3, 5, 7, 9\}$

28. If $P = \{$Factors of 36$\}$ and $Q = \{$Factors of 48$\}$, then find $n(P \cap Q)$.

    (a) 6
    (b) 5
    (c) 7
    (d) 8

29. $S = abcdef.....z$, $\mu = \{$Vowels in $S\}$ and $B = \{$Vowels in even positions of $S\}$. Find $n(B')$.
    (a) 4
    (b) 5
    (c) 3
    (d) 2

30. If N is a natural number, $A = \{$Factors of $N\}$ and $B = \{$Multiples of $N\}$, then $n(A \cap B) = $ _____.
    (a) 2
    (b) 3
    (c) 4
    (d) 1

31. If $\mu = \{$Natural numbers up to 32$\}$, $C = \{2, 5, 8, 11,....32\}$ and $D = \{2, 4, 6, 8, ...32\}$, then find $n(C \cap D')$.
    (a) 3
    (b) 4
    (c) 6
    (d) 5

Directions for questions 32 to 65: Select the correct alternative from the given choices.

32. $E = \{$Natural numbers up to 30$\}$
    $X = \{$Multiples of 2 up to 30$\}$
    $Y = \{x/x = 4y + 2, y \in E\}$
    Find $n(X \cap Y)$.
    (a) 2
    (b) 5
    (c) 7
    (d) 9

33. $X = \{$The units digit of the sum of 10 consecutive natural numbers$\}$. Find X.
    (a) $\{5\}$
    (b) $\{2\}$
    (c) $\{3\}$
    (d) $\{0\}$

34. If $A = \{1, 2, 3, 4, 8\}$, then which of the following can be concluded?
    (a) $8 \in A$
    (b) $9 \notin A$
    (c) $\{2, 3\} \subset A$
    (d) All of these

35. If A and B are two disjoint sets; $n(A) + n(B) = 24$, then find $n(A \cup B)$.
    (a) 16   (b) 18
    (c) 24   (d) Cannot say

36. If $A = \{1, 2, 3, 4\}$ and $B = \{2, 4, 8, 9\}$, then $(A - B) \cup (B - A) = $ _____.
    (a) $\{1, 3, 8, 9\}$ (b) $\{2, 4\}$
    (c) A (d) B

37. In the question above, find $(A - B) \cap ((B - A)$.
    (a) $\{2, 4\}$ (b) A
    (c) B (d) ϕ

38. Which of the following is/are false?
    (a) $A - A = \phi$
    (b) $A \cup A' = \mu$
    (c) Both (a) and (b)
    (d) None of these

39. Which of the following is/are true?
    (a) $P \cup P' = \mu$
    (b) $P \cap P' = \phi$
    (c) Both (a) and (b)
    (d) None of these

40. If two sets are disjoint, then _____.
    (a) They have one element in common
    (b) They have 0 as the common element
    (c) They have no element in common
    (d) None of these

41. If $Y = \{a, e, \{i, o\}, u\}$, then which of the following is a correct statement?
    (a) $\{i, o\} \subset Y$
    (b) $\{i, o\} \in Y$
    (c) $\{\{i, o\}\} \in Y$
    (d) $\{i, o\} \notin Y$

42. If $U = \{x : x$ is an alphabet$\}$ and $C = \{x : x$ is a consonant$\}$, then $C' = $ _____.
    (a) $\{a, e, i, o\}$ (b) $\{a, e, i\}$
    (c) $\{i, o, u\}$ (d) $\{a, e, i, o, u\}$

43. If $O = \{o\}$, then the number of all possible subsets of O is _____.
    (a) 2   (b) 3
    (c) 1   (d) 4

44. If $n(A) = 10$, $n(A \cap B) = 5$ and $n(A \cup B) = 35$, then $n(B) = $ _____.
    (a) 30   (b) 10
    (c) 40   (d) None of these

45. If $X = \{0, 1, 2, 3, 4, 5, 6, 7, 8, 9, 10\}$ and $Y = \{1, 3, 5, 7, 9\}$, then $X - Y = $ _____.
    (a) $\{1, 2, 3, 4, 5\}$
    (b) $\{1, 3, 5, 7, 9\}$
    (c) $\{0, 2, 4, 6, 8, 10\}$
    (d) None of these

46. Given that $A = \{$Perfect cubes between 10 and 100$\}$ and $B = \{$Perfect squares between 10 and 100$\}$.
    Find $n(A \cap B)$.
    (a) 2   (b) 1
    (c) 5   (d) 3

Level 3

47. If $\mu = \{$Natural numbers up to 30$\}$, $Q = \{$Multiples of 4 less than 30$\}$ and $R = \{$Multiples of 6 less than 30$\}$, then find $n[(Q \cap R)']$.
    (a) 27   (b) 26
    (c) 29   (d) 28

48. $A = \{$Natural numbers less than 200 divisible by 9$\}$
    $B = \{$Natural numbers less than 200 divisible by 12$\}$
    $C = \{$Natural numbers less than 200 divisible by 15$\}$. Then $(A \cap B \cap C) = $ _____.
    (a) $\{180\}$
    (b) $\{120\}$
    (c) $\{105, 150\}$
    (d) $\{120, 180\}$

10.16 Chapter 10

49. In the previous question, find $(A \cap B)$.
    (a) {60, 120, 180}
    (b) {45, 90, 135, 180}
    (c) {36, 72, 108, 144, 180}
    (d) {36, 60, 84}

50. If P = {Factors of 48} and Q = {Factors of 60}, then find $n[(P - Q) \cup (Q - P)]$.
    (a) 5
    (b) 7
    (c) 8
    (d) 10

51. Given that E = {Natural numbers up to 30}, $P = \{x/x = 4y + 1, x \in E\}$ and $Q = \{x/x = 6y + 1, x \in E\}$. Find $n(P \cap Q)$.
    (a) 0
    (b) 2
    (c) 1
    (d) 3

52. A class has 50 students, each student likes either cricket or football or both. Sixteen students like both the games. Find the number of students who like exactly one game.
    (a) 34
    (b) 32
    (c) 38
    (d) 36

53. In a group of 36 persons, 20 take coffee but not tea. 16 take tea but not coffee. Find the number of persons who take neither tea nor coffee.
    (a) 2
    (b) 1
    (c) 0
    (d) Cannot be determined

54. In a locality, there are 100 residents sixty of them read 'The Hindu' and 40 of them read 'The Times of India'. The number of residents who read both newspapers must be
    (a) More than the number of those who read neither newspaper.
    (b) Less than the number of those who read neither newspaper.
    (c) Equal to the number of those who read neither newspaper.
    (d) Cannot say

55. In a class, there are 80 students. The ratio of the number of those who like only chocolates, only ice creams, both of these and neither of these is 4 : 3 : 2 : 1. How many like utmost one of the chocolates and ice creams?

    (a) 72
    (b) 56
    (c) 64
    (d) 60

56. In the question above, how many do not like ice creams?
    (a) 40
    (b) 44
    (c) 148
    (d) 36

57. In a class, each student plays chess or carom or both. The number of students who play chess, carom and both are 11, 12 and 3 respectively. Find the percentage of those who play only chess.
    (a) 36%
    (b) 40%
    (c) 44%
    (d) 48%

58. In the question above, find the percentage of those who play only caroms.
    (a) 45%
    (b) 40%
    (c) 50%
    (d) 55%

59.

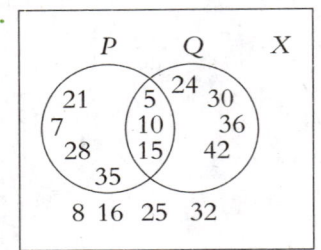

```
    P         Q        X
       _____
      /   \  24 /   \
     | 21  \  / 30   |
     |  7   \/  36   |
     | 28  5\ 42     |
     |     10\       |
     |    15 /       |
     |   35 /        |
      \____/_____/
     8  16  25  32
```

From the above Venn diagram, find $n(P - Q) + n(Q - P) = $ _____.
    (a) 10
    (b) 4
    (c) 6
    (d) 8

60. In a class of 50 students, each student passed Maths or English. If 10 students passed both Maths and English and 30 students passed Maths, then find the number of students who passed English.
    (a) 30
    (b) 20
    (c) 10
    (d) None of these

61. If $A = \{x/x$ is a factor of 4$\}$, $B = \{x/x$ is a factor of 8$\}$ and $C = \{x/x$ is a factor of 10$\}$, then which of the following statements is true?
    (a) $A \subset C$
    (b) $C \subset A$
    (c) $A \subset B$
    (d) $B \subset A$

62. Given that X = {Natural numbers less than 100 divisible by 6}, Y = {Natural numbers less than 100 divisible by 8} and Z = {Natural numbers less than 100 divisible by 18}.

    Find $n(X \cap Y \cap Z)$.

    (a) 4   (b) 1
    (c) 3   (d) 2

63. The strength of a class is 96. In it, 56 students like cricket and 40 students like football. Which of the following can be concluded?

    (a) No student likes either cricket or football
    (b) Each student likes either cricket or football
    (c) Neither (a) nor (b)
    (d) Both (a) and (b)

64. In an office, the ratio of the percentage of employees who like only tea, percentage of employees who like only coffee, percentage of employees who like both the drinks and percentage of employees who like neither of the drinks is 8 : 7 : 6 : 4. Find the percentage of employees who like neither of the drinks.

    (a) 12%   (b) 8%
    (c) 20%   (d) 16%

65. In the previous question, if the number of employees in the office is 150, then find the number of employees who like only tea.

    (a) 48   (b) 54
    (c) 120   (d) 66

ASSESSMENT TEST

Test 1

Directions for questions 1 to 11: Select the correct alternative from the given choices.

1. The mean of 36 observations is 22. If one observation 22 is deleted, then find the new mean.

   The following are the steps involved in solving the above problem. Arrange them in sequential order.

   (A) ∴ New arithmetic mean = $\frac{770}{35}$ = 22.

   (B) Arithmetic mean = $\frac{\text{The sum of the observations}}{\text{Total number of the observations}}$.

   ⇒ 22 = $\frac{\text{The sum of the observations}}{36}$.

   (C) The sum of the observations = 36 × 22 = 792.

   (D) Since one observation 22 is deleted, the new sum = 792 − 22 = 770 and the number of observations is 36 − 1, i.e., 35.

   (a) BDCA   (b) BCDA
   (c) DBAC   (d) CBDA

2. If $A = \{x : x \in W, x \leq 10\}$ and $B = \{x : x \in N, x \leq 10\}$, then find $n(A \cup B)$.

   The following are the steps involved in solving the above problem. Arrange them in sequential order.

   (A) ⇒ $n(A \cup B) = 11$

   (B) $A \cup B = \{0, 1, 2, 3, 4, 5, 6, 7, 8, 9, 10\} \cup \{1, 2, 3, 4, 5, 6, 7, 8, 9, 10\}$

   (C) From the given data $A = \{0, 1, 2, 3, 4, 5, 6, 7, 8, 9, 10\}$ and $B = \{1, 2, 3, 4, 5, 6, 7, 8, 9, 10\}$

   (D) ∴ $A \cup B = \{0, 1, 2, 3, 4, 5, 6, 7, 8, 9, 10\}$

   (a) CABD   (b) CBDA
   (c) DCBA   (d) BCDA

3. If the mean of 4, x and y is 6, then the mean of x, y and 10 is _____.

   (a) 8   (b) 9
   (c) 12   (d) 10

4. A bar graph is drawn to the scale, 1 cm = 4x units. The length of the bar representing a quantity 1000 units is 1.25 cm. Find x.

   (a) 200   (b) 175
   (c) 250   (d) 275

10.18 Chapter 10

5. In a pie graph, a component is represented as a sector with sector angle 108°. Find the percentage of the component value in total.
   (a) 28% (b) 30%
   (c) 32% (d) 35%

6. Find the mode of the data 2, 4, 6, 4, 6, 7, 6, 7 and 8.
   (a) 4 (b) 6
   (c) 7 (d) All of these

7. The mean height of a group of 30 students is 150 cm. If a 150 cm tall student is included in the group, then the mean height of the new group is _____.
   (a) 151 cm (b) 149 cm
   (c) 150 cm (d) None of these

8. If the mean of 2, 3, x, 7, 8 is x, then find the value of x.
   (a) 3 (b) 5
   (c) 4 (d) 6

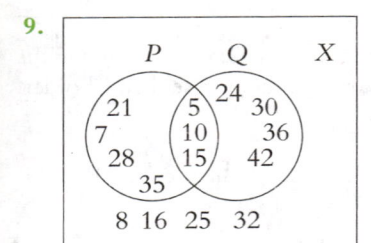

9.

   From the above Venn diagram, find $n(P - Q) + n(Q - P) =$ _____.
   (a) 10 (b) 4
   (c) 6 (d) 8

10. In a class of 50 students, each student passed Maths or English. If 10 students passed both Maths and English, and 30 students passed Maths, then find the number of students who passed English.
    (a) 30 (b) 20
    (c) 10 (d) None of these

11. There are 100 children in a colony. Of them, 70 watches Disney channel, 50 watch both Cartoon network and Disney channels and 20 watch none of these. Find the number of children who watch only cartoon network.
    (a) 40 (b) 30
    (c) 20 (d) 10

Directions for questions 12 to 15: Match the Column A with Column B.

Column A		Column B
12. If $\mu = \{0, 1, 2, 3, 4, 5, 6\}$ and $A = \{0, 1, 4\}$, then $n(A') =$ _____	( )	(a) 4
13. If $A = \{b, c, y, a, q, r\}$ and $B = \{a, x, p\}$, then $n(A - B) =$ _____	( )	(b) 6
14. If $K = \{0, 1, 2, 3\}$, then the number of subsets of K is _____	( )	(c) 8
15. If $X = \{x: x \text{ is even}, x \in N \text{ and } x \leq 12\}$ and $Y = \{x: x \text{ is a prime}, x \in N \text{ and } x \leq 12\}$, then $n(X \cup Y)$ is _____	( )	(d) 10
		(e) 16

Test 2

Directions for questions 16 to 26: Select the correct alternative from the given choices.

16. The mean of 2, 12, x, 15, 20 and 17 is 16, then find the value of x.

    The following are the steps involved in solving the above problem. Arrange them in sequential order.

    (A) $16 = \dfrac{2 + 12 + x + 15 + 20 + 17}{6}$

    (B) $96 = 66 + x$

    (C) $\Rightarrow x = 96 - 66 = 30.$

    (D) We have, arithmetic mean
    $= \dfrac{\text{The sum of observations}}{\text{Total number of observations}}$

    (a) ADBC (b) DABC
    (c) CABD (d) DBAC

17. In a class, there are 100 students. Of them, 60 students attend music classes, 40 students attend dance classes and 20 students attend both the classes. Find the number of students who attend neither of the classes.

The following are the steps involved in solving the above problem. Arrange them in sequential order.

(A) $n(M \cup D) = n(M) + n(D) - n(M \cap D) \Rightarrow n(M \cup D) = 60 + 40 - 20 = 100 - 20 = 80$.

(B) ∴ Number of students who attend neither of the classes = 100 − 80 = 20.

(C) $n(M) = 60$, $n(D) = 40$ and $n(M \cap D) = 20$ (given).

(D) Let $n(M)$ be the number of students who attend music classes and $n(D)$ be the number of students who attend dance classes.

(a) DACB (b) DCBA
(c) ACDB (d) DCAB

18. The mean of p, q and r is same as the mean of q, $2r$ and s. Then which of the following is correct?

(a) $p = q = r$ (b) $q = r = s$
(c) $q = r$ (d) $p = r + s$

19. A bar graph is drawn to the scale of 1 cm = 2 m units. The length of the bar representing a quantity of 875 units is 1.75 cm. Find m.

(a) 125 (b) 225
(c) 250 (d) 375

20. In a pie graph, a component is represented as a sector with sector angle 72°. Find the percentage of the component value in total.

(a) 21% (b) 27.5%
(c) 22.5% (d) 20%

21. The mode of the unimodal data 7, 8, 9, 8, 9, 10, 9, 10, 11, 10, 11, 12 and x is 10. Find the value of x.

(a) 10 (b) 9
(c) 8 (d) 11

22. The mean weight of 21 students is 21 kg. If a student weighing 21 kg is removed from the group, then what is the mean weight of the remaining students?

(a) 20 kg (b) 21 kg
(c) 19 kg (d) None of these

23. If the mean of 2, 4, p, 8 and 10 is p, then find the value of p?

(a) 4 (b) 5
(c) 8 (d) 6

24.

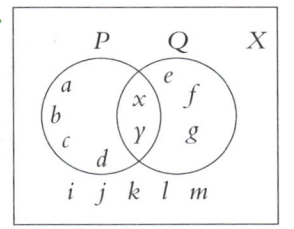

From the above Venn diagram, find $n(P - Q) + n(Q - P)$.

(a) 11 (2) 9
(c) 6 (4) 7

25. In a class of 70 students, each student passed Hindi or English. Of them, 15 students passed both Hindi and English, and 40 students passed Hindi. Find the number of students who passed English.

(a) 35 (b) 25
(c) 55 (d) 45

26. In a school, there are 150 students. If 90 of them play chess, 70 of them play both chess and carrom and 40 play none of these games, then find the number of children who play only carrom.

(a) 90 (b) 10
(c) 20 (d) None of these

Directions for questions 27 to 30: Match the Column A with Column B.

Column A		Column B
27. If $\mu = \{p, q, r,\}$ and $A = \{p\}$, then $n(A') = $ _____	( )	(a) 0
28. If $A = \{1, 2, 3, 4, 5, 6, 7, 8, 9, 10\}$ and $B = \{2, 4, 6, 8, 10\}$, then $n(A - B)$ is _____	( )	(b) 1
29. If $X = \{a, e, I, o, u\}$, then the number of improper subsets of X is _____	( )	(c) 2
30. If $A = \{x: x$ is an even prime, $x \in N\}$ and $B = \{x: x$ is an odd natural number, $x < 10\}$, then $n(A \cap B) = $ _____	( )	(d) 5
		(e) 31

10.20 Chapter 10

TEST YOUR CONCEPTS

Very Short Answer Type Questions

1. True
2. True
3. False
4. False
5. True
6. disjoint
7. infinite
8. 5
9. 15
10. 7
11. Not a set
12. Not a set
13. Not a set
14. Set
15. (b)
16. (d)
17. (d)
18. (b)
19. (c)
20. (a)
21. (b)
22. (c)
23. (c)
24. (a)
25. (c)
26. (b)
27. (b)
28. (d)

Short Answer Type Questions

29. 1
30. $W = \{x : x$ is an woman prime minister of India$\}$
31. 8
32. $A - B = \{S\}$ and $B - A = \{R, H, M, E\}$
33. Disjoint sets.
34. $\mu = \{1, 2, 3, 4, 5, 6, 7, 8, 9, 10, 12\ \text{--------}, 99\}$
35. $\{x : x$ is a natural number and $x < 100\}$
36. $K' = \{x : x$ is a natural number and $x \neq 2\}$
37. $\{x\}, \{y\}, \{z\}, \{x, y\}, \{x, z\}, \{y, z\}, \{x, y, z\}, \phi.$
38. $\{p\}, \{q\}, \{p, q\},$ and $\phi.$
39. $\phi, \{0\}, \{-1\}, \{1\}, \{0, -1\}, \{0, 1\}$ and $\{-1, 1\}$
40. $\phi, \left\{\dfrac{-1}{2}\right\}$ and $\left\{\dfrac{1}{2}\right\}$
41. 1
42. 80
43. $\{4, 9, 16, 25, 36, 49, 64, 81, 100\}$
44. 8
45. $\{0, 1, 4, 9, 16, 25\}$
46. $\{0, 1, 4, 9, 16, 25\}.$
47. $A - B = \{2, 11\}$ and $B - A = \{1, 9\}$
48. $\{5, 7, 9, 11\}$
49. 1
50. 2^{26}
51. 60
52. $\{0, 1, 3, 5, 7, 9\}$

Essay Type Questions

53. (a) $\{1, 2, 3, 4, 5, 6, 7, 8, 9\}$
    (b) $\{4, 5\}$
    (c) $\{1, 2, 3, 6, 7, 8, 9, 10, 11, 12, 13, 14, 15\}$
54. (a) $\{f, g, h, i, j, k, l, m\}$
    (b) $\{f, g, h, i, j, k, l, m\}$
    (c) $\{c, d, e, f, g, h, i\}$

Set Theory **10.21**

55. 8
56. 18, 24
57. 40
58. 30

59. 7
60. 45
61. 20

CONCEPT APPLICATION

Level 1

1. (b)   2. (c)   3. (d)   4. (c)   5. (a)   6. (d)   7. (a)   8. (b)   9. (d)   10. (d)
11. (c)   12. (b)   13. (a)   14. (d)   15. (a)   16. (d)   17. (c)   18. (f)   19. (a)   20. True
21. True   22. False   23. False

Level 2

24. (b)   25. (a)   26. (d)   27. (a)   28. (a)   29. (b)   30. (d)   31. (d)   32. (c)   33. (a)
34. (d)   35. (c)   36. (a)   37. (d)   38. (d)   39. (c)   40. (c)   41. (b)   42. (d)   43. (a)
44. (a)   45. (c)   46. (b)

Level 3

47. (d)   48. (a)   49. (c)   50. (d)   51. (b)   52. (a)   53. (c)   54. (c)   55. (c)   56. (a)
57. (b)   58. (a)   59. (d)   60. (a)   61. (c)   62. (b)   63. (c)   64. (d)   65. (a)

ASSESSMENT TESTS

Test 1

1. (b)   2. (b)   3. (a)   4. (a)   5. (b)   6. (b)   7. (c)   8. (b)   9. (d)   10. (a)
11. (d)   12. (a)   13. (b)   14. (e)   15. (d)

Test 2

16. (b)   17. (d)   18. (d)   19. (c)   20. (d)   21. (a)   22. (b)   23. (d)   24. (d)   25. (d)
26. (c)   27. (c)   28. (d)   29. (b)   30. (a)

10.22 Chapter 10

CONCEPT APPLICATION

Level 1

1. $A' = \mu - A$
   $A' = \{1, 3, 4, 6, 7\}$
   $\therefore n(A') = 5.$
   Hence, the correct option is (b)

2. 1 is the only number which is neither prime nor composite.
   $\therefore n(X \cap Y) = 1$
   Hence, the correct option is (c)

3. All the factors of 6 are factors of 12.
   $\therefore P$ is a subset of $Q. \Rightarrow P \cup Q = Q$
   $(Q = \{1, 2, 3, 4, 6, 12\}$
   $\therefore n(P \cup Q) = n(Q) = 6.$ Choice (4)

4. Let U be the universal set.
   $U = M \cup M' = N \cup N'$
   If $M = N$, then $M' = N'$
   If $M' = N'$, then $M = N$
   $\therefore$ Choices (a) and (b) follow, i.e., Choice (c) follows.
   Hence, the correct option is (c)

5. $n(A) + n(B) = n(A \cup B) + n(A \cap B)$
   $10 + 20 = 26 + n(A \cap B)$
   $n(A \cap B) = 4.$
   Hence, the correct option is (a)

6. $B \cup C = \{1, 2, 3, 4, 5, 6, 7, 8, 9\}$ (1)
   $B \subset A$ and $C \subset A$
   $\therefore A \cup (B \cup C) = A$ (2)
   From (1) and (2), $A \cup (B \cup C) = A = B \cup C.$
   Hence, the correct option is (d)

7. $B \subset A \Rightarrow A \cap B = B$
   Hence, the correct option is (a)

8. $C \subset A \Rightarrow A \cap C = C$
   Hence, the correct option is (b)

9. $A \cap B = B; A \cap C = C$
   $((A \cap B) \cup (A \cap C)) = B \cup C$

   $A = A \cup B = A \cup C = B \cup C$
   Hence, the correct option is (d)

10. $\{0\}$ contains one element, i.e., 0
    $\{\phi\}$ contains one element, i.e., ϕ
    $\{x: x$ is an even prime number$\} = \{2\}$ contains one element, i.e., 2
    All the above are singleton sets.
    Hence, the correct option is (d)

11. $n(A) = $ The number of distinct elements in A.
    $= 10$
    Hence, the correct option is (c)

12. $A = \{x/x + 5 = 5\} \Rightarrow A = \{0\}$
    $(\because x + 5 = 5 \Rightarrow x = 0)$
    $\therefore n(A) = 1.$
    Hence, the correct option is (b)

13. $4 \in X.$ (since i is an element of X)
    Hence, the correct option is (a)

14. (A), (B) and (C) is the required sequential order.
    Hence, the correct option is (d)

15. (C), (D), (B) and (A) is the required sequential order.
    Hence, the correct option is (a)

16. $\rightarrow d: A = \{0, 1, 2, 3, 4, 5, 6, 7, 8, 9, 10\}$ and
    $B = \{1, 2, 3, 4, 5, 6, 7, 8, 9\}$
    $A - B = \{0, 10\}$ and $n(A - B) = 2.$

17. $\rightarrow c: n(X) = 0$ $(\because X$ is null set$)$
    $\therefore n(P(X)) = 2^{n(x)} = 2^0 = 1$

18. $\rightarrow f: A = \{G, O, L, E\}$ is the set containing the letters of the word 'GOOGLE'.
    $\therefore n(A) = 4.$

19. $\rightarrow a: \because$ Given $A \subset B$

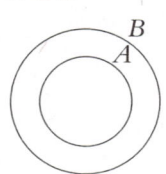

    $\Rightarrow A \cap B = A$

20. Given $A = \{T, I, G, E, R\}$ and $B = \{G, I, N, T, E, R\}$

$\Rightarrow A \cup B = \{T, I, G, E, R, N\}$

$\Rightarrow n(A \cup B) = 6$

∴ The given statement is true.

21. $x = \{G, I, N, E, R, C, O, K\}$ is the set containing the letters of the word 'GINGERCOOK'.

∴ $n(x) = 8$

∴ The given statement is true.

22. Given $A \subset B$

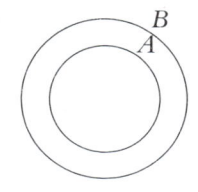

[Venn diagram: B containing A]

$\Rightarrow A \cup B = B$

∴ The given statement is false.

23. Since $\{4, 5\}$ is an element of set X and it is not a subset.

i.e., $\{4, 5\}$ belongs to set X.

∴ $\{4, 5\}$ is not a subset of X.

∴ The given statement is false.

Level 2

24. $A = \{1, 4, 9, 16, 25, 36, 49, 64, 81\}$

$B = \{1, 8, 27, 64\}$

$A \cap B = \{1, 64\}$ ∴ $n(A \cap B) = 2$

Hence, the correct option is (b).

25. $O' = \{\text{Set of all even primes}\}$

∴ $O' = \{2\}$ (∵ 2 is the only even prime)

∴ $n(O') = 1$.

Hence, the correct option is (a).

26. $\mu = \{16, 25, 36, 49, 64, 81\}$

$X = \{36, 81\}$

$Y = \{16, 25, 49\}$

$X \cup Y = \{16, 25, 36, 49, 81\}$

$(X \cup Y)' = \{64\}$

Hence, the correct option is (d).

27. Given, $\mu = \{0, 1, 2, 3, 4 \ldots \ldots 9\}$,

$X = \{2, 3, 5, 7\}$ and $Y = \{2, 5, 8\}$

$X' = \{0, 1, 4, 6, 8, 9\}$

$Y' = \{0, 1, 3, 4, 6, 7, 9\}$

$X' \cup Y' = \{0, 1, 3, 4, 6, 7, 8, 9\}$

$n(X' \cup Y') = 8$.

Hence, the correct option is (a)

28. $P = \{1, 2, 3, 4, 6, 9, 12, 18, 36\}$

$Q = \{1, 2, 3, 4, 6, 8, 12, 16, 24, 48\}$

$P \cap Q = \{1, 2, 3, 4, 6, 12\}$

$n(P \cap Q) = 6$.

Hence, the correct option is (a)

29. $E = \{a, e, i, o, u\}$

The positions of a, e, i, o and u in S are 1, 5, 9, 15 and 21 respectively.

∴ Every vowel has an odd position in S.

∴ $B = \phi$

∴ $B' = \mu$

$n(B') = n(\mu) = 5$.

Hence, the correct option is (b)

30. Every number is a factor and a multiple to itself.

∴ $n(A \cap B) = 1$.

Hence, the correct option is (d)

31. $C = \{2, 5, 8, 11, 14, 17, 20, 23, 26, 29, 32\}$

$D = \{2, 4, 6, 8, \ldots \ldots 32\}$

10.24 Chapter 10

$D' = $ {All odd natural numbers up to 32}

$C \cap D' = \{5, 11, 17, 23, 29\}$

$n(C \cap D)' = 5$

Hence, the correct option is (d)

32. $E = \{1, 2, 3, \ldots\ldots 30\}$

    $X = \{2, 4, 6, \ldots\ldots 30\}$

    $Y = \{6, 10, 14, 18, 22, 26, 30\}$

    $X \cap Y = Y (\because Y \subset X)$

    $\therefore n(X \cap Y) = 7$.

    Hence, the correct option is (c)

33. 10 consecutive natural numbers will have their units digits as 0, 1, 2, 3, 4, 6, 7, 8, 9.

    $\therefore$ Units digit of their sum = 5.

    $\therefore X = \{5\}$

    Hence, the correct option is (a)

34. All the given statements with respect to the set A are true.

    Hence, the correct option is (d)

35. $n(A) + n(B) = 24$

    A and B are disjoint.

    $\therefore n(A \cap B) = 0$

    $n(A \cup B) + n(A \cap B) = n(A) + n(B)$

    $\Rightarrow n(A \cup B) = 24$.

    Hence, the correct option is (c)

36. $A = \{1, 2, 3, 4\}$

    $B = \{2, 4, 8, 9\}$

    $A - B = \{1, 3\}$

    $B - A = \{8, 9\}$

    $(A - B) \cup (B - A) = \{1, 3, 8, 9\}$

    Hence, the correct option is (a)

37. $(A - B) \cap (B - A) = \phi$

    Hence, the correct option is (d)

38. None of the given statements are false.

    Hence, the correct option is (d)

39. Both the given statements are true.

    Hence, the correct option is (c)

40. Disjoint sets have no elements in common.

    Hence, the correct option is (c)

41. Since $\{i, o\}$ is an element of Y.

    $\therefore \{i, o\} \in Y$.

    Hence, the correct option is (b)

42. $U = \{a, b, c, d, e, f, g, h, i, j, \ldots\ldots\ldots z\}$

    $C = \{b, c, d, f, g, h, j, k, l, m, n, p, q, r, s, t, v, w, x, y, z\}$

    $C' = U - C = \{a, e, i, o, u\}$.

    Hence, the correct option is (d)

43. If a set has n elements, then number of all possible subsets = 2^n

    Since O contains only one element.

    $\Rightarrow$ Number of subsets = $2^1 = 2$.

    Hence, the correct option is (a)

44. $n(A \cup B) = n(A) + n(B) - n(A \cap B)$

    $35 = 10 + n(B) - 5$

    $\Rightarrow n(B) = 35 - 10 + 5 = 30$

    $\therefore n(B) = 30$.

    Hence, the correct option is (a)

45. $X - Y = \{0, 2, 4, 6, 8, 10\}$

    Hence, the correct option is (c)

46. $A = \{27, 64\}$

    $B = \{16, 25, 36, 49, 64, 81\}$

    $A \cap B = \{64\}$

    $n(A \cap B) = 1$.

    Hence, the correct option is (b)

Set Theory

Level 3

47. $\mu = \{1, 2, 3, \ldots 30\}$

    $Q = \{4, 8, 12, 16, 20, 24, 28\}$

    $R = \{6, 12, 18, 24\}$

    $Q \cap R = \{12, 24\}$

    $\therefore n(Q \cap R) = 2$

    $n[(Q \cap R)'] = n(\mu) - n(Q \cap R) = 30 - 2 = 28.$

    Hence, the correct option is (d)

48. $A = \{9, 18, 27, \ldots 198\}$

    $B = \{12, 24, 36, \ldots 192\}$

    $C = \{15, 30, 45, \ldots 195\}$

    $A \cap B \cap C =$ [Natural numbers less than 200 divisible by 9, 12 and 15, i.e., by LCM of $\{9, 12, 15\}$]

    But LCM of 9, 12, 15 is 180.

    $\therefore A \cap B \cap C = \{180\}.$

    Hence, the correct option is (a)

49. $A \cap B = \{$Natural numbers less than 200 divisible by 9 and 12, i.e., by LCM of $(9, 12)\}$.

    But LCM of 9 and 12 is 36.

    $\therefore A \cap B = \{36, 72, 108, 144, 180\}.$

    Hence, the correct option is (c)

50. $P = \{1, 2, 3, 4, 6, 8, 12, 16, 24, 48\}$

    $Q = \{1, 2, 3, 4, 5, 6, 10, 12, 15, 20, 30, 60\}$

    $P - Q = \{8, 16, 24, 48\}$

    $Q - P = \{5, 10, 15, 20, 30, 60\}$

    $n[(P - Q) \cup (Q - P)] = n(P - Q) + n(Q - P)$

    $= 4 + 6 = 10$

    Hence, the correct option is (d)

51. $P = \{5, 9, 13, 17, 21, 25, 29\}$

    $Q = \{7, 13, 19, 25\}$

    $P \cap Q = \{13, 25\}$

    $n(P \cap Q) = 2.$

    Hence, the correct option is (b)

52. There are 50 students. Each student likes at least one game. 16 students like both games.

    $\therefore$ The remaining 34 students like exactly one game.

    Hence, the correct option is (a)

53. 20 take only coffee.

    16 take only tea.

    $\therefore$ Number of persons who take either coffee or tea = 36

    = Number of persons in the group.

    $\therefore$ Required number = 0.

    Hence, the correct option is (c)

54. Let the number of residents who read both newspapers be b.

    Let the number of residents who read neither newspaper be n.

    The number of residents who read either of the newspapers

    $= 60 + 40 - b$

    $\therefore 60 + 40 - b + n = 100$

    $b = n.$

    $\therefore$ Choice (c) follows.

    Hence, the correct option is (c)

55. Let the number of those who like neither chocolates nor ice creams be x.

    Number of those who like both of these = $2x$.

    Number of those who like only ice creams = $3x$.

    Number of those who like only chocolates = $4x$.

    $x + 2x + 3x + 4x = 80$

    $10x = 80, x = 8$

    Atmost means maximum of one.

    Required number = 80 − Number of students who like both

    $= 80 - 2x = 80 - 16 = 64.$

    Hence, the correct option is (c)

56. Number of those who like ice creams = $3x + 2x = 5x = 40.$

    Required number = $80 - 5x = 40.$

    Hence, the correct option is (a)

10.26 Chapter 10

57. The number of students who like only chess = 11 − 3 = 8.

The number of students who like only carom = 12 − 3 = 9.

The number of students who like either chess or carom = 8 + 9 + 3 = 20.

The required percentage = $\left(\dfrac{8}{20} \times 100\right)\%$

= 40%

Hence, the correct option is (b)

58. The required percentage = $\left(\dfrac{9}{20} \times 100\right)\% = 45\%$

Hence, the correct option is (a)

59. P − Q = {21, 7, 28, 35}

⇒ n(P − Q) = 4

Q − P = [24, 30, 36, 42}

⇒ n(Q − P) = 4

∴ n(P − Q) + n(Q − P) = 4 + 4 = 8.

Hence, the correct option is (d)

60.

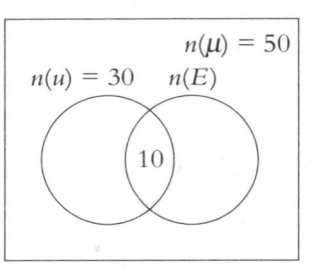

Given, $n(M \cap E) = 10$ and $n(M) = 30$ and $n(M \cup E) = 50$.

$n(M \cup E) = n(M) + n(E) − n(M \cap E)$

50 = 30 + n(E) − 10

⇒ n(E) = 30.

Hence, the correct option is (a)

61. A = {1, 2, 4}

B = {1, 2, 4, 8}

C = {1, 2, 5, 10}

⇒ A ⊂ B

Hence, the correct option is (c)

62. $n(X \cap Y \cap Z) = n$ (Natural numbers less than 100 and divisible by 6, 8 and 18 are multiples of LCM of 6, 8, 18, i.e., 72)

$X \cap Y \cap Z = \{72\}$

$n(X \cap Y \cap Z) = 1$.

Hence, the correct option is (b)

63.

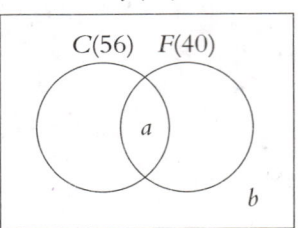

The number of students who like either cricket or football

= 56 + 40 − a = 96 − b.

96 − a + b = 96

b = a.

Choice (a) ⇒ a = 0 and Choice (b) ⇒ b = 0.

Choice (a) and (b) need not be true.

Hence, the correct option is (c)

64. Let the percentage of employees who like only tea be 8x%.

The percentage of employees who like only coffee = 7x%.

The percentage of employees who like both the drinks = 6x%.

The percentage of employees who like neither of the drinks = 4x%.

$n(\mu) − n(A \cap B)$

= 8x% + 7x% + 6x% + 4x% = 100%

25x = 100

x = 4

∴ 4x = 16.

Hence, the correct option is (d)

65. The percentage of employees who like only tea = 8x% = 8 × 4% = 32%.

The required number of employees who like only tea

= $\dfrac{32}{100} \times 150 = 48$.

Hence, the correct option is (a)

ASSESSMENT TESTS

Test 1

Solutions for questions 1 to 11:

1. (B), (C), (D) and (A) is the required sequential order.
   Hence, the correct option is (b)

2. (C), (B), (D) and (A) is the required sequential order.
   Hence, the correct option is (b)

3. Given, $\dfrac{4 + x + y}{3} = 6$
   $x + y = 6 \times 3 - 4$
   $x + y = 14$
   $\dfrac{x + y + 10}{3} = \dfrac{14 + 10}{3} = 8.$
   Hence, the correct option is (a)

4. Given, 1 cm = 4x units
   1.25 cm = 1000 units
   $1 \text{cm} = \dfrac{1000}{1.25}$ units
   1 cm = 800 units
   $\Rightarrow 4x = 800 \Rightarrow x = 200.$
   Hence, the correct option is (a)

5. Percentage of the required component
   $= \dfrac{108°}{360°} \times 100\% = 30\%.$
   Hence, the correct option is (b)

6.
Observations	Frequency
2	1
4	2
6	3
7	2
8	1

   6 has the highest frequency.
   ∴ Mode = 6.
   Hence, the correct option is (b)

7. Given that the mean height of 30 students = 150 cm.
   ∴ The total height of 30 students
   = (150 cm) 30 = 4500 cm.

   If 150 cm tall student is included in the group, then the total height of 31 students = 4500 + 150 = 4650 cm.
   ∴ The mean height of 31 students
   $= \dfrac{4650}{31} = 150$ cm.
   Hence, the correct option is (c)

8. Given, mean = x
   $\dfrac{2 + 3 + x + 7 + 8}{5} = x$
   $\Rightarrow 4x = 20 \Rightarrow x = 5.$
   Hence, the correct option is (b)

9. $P - Q = \{21, 7, 28, 35\}$
   $\Rightarrow n(P - Q) = 4$
   $Q - P = \{24, 30, 36, 42\}$
   $\Rightarrow n(Q - P) = 4$
   $\therefore n(P - Q) + n(Q - P) = 4 + 4 = 8.$
   Hence, the correct option is (d)

10. 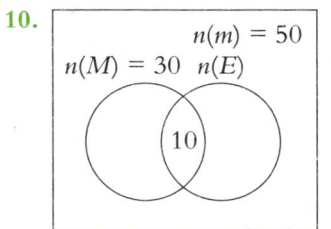
    $n(m) = 50$
    $n(M) = 30 \quad n(E)$
    10

    Given, $n(M \cap E) = 10$ and $n(M) = 30$ and $n(M \cup E) = 50.$
    $n(M \cup E) = n(M) + n(E) - n(M \cap E)$
    $50 = 30 + n(E) - 10 \Rightarrow n(E) = 30.$
    Hence, the correct option is (a)

11. 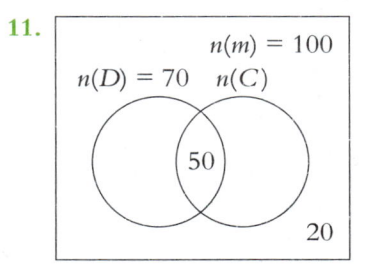
    $n(m) = 100$
    $n(D) = 70 \quad n(C)$
    50
    20

Chapter 10

Given $n(D) = 70$

$n(D \cup C) = 100 - 20 = 80$

$n(D \cap C) = 50$

$n(D \cup C) = n(D) + n(C) - n(D \cap C)$

$80 = 70 + n(C) - 50$

$\Rightarrow n(C) = 60$.

∴ The number of children who watch only cartoon network = $60 - 50 = 10$.

Hence, the correct option is (d)

Solutions for questions 12 to 15:

12. → a: $\mu = \{0, 1, 2, 3, 4, 5, 6\}$ and $A = \{0, 1, 4\}$

    $\Rightarrow A' = \{3, 4, 5, 6\}$

    ∴ $n(A') = 4$.

13. → b: $A - B = \{b, c, y, z, q, r\} - \{a, x, p\} = \{b, c, y, z, q, r\}$

    ∴ $n(A - B) = 6$.

14. → e: $K = \{0, 1, 2, 3\} \Rightarrow n(K) = 4$

    ∴ The number of subsets = $2^4 = 16$.

15. → d: $X = \{2, 4, 6, 8, 10, 12\}$ and $Y = \{2, 3, 5, 7, 11\}$

    $\Rightarrow X \cup Y = \{2, 3, 4, 5, 6, 7, 8, 10, 11, 12\}$

    ∴ $n(X \cup Y) = 10$.

Test 2

Solutions for questions 16 to 26:

16. dabc is the required sequential order.

    Hence, the correct option is (b)

17. dcab is the required sequential order.

    Hence, the correct option is (d)

18. Given, $\dfrac{p + q + r}{3} = \dfrac{q + 2r + s}{3}$

    $\Rightarrow p = r + s$

    Hence, the correct option is (d)

19. Given, 1 cm = 2 m units → (1)

    And 1.75 cm = 875 units

    $\Rightarrow$ 1 cm = $\dfrac{875}{1.75}$ units

    $\Rightarrow$ 1 cm = 500 units → (2)

    From (1) and (2), we get,

    2 m = 500

    m = 250.

    Hence, the correct option is (c)

20. Required percentage = $\dfrac{72°}{360°} \times 100 = 20\%$

    Hence, the correct option is (d)

21. The given observations are 7, 8, 9, 8, 9, 10, 9, 10, 11, 10, 11, 12 and x.

Observations	Frequency
7	1
8	2
9	3
10	3
11	2
12	1
x	1

    And also given that the mode is 10 and the given data is unimodal.

    ∴ $x = 10$

    Hence, the correct option is (a)

22. Given that the mean weight of 21 students = 21 kg.

    ∴ The total weight of 21 students = 21 (21 kg).

    If a student weighing 21 kg is removed from the group, then the total weight of 20 students = 21(21 kg) − 21 kg = 420 kg.

    ∴ The mean weight of 20 students = $\dfrac{420}{20}$ = 21 kg.

    Hence, the correct option is (b)

23. Given, mean = p

$\dfrac{2+4+p+8+10}{5} = p$

$\Rightarrow 4p = 24$

$\Rightarrow p = 6$.

Hence, the correct option is (d)

24. $P - Q = \{a, b, c, d\}$

$Q - P = \{e, f, g\}$

$\Rightarrow n(P - Q) = 4$ and $n(Q - P) = 3$

$\therefore n(P - Q) + n(Q - P) = 4 + 3 = 7$.

Hence, the correct option is (d)

25.

$n(H) = 30 \quad n(E)$

15

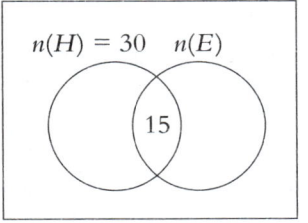

Given, $n(H) = 40$, $n(H \cap E) = 15$ and $n(H \cup E) = 70$

$n(H \cup E) = n(H) + n(E) - n(H \cap E)$

$70 = 40 + n(E) - 15 \Rightarrow n(E) = 45$.

Hence, the correct option is (d)

26.

$n(\mu) = 150$
$n(A) = 90 \quad n(B)$

70

40

Given, $n(A) = 90$, $n(A \cap B) = 70$

and $n(A \cup B) = 150 - 40 = 110$.

$n(A \cup B) = n(A) + n(B) - n(A \cap B)$

$110 = 90 + n(B) - 70$

$\Rightarrow n(B) = 90$

$\therefore$ The number of children who play only carrom.

$= 90 - 70 = 20$.

Hence, the correct option is (c)

Solutions for questions 27 to 30:

27. → c: Given $\mu = \{p, q, r\}$ and $A = \{p\}$

$A' = \{q, r\}$

$(A') = 2$

28. → d: $A = \{1, 2, 3, 4, 5, 6, 7, 8, 9, 10\}$ and $B = \{2, 4, 6, 8, 10\}$

$A - B \{1, 3, 5, 7, 9\}$

$n(A - B) = 5$.

29. → d: Number of improper subsets is 1

30. → a: Given $A = \{2\}$ and $B = \{1, 3, 5, 7, 9\}$

$\Rightarrow A \cap B = \{\ \} = \varnothing$

$n(A \cap B) = 0$

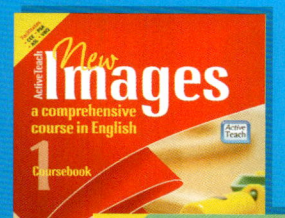

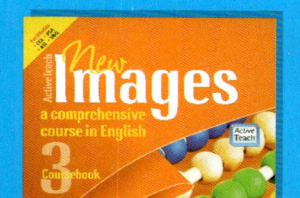

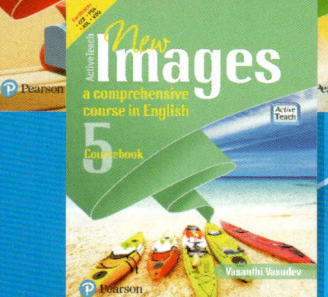

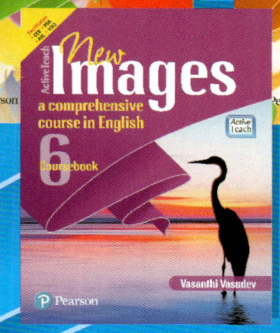

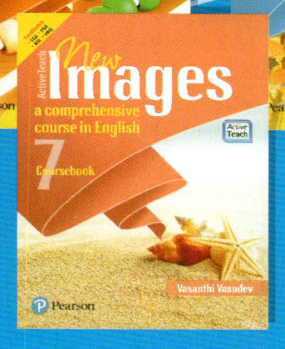

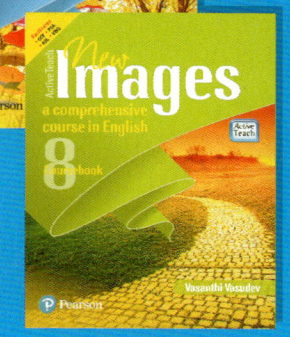

ActiveTeach New Images is a unique learning solution that combines a pedagogically sound and comprehensive English language course with innovative teaching and learning digital resources—seamlessly integrating technology with the syllabus to ensure effective learning in a learner-friendly classroom environment.

This series has been designed according to the latest teaching-learning pedagogies and adheres to the **National Curriculum Framework, 2005**. A variety of easily navigable and accessible media and digital resources are strategically embedded within the text of the digital book.

ActiveTeach New Images also takes into consideration the recent reforms in the educational system, and facilitates Continuous and Comprehensive Evaluation **(CCE)** and Assessment of Speaking and Listening **(ASL)** in the classroom. Besides covering tasks for Problem Solving Assessments **(PSA)**, it includes tasks for developing critical and creative thinking. The units also cover value-based questions **(VBQ)** for developing twenty-first-century life skills.

The **workbooks**, which are thematically linked to the **coursebooks**, contain two model test papers for **summative assessment**. Workbooks 5 to 8 contain material for practising Open Text-Based Assessment **(OTBA)**.

Salient Features of *ActiveTeach New Images*

- content-mapped digital resources
- a range of interactive activities and games
- educational puzzles and activities in the form of Active Widgets
- engaging animations and audio inputs
- differentiated worksheets augmenting lessons
- captivating videos with video worksheets
- detailed glossary
- informative slideshows
- annotation tools to aid teaching and learning
- **New** interactive Language Lab for practice of LSRW, vocabulary & grammar

FREE SCHOOL SUPPORT MATERIALS

This book is supplemented by a *Teacher's Resource Book* and a *Teacher's CD-ROM* which contain

- **E-book** with interactivities and audio recordings of the poems, stories (model reading), plays, pronunciation drills and listening inputs
- **Animations** and **Worksheets** for use in the classroom
- **Lesson plans** with teaching guidelines for all the lessons
- A **Dynamic Question Bank** that allows teachers to generate question papers for various assessment purposes

For further information, please contact us at schoolmarketing@pearson.com.

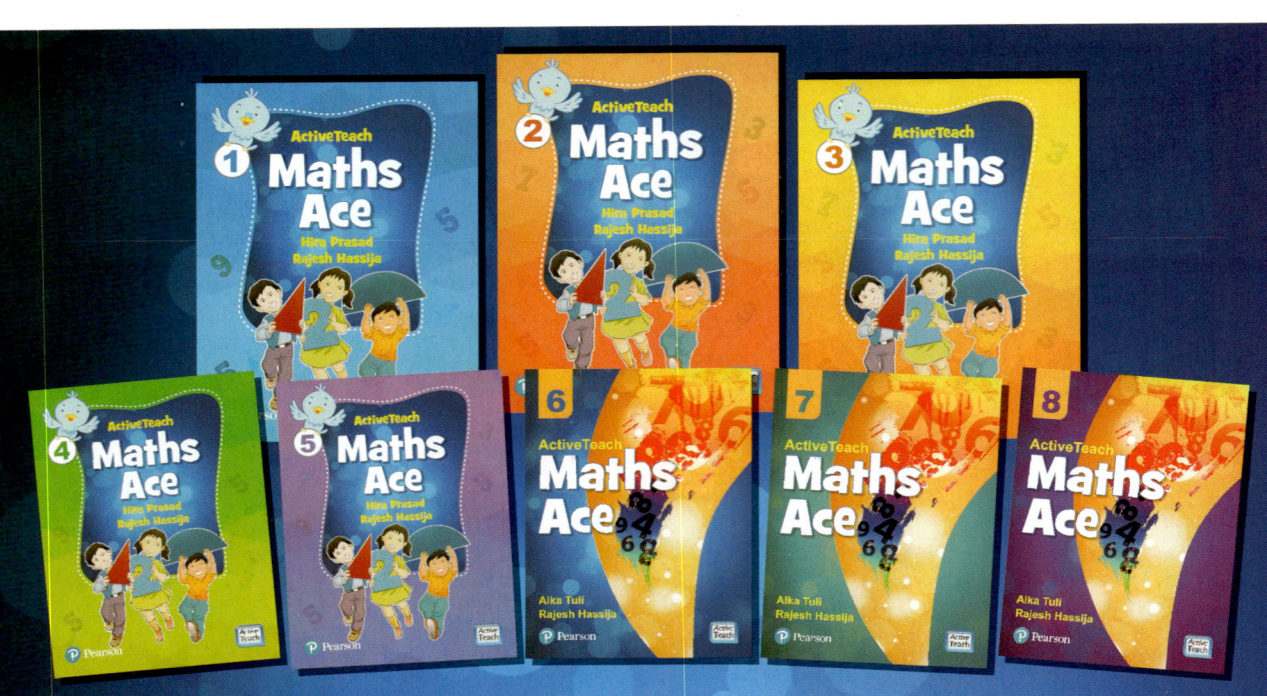

ActiveTeach Maths Ace is a blended-learning course in mathematics for classes 1 to 8. The series adheres to National Curriculum Framework 2005 and the books have been designed in accordance with the latest guidelines laid down by the National Council of Educational Research and Training (NCERT).

The series is based on the extensive feedback received from teachers. It employs dynamic and innovative teaching-learning techniques and resources that stimulate and motivate young students to develop a passion for mathematics.

Key Features

- The platform can project, zoom, highlight or annotate every area of the digital book.
- The easy navigation buttons help one to move quickly through one part of the digital content to the other.
- The various resource icons indicate what lies embeddead in the ActiveTeach digital book.
- The series provides innumerable examples, illustrations, engaging animations, captivating videos, video-based tutorials, concept-enhancing audios, interesting slideshows, differentiated worksheets and a host of interactive activities.
- Interesting games will help students learn in the play-way method.
- **Glossary** section is equipped with illustrations and voice-overs that will help in building a strong foundation in mathematics.
- **Lesson** and **study plan** will give teachers and students the option to explore beyond what is given and help them design their own plans.

- **Find Resources** will help teachers get a list of available resources.
- **Dynamic Question Zone** is an exclusive resource provided to teachers to help them create question papers.
- A special tab for **CCE** will also equip teachers to select appropriate tasks for formative and summative assessment throughout the year.
- The **hotspot** and **curtain** tools can be used to mask the page selectively and offer emphasised learning.
- **BTW** section includes the Bloom's Taxonomy Worksheets for every chapter covering the six levels—**Remembering**, **Understanding**, **Applying**, **Analyzing**, **Evaluating** and **Creating**, along with answers.
- **Slideshows** include Chapter Openings, PSAs and Life Skills.
- A special section on **Active Widgets** contain interesting mathematics resources such as well-known mathematicians, formula chart, find the operator, plot it, 2D and 3D objects, virtual geometry box, convertor, calculator and stopwatch.

FREE SCHOOL SUPPORT MATERIALS

This book is supplemented by a **Teacher's Resource Book** and a **Teacher's CD-ROM** which contain

- **E-book** with interactivities and audio recordings of the poems, stories (model reading), plays, pronunciation drills and listening inputs
- **Animations** and **Worksheets** for use in the classroom
- **Lesson plans** with teaching guidelines for all the lessons
- A **Dynamic Question Bank** that allows teachers to generate question papers for various assessment purposes

For further information, please contact us at schoolmarketing@pearson.com.

ActiveTeach Universal Science is a unique learning solution that combines a pedagogically sound and comprehensive Science course with innovative teaching and learning resources—seamlessly integrating technology with the syllabus to ensure effective learning in a learner-friendly classroom environment.

Recommended for primary- and middle-school students, **Universal Science** is a series of books that adheres to the National Curriculum Framework (2005). The books have been designed in accordance with the latest guidelines laid down by the National Council of Educational Research and Training.

The series is based on extensive feedback received from teachers and education consultants experienced in teaching and interacting with students in this age group. All the books present concepts and provide exercises with the view to nurturing the scientific temperament in young learners. The well-structured chapters interspersed with interesting information and questions make learning almost effortless. Together with the activities which instil the spirit of experimentation, the detailed coverage of topics and the variety of exercises lend the textbooks the right balance between the theoretical and practical aspects of science.

The series is complemented by a set of **Comprehensive Teacher's Resource Books**, which further equips the teacher to apply CCE to classroom teaching and make learning interactive and enjoyable.

ActiveTeach provides ample scope for students to use digital content at home to revise lessons and prepare for tests while facilitating self-assessment.

Salient Features of ActiveTeach Universal Science

- Content-mapped digital resources
- Engaging animations and audio inputs
- Captivating videos
- Slide shows with quizzes and PSAs
- A range of interactive activities
- Worksheets augmenting lessons
- Detailed glossary
- Zoom images
- Innovative widgets and tools to aid teaching and learning
- Information on scientists

FREE SCHOOL SUPPORT MATERIALS

This book is supplemented by a **Teacher's Resource Book** and a **Teacher's CD-ROM** which contain

- **E-book** with interactivities and audio recordings of the poems, stories (model reading), plays, pronunciation drills and listening inputs
- **Animations** and **Worksheets** for use in the classroom
- **Lesson plans** with teaching guidelines for all the lessons
- A **Dynamic Question Bank** that allows teachers to generate question papers for various assessment purposes

For further information, please contact us at **schoolmarketing@pearson.com**.

As per the latest ICSE 2018 syllabus

The *Unravelling Science (ICSE)* series, consisting of a set of nine coursebooks and nine workbooks for middle school, is based on the latest syllabi of the Inter-State Board for Anglo-Indian Education. The content has been presented in a manner that provides a clear insight of the basic concepts and principles of science.

KEY FEATURES

- **Learning Objectives** at the beginning of each chapter clearly define the goals of study.
- **Activities**, wherever appropriate within the text, are easy to do for practical verification.
- **Quick Check** questions are interspersed within the text for quick self-testing.
- **Think Through** questions present within the text prompt the students to think out of the box.
- **Know Your Scientist** introduces the students to the works of some renowned scientists in the various fields of science.
- **Did You Know?** in each chapter provides interesting facts to enhance learning.
- **Connecting the Dots** is a recapitulation of a chapter through mind maps.
- **Keywords** presents all the key terms along with their definitions.
- **Assessment Time** consists of a mix of objective- and subjective-type questions for the purpose of recall and revision.
- **Application-based Questions** test the conceptual understanding of students through thought-provoking problems.
- **Let Us Explore** contains interesting activities or projects meant for the reinforcement of concepts.

FREE SCHOOL SUPPORT MATERIALS

THIS BOOK IS SUPPLEMENTED BY A CD-ROM WHICH CONTAINS

- Answer key
- Lesson plans
- E-book
- Animations
- Dynamic question bank with answer key

For further information, please contact us at schoolmarketing@pearson.com.